PLATE TECTONICS

How It Works

PLATE TECTONICS

How It Works

Allan Cox

Stanford University

Robert Brian Hart

b

Blackwell Science

Editorial Offices

238 Main Street, Cambridge, Massachusetts 02142
Osney Mead, Oxford, OX2 0EL, England
25 John Street, London WC1N 2ES, England
23 Ainslie Place, Edinburgh, EH3 6AJ, Scotland
54 University Street, Carlton, Victoria 3053, Australia
Arnette Blackwell SA, 1 rue de Lille, 75007 Paris, France
Blackwell Wissenschafts-Verlag GmbH
Kurfürstendamm 57, 10707 Berlin, Germany
Blackwell MZV, Feldgasse 13, A-1238 Vienna, Austria

Distributors

USA
Blackwell Science, Inc.
238 Main Street
Cambridge, MA 02142
(Telephone orders: 800-215-1000)

Canada
Oxford University Press
70 Wynford Drive
Don Mills, Ontario M3C 1J9
(Telephone orders: 416-441-2941)

Australia
Blackwell Science Pty Ltd
54 University Street
Carlton, Victoria 3053
(Telephone orders: 03-347-5552)

Outside North America and Australia

Blackwell Science, Ltd
% Marston Book Services, Ltd.
P.O. Box 87
Oxford OX2 0EL
England
(Telephone orders: 44-865-791155)

Sponsoring Editor: John Staples

Manuscript Editor: Andrew Alden

Production Coordinator: Robin Mitchell

Interior and Cover Design: Gary Head

Composition: Graphic Typesetting Service

©1986 Blackwell Science, Inc.

00 99 98   7 8 9

Library of Congress Cataloging in Publication Data

Cox, Alan, 1926–
  Plate Tectonics.
  Includes index.   1. Plate tectonics.   I. Hart, R. Brian.   II Title.
  QE511.4.C683   1986   551.1'36   86-6138
  ISBN 0-86542-313-X

British Library Cataloguing in Publication Data

Cox, Alan
  Plate tectonics: how it works.
  1. Plate tectonics
  I. Title   II. Hart, R. Brian
  551.1'36   QE511.4

  ISBN 0-86542-313-X

Contents

4   Wrapping Plate Tectonics Around a Globe   127

5   Plotting Planes and Vectors in Local Coordinates   159

Preface

This book is intended for the reader whose imagination has been captured by reading a popular account about plate tectonics and would like to know more. It concentrates on the quantitative side of plate tectonics because most scientifically literate people are already familiar with the qualitative side. The book will enable the reader to answer questions like the following:

How fast is London moving away from New York?

How fast was Los Angeles moving toward San Francisco 50 million years ago?

How are the motions of plates described in mathematical terms?

What geophysical observations are used to determine plate motions?

How are earthquakes related to plate motions?

How are the magnetic poles related to plate motions?

What drives the plates?

The guiding philosophy of this book is that in plate tectonics, as in chess, more insight comes from playing the game than from talking or reading about it. This is a hands-on, how-to-do-it book. Most students find that through learning the nuts and bolts of plate tectonics, they gain new insight into its power and its limitations.

The basics of quantitative plate tectonics can be mastered in a few months. Even less time is required to understand how plates would move if the surface of the earth was a plane instead of a sphere. All the reader needs to get started is a little knowledge of geometry, a piece of paper, a pair of scissors, and a logical mind.

We begin in Chapter 1 by representing the Earth's surface as a plane and by representing plates as pieces of paper with simple boundaries made up of straight lines and circular arcs. These simple pieces of paper are used to present the main elements of plate tectonics. In Chapter 2 we look at the velocities of these pieces of paper as they move over the surface of a plane. By the end of Chapter 2, the reader will have mastered most of the key ideas of plate tectonics on a two-dimensional, planar earth.

Doing plate tectonics on the spherical earth is a little more complicated and a lot more interesting. To move from the plane to the sphere, the student will need to know something about drawing and moving circles on a sphere. Techniques for doing this using an intuitive, graphical approach are introduced in Chapter 3. No prior knkowledge of spherical geometry or stereographic projections is required.

In Chapter 4, plate tectonics is moved from a plane onto the spherical earth using the geometrical techniques of Chapter 3.

In Chapter 5, as a prelude to seismology, techniques are developed for plotting planes and lines on projections. The approach taken is a direct extension of what was learned in Chapter 3, where a sphere was used to represent the Earth. The same sphere is now used to describe the space around some local point of interest—an earthquake epicenter, for example, or a point on a fault.

Having mastered the techniques of Chapter 3 and 5, students are generally pleased to discover that the same basic set operations can be used to find the great circle distance from San Francisco to Tokyo, to rotate continents, to calculate plate velocities, to locate the epicenters of earthquakes, to interpret the stress fields of earthquakes, and to find paleomagnetic poles. The same graphical techniques can be used to solve many of the problems the student will encounter in courses in structural geology, crystallography, and observational astronomy.

Chapter 6 develops the strong link that exists between plate motions and earthquakes. Plate tectonics provides a conceptual framework for understanding earthquakes. Conversely, earthquakes are a primary source of information about plate motions. After finishing this chapter, the reader will understand the familiar black-and-white "beach balls' that are used to describe the orientation of fault planes and the directions of slip of earthquakes along plate boundaries. He or she will also understand how this information is used to determine the relative motions between pairs of plates.

In Chapter 7 we show how to move a point on a sphere along a small circle. This simple operation of "finite rotation," which is at the geometrical heart of plate tectonics, is described from several viewpoints with lots of examples. Handling a sequence of finite rotations is quite tricky—so tricky, in fact, that mistakes in performing this operation have produced a number of errors in the plate tectonic literature. The goal of Chapter 7 is to help the reader develop insights and techniques that will make such errors less likely.

Chapters 8 and 9 are a mini-text in paleomagnetism. Chapter 8 shows the scientific basis for interpreting the famous marine magnetic anomalies, which provided the magnetic key that unlocked plate tectonics. Chapter 9 describes how rocks get magnetized, how the magnetism of rocks is used to find paleomagnetic poles, and how paleomagnetic poles are connected to plate motions.

In Chapter 10 we turn from the techniques of plate tectonics to some broader issues of current interest in plate tectonic research: the cause of plate motions, hotspots, absolute plate motion, and true polar wander.

The intended reader of the book is a college undergraduate whose appetite for plate tectonics was whetted by an introductory course in geology. However, to make the book accessible to a larger, scientifically literate audience, we have defined expressions like "sinistral faulting" covered in beginning geology courses—the knowledgeable geology student will quickly skip over these with a superior smile.

The required mathematical background is minimal: some trigonometry, elementary calculus on the level of knowing what is meant by *"dx/dt,"* and familiarity with vectors. The latter are defined when they are introduced. As a reminder, examples are given showing how vectors add.

For students interested in computers, we show how to translate the basic geometrical operations used throughout the book into algebraic operations suitable for programming. In Chapter 7, for example, we show how to do finite rotations using standard matrix equations. These will be of interest mainly to students who wish to go on to advanced work in plate tectonics. We hope that even hackers will first do problems by hand so they will at least know whether their computer programs are working. As a fringe benefit, doing a few problems the old-fashioned way helps develop an intuitive understanding of how plate tectonics works on a globe. Armed with this insight, hackers have our blessing if they want to go on to create the perfect, all-purpose plate tectonic program.

The hand-drawn figures and cartoons of this work book and its informal style betray its origin as a set of class notes. These notes improved through a decade of use at Stanford, thanks to the criticism and advice of students, especially that of Douglas Wilson, who was first a student and then a teaching assistant in the plate tectonics course. Eli Silver at the University of California at Santa Cruz and Walter Alvarez at the University of California at Berkeley and their students used early versions of the notes and provided useful feedback. We thank Gary Head, the designer, for encouraging us to use our original handmade figures, both in order to reduce the final cost for the student and also to retain the work-in progress feeling of the original notes. The editor, Andrew Alden, helped us on many levels from elements of style to the flow of logic; he even checked some of the math. The catalyst who brought the book together was John Staples, whose love of books and patience with authors makes him more than a publisher's agent.

Introduction

Plate tectonics is a major new paradigm, or scientific world-view, that profoundly changes our ideas about how the earth works. It has been compared to the Bohr theory of the atom in its simplicity, its elegance, and its ability to explain a wide range of experiments and observations.

Tectonics is the study of the forces within the earth that give rise to continents, ocean basins, mountain ranges, earthquake belts, and other large-scale features of the earth's surface. A revolution in tectonic thinking was brought about by plate tectonics and two closely related ideas, seafloor spreading and the use of geomagnetic reversals. The latter is a method for clocking plate tectonic processes. These three ideas were advanced and substantiated between 1962 and 1968 by a handful of scientists working on problems that at first seemed unrelated but which suddenly came together to form a tightly knit fabric. This major revolution was triggered by no more than a dozen key articles that were published during these few years.

Like all revolutions, plate tectonics started with something as fragile as a nest of wild birds' eggs: some tentative ideas in the minds of a few scientists. Today most of these ideas seem obvious. In fact, students sometimes ask their teachers, "Why did it take your generation so long to tumble to the idea of plate tectonics?"

The question is a good one, for ideas closely related to plate tectonics were known long before the mid-1960s. For example, the theory of continental drift, which can be regarded as the grandfather of plate tectonics, was put forward by Alfred Wegener in 1912. The theory of seafloor spreading, which surely is the intellectual father of plate tectonics, was proposed by Arthur Holmes in 1929. A convection current rises up through the earth's mantle, said Holmes, to form the large mountain range or ridge running down the middle of the Atlantic Ocean basin. The rising current then spreads to either side of the ridge, pushing aside the continents and forming the Atlantic basin. David Griggs came even closer to the core idea of plate tectonics in 1939. The mountain ranges and earthquake belts that ring the Pacific basin, said Griggs, are due to convection currents rising in the center of the basin and sinking along its margin. "Such an interpretation," Griggs wrote, "would partially explain the sweeping of the Pacific basin clear of continental material. The seismologists all agree that the foci of deep earthquakes in the circum-Pacific region seem to be on planes inclined about 45° toward the continents. It might be possible that these quakes were caused by slipping along the convection-current surfaces."

These ideas are so close to plate tectonics that, reading them today, it is difficult to imagine that they were not taken seriously at the time they were put forward. Yet until the mid-1960s, few North American geologists accepted any of them. Most regarded the ideas of continental drift and mantle convection as unproven, untestable, or wrong. Some rejected continental drift because they were unconvinced by the argument of the "drifters" that the geology on opposite sides of the Atlantic and Indian oceans matched better if the oceans were closed. Geologists who were impressed by the match across ocean basins still tended to reject Wegener's idea because no known mechanism was capable of forcing continents to move like rafts through the strong rocks that make up the ocean floor.

Students today may find it hard to imagine an intellectual landscape in which almost all geologists and geophysists in the United States were dead set against continental drift. The articles about continental drift and seafloor spreading that we have just quoted were rarely cited before 1962. In North America, these ideas had not entered the mainstream of scientific thought, whereas in Europe and Africa, most geologists were open to the idea of continental drift. Textbooks naturally reflected the view of the profession. In those written between 1930 and 1960 in North America, continental drift either was not mentioned or was dismissed as speculation. For example, the most advanced and influential textbook used in the United States during the 1950s summarized a discussion of continental drift with the statement: "Though the theory is a brilliant tour-de-force, its support does not seem substantial." Being *for* continental drift in North America was as unpopular as being *against it* is today. Recent history has taught geologists that in science, the majority is sometimes wrong—as are the experts and the textbooks.

Plate tectonics was more persuasive than earlier tectonic theories because it was able to make predictions that could be tested against observations. The link to observations was provided by two quantitative elements of the hypothesis, **geometrical precision,** and **accurate timing.** Geometrical precision was the result of several key geometrical ideas that lie at the core of the hypothesis. The first of these is that the earth is distinctly layered. Its outer layer is the lithosphere, a spherical shell about 80 km thick, so rigid and strong that little deformation occurs within it. Beneath the strong lithosphere is the weak, ductile asthenosphere with a viscosity much lower than that of the lithosphere. As the lithosphere moves laterally, little stress is generated in the asthenosphere because the latter is so ductile. The strong contrast in rheology (flow behavior) between the lithosphere and asthenosphere allows stress to be transmitted long distances through the lithosphere. The resulting pattern of motion is quite different from what it would be if the earth were a planet with a single-layer mantle having uniform or nearly uniform rheology. If that were the case, the flow pattern would be an irregular and diffuse pattern similar to the motion of water in a kettle heated from below. On planet Earth with its highly layered structure, the motion of the lithosphere is analogous to the motion of drifting sheets of ice on a pond.

The second geometrical idea is that the earth's lithospheric shell is divided into about a dozen pieces, each of which is rigid and all of which are moving relative to each other. A key step in plate tectonics was to look closely at the boundaries between the plates and to recognize that boundaries can be divided into exactly three classes. **Trenches** are boundaries where two plates are converging. **Ridges** are boundaries where two plates are diverging. **Transforms** are boundaries where two plates are moving tangentially past each other. Each type of boundary turns out to be a major geological feature, the origin of which was poorly understood before plate tectonics.

The third geometrical idea, explored in Chapter 4, stems from recognizing an analogy between lines on a plane and circles on a globe. On a plane an object propelled by a constant force moves along a line. On a globe an object propelled by a constant torque moves along a circle. Because plates move under the influence of nearly constant torques for tens of millions of years, their motions are along circles the locations of which can be deduced from geological and geophysical data. These circles can be described efficiently by specifying the coordinates of the so-called **Euler poles** that lie at the center of the circles.

The other major element of the plate tectonic revolution was based on measuring time. Determining the age of rocks has been a central theme in geology since the beginning of the science. In the early days of geology the first great advance was to use fossils, which are still important for dating today. The next quantum jump in dating came with the use of radiogenic isotopes. By the early 1960s the ages of most parts of most continents had been determined by geologists using these two dating techniques. However, surprisingly little was known about the age of the oldest ocean floor. Estimates ranged from Precambrian (about 600 million years) to late Mesozoic (about 70 million years).

The third quantum jump in determining the age of the earth's surface came in the early 1960s, when geophysicists discovered that they could determine the age of the seafloor through studying the magnetic field at the sea surface. This new dating technique is described in Chapter 8. Its use was a key element in plate tectonics because it provided a clear geometrical picture of the way rocks of different age were created by a process of seafloor spreading. The new dating technique also permitted the rates of plate tectonic processes to be determined much more accurately than had been possible in earlier studies of continental tectonics.

The state of tectonics before the introduction of these four quantitative concepts can be imagined by considering what physics would be like without an appropriate mathematical framework. It would have the quality of the traditional "Physics for Poets" course in a liberal arts curriculum: interesting and stimulating, but not quantitative enough to be tested against observation. Tectonics was like that prior to plate tectonics. In fact, in his classic paper on seafloor spreading, which appeared in 1961 on the eve of plate tectonics, Harry Hess described his study somewhat apologetically as "an essay in geopoetry" (a self-assessment with which many would now disagree). It was only after the introduction of new geometrical concepts and accurate timing that tectonics became a quantitative science characterized by interplay between theory and observation. The main goal of this book is to provide students with the quantitative tools of the plate tectonic trade.

A word of caution may be in order because it is easy to claim too much in the flush of excitement over a new paradigm. Plate tectonics won't tell you everything about the history of the earth any more than evolution explains all of biology. What it does explain is many of the large-scale processes that shape the surface of our planet. Left unexplained are many equally exciting facets of earth science, including the origin and evolution of the earth's atmosphere, the chemical evolution of the crust and mantle, the geology of other planets and their moons, and the evolution of life. Plate tectonics provides an intellectual framework to use in attacking some of these problems, but it doesn't provide the answers. So geology students need not fear that all of the important problems of earth sciences were solved with the advent of plate tectonics. Like other new paradigms, plate tectonics has not produced a winding down of a discipline because all problems are solved. On the contrary, it has provided a solid foundation for attacking a new set of problems.

A subliminal goal of this book is to convey some of the playfulness and lightness that characterized early research in plate tectonics. The originator of the idea, J. Tuzo Wilson, took great delight in using paper cutouts to demonstrate the theory of plate tectonics. Some of us still remember watching Wilson play his embarrassingly child-like games before large audiences of scientists. Now that plate tectonics has become a serious subject, it's hard to remember why we had so much fun doing plate tectonics in those early days. Perhaps it recalled the fun we had as kids making model airplanes and other toys that really worked. In retrospect it seems almost immoral that such enjoyable, childish games should have provided the answer to questions that had puzzled geologists since the beginning of science.

As you shuffle pieces of paper around in the first few chapters, we hope that you can recapture some of Wilson's playful spirit and also, perhaps, some of his satisfaction when it dawned on him that these simple games were, for the first time, explaining the origin of mountain chains, volcanoes, major faults, and earthquakes.

PLATE TECTONICS

How It Works

1

Basics of a Revolution

This chapter has two goals. The first is to develop two of the key ideas of plate tectonics, layering and plate geometry. The second is to give the reader a glimpse of what tectonics was like prior to plate tectonics. A comparison of pre- and post-plate tectonic theories will show how profoundly our views changed. This comparison will also interest those who are curious why so obvious a theory wasn't discovered much sooner. We will see that many of the key ideas of plate tectonics were in the air for a long time before plate tectonics. All that was missing was a few key pieces of data and the imagination and insight needed to put the ideas and the data together.

Earth's Layers

The earth was known to be layered long before the advent of plate tectonics. The layers consist of three concentric shells, the core, the mantle, and the crust, each with a different chemical composition (Figure 1-1). In plate tectonics this model is retained and a new pair of layers is added, the lithosphere and asthenosphere, based not on composition but on rheology, that is, how easily rocks flow. The boundary between the two new layers lies within the mantle.

Core, Mantle, and Crust

The fact that the earth is layered was deduced from two sets of geophysical data. The first clue came from acoustic waves

Figure 1-1.
Pre-plate tectonic cross section of the earth.

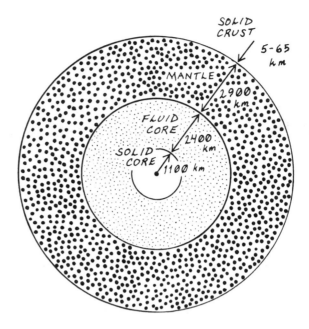

generated by earthquakes. Because the earth is transparent to these waves, its spherical shells act like lenses that reflect and refract the waves, producing patterns that reveal the earth's internal structure. The earth's layering can be seen in these patterns. The second clue was provided by the earth's gravity field. If the earth's density were uniform and equal to that of the rocks at the surface, the force of gravity would be only half the force observed. The observed strength of the gravity field shows that the earth's density increases with depth.

In trying to come up with a model consistent with both the seismological and the gravitational data, geophysicists were driven to the conclusion that the earth must have an extremely dense **core,** so dense, in fact, that the core must be made of a heavy metal. The best candidate is iron, an element that is abundant in the cosmos. At the very center the inner core is solid metal. The next layer, the outer core, consists of liquid metal. In Chapter 8 we will learn that the earth's magnetic field originates in this liquid layer.

Above the core lies the thick shell called the **mantle.** Pieces of the mantle are sometimes torn loose at depth and blasted from the throats of volcanoes. Hold one of these pieces of peridotite in your hand. You will notice that it is dark green in color, textured like an ordinary rock with transparent crystals, and a little heavier than most rocks you

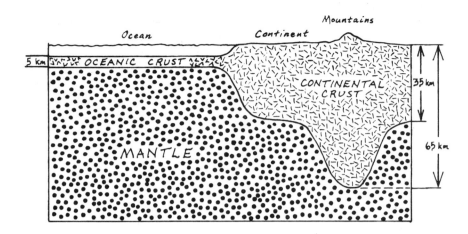

are familiar with. Its minerals incorporate into their crystal lattices silicon and oxygen, light elements that lower the density of the mantle below that of the metallic core.

The **crust** consists of the rocks we walk on every day. Its density is less than that of the mantle because its constituent minerals incorporate even more light elements than do mantle minerals. The most abundant of these elements are silicon, oxygen, aluminum, potassium, and sodium.

The thickness of the crust is not uniform. Typical thicknesses beneath continents are 30 to 50 km, increasing to 65 km beneath mountain ranges. The thickness of the crust beneath ocean basins is typically 5 km.

Deducing the presence of the core, mantle, and crust from earthquake waves and gravity was one of the great achievements of earth science. The layering tells us that early in its history, planet Earth was hot and soft enough for dense material to sink and for light material to rise. Important as it was, however, knowledge of this layering did not lead to the discovery of plate tectonics.

Strength of the Mantle

The stage was set for plate tectonics during the 1950s by a growing interest in the strength of the mantle. Wegener thought that continents move *through* the mantle, much like a raft moves through water. If Figure 1-2 is viewed as a cross section through western South America looking northward, the continent is moving like a ship to the left through the mantle. Wegener thought that the rocks of the oceanic crust and mantle in front of the moving continent are deformed

Figure 1-2.
Cross section through thin oceanic crust and thick continental crust—note crustal root beneath mountain range. In classical continental drift, the continent plows through the mantle, deforming and displacing it.

and displaced. Geologists had difficulty with this idea. The driving forces Wegener proposed were too small to produce stresses as large as those needed to break or deform mantle rocks in the laboratory. This conclusion was reinforced by the observation that the upper layers of the earth are strong enough to hold up mountain ranges, withstanding stresses much larger than those produced by Wegener's driving mechanisms. The mantle appears to be much too strong for continents to plow through it. Therefore continental drift was widely rejected, despite much evidence in its favor, because it seemed to lack a viable mechanism.

A second question closely related to the strength of the mantle is whether the earth's rotation axis is capable of shifting, a process known as polar wander. In a rigid earth, the equatorial bulge provides a formidable element of stability. In a soft earth, the equatorial bulge does not prevent polar wander. By the 1950s it was widely recognized that the question of whether polar wander is possible rests critically upon the question of whether the mantle possesses "finite strength." This refers to a rheology in which rocks deform and flow only above a threshold stress, termed the finite strength. If the mantle has finite strength greater than the driving stresses of continental drift and polar wander, neither is possible. On the other hand, if the mantle flows under an arbitrarily small stress, both continental drift and polar wander are inevitable.

The following quote from an influential 1960 monograph on the subject of the earth's rotation conveys something of the spirit of the times prior to plate tectonics. After a thorough review of relevant theories and observational data, the authors concluded that although the problem of polar wander was unsolved, they favored an earth model with finite strength and no wander: "From the viewpoint of dynamic considerations and rheology, the easiest way out is to assign sufficient strength to the Earth to prevent polar wandering, and the empirical evidence, in our view, does not compel us to think otherwise."

When the importance of the finite strength of the mantle was recognized, geophysicists were eager to try to settle the matter by squeezing mantle rocks in the laboratory. The challenge was (and still is) to reproduce in the laboratory conditions occurring in the mantle, where temperatures and pressures are extremely high, stresses are low, and rates of deformation are excruciatingly slow. By the early 1960s when plate tectonics appeared on the scene, the issue was still in doubt. In the end it was not laboratory measurements that persuaded many geologists that rocks in the upper mantle are soft enough to be deformed. It was the internal consistency of plate tectonics itself, including a new theory showing how South America could move westward without plowing through the mantle beneath the Pacific seafloor.

Plate Tectonic Layering

In plate tectonics, the mantle is divided into two and possibly three layers with different deformational properties. The upper layer is highly resistant to deformation, indicating that it has either high viscosity or finite strength. If the entire mantle were like this, plate tectonics and polar wander would not occur. The **lithosphere** consists of this rigid upper layer of mantle and the overlying layer of rigid crust. Beneath the lithosphere is the soft, easily deformable layer of mantle called the **asthenosphere.** The plates glide as nearly rigid bodies over the soft asthenosphere. The lithosphere is about 80 km thick and the asthenosphere is at least several hundred kilometers thick. Beneath the asthenosphere lies the mesosphere, the innermost shell of the mantle. Its physical properties and the location of its upper boundary are not well known, although it appears to be less deformable than the asthenosphere and more deformable than the lithosphere.

The theory of plate tectonics is largely based upon the presumed contrast in the rheological (deformational) properties of the two outer layers. Because of this contrast in rheology, stresses exerted along one part of a plate can be transmitted to distant parts of the plate much as a force exerted on one side of a floating raft is transmitted across the raft. How this works may be seen in the following thought experiment. Start with two ponds, one frozen solid to the bottom and one with a layer of ice 10 centimeters thick floating on water. Across both ponds make two parallel cuts 1 meter apart and 10 centimeters deep. Now stand on the bank at one end of the cuts and push horizontally on the ice between the two cuts. On the completely frozen pond, the shallow cuts have little effect and the stress is transmitted from the spot where you are pushing through the entire body of ice to the sides and bottom of the pond. Now go to the other pond and push on the ice between the cuts. This time the 1-meter-wide strip of ice floating on water moves easily because there is little friction on the sides or along the bottom of the strip. Most of the stress is transmitted to

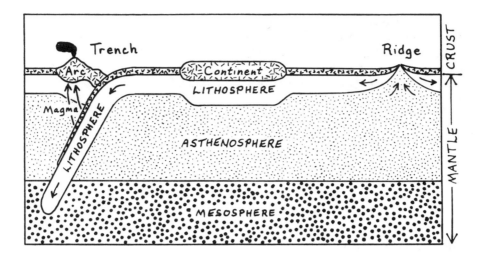

Figure 1-3.
Rigid, cold lithosphere slides over the soft asthenosphere until it encounters a trench, where it sinks. Plate tectonic cross section: continent does not plow through the lithosphere but rides with it.

a region 1 meter wide on the opposite bank, where the strip of ice will probably be pushed up onto the bank. Almost no stress is transmitted through the water to the bottom and sides of the pond because of the low viscosity of the water. In these two experiments exactly the same driving forces applied to bodies of exactly the same geometrical shape produce vastly different styles of deformation because of the difference in the rheology of the two bodies.

Before plate tectonics, most tectonic models were based on a rheologically homogeneous mantle analogous to the completely frozen pond. After plate tectonics, the pond with ice over water became the proper analogy. This change in rheological models brought about a profound change in our ideas about how the earth works.

Instead of continents plowing through the mantle, as in Wegenerian tectonics, continents are embedded in the lithosphere (Figure 1-3) and move with the lithospheric plates in which they are embedded. Where two plates converge, one plate usually slips beneath the other and sinks into the soft asthenosphere without large-scale deformation of either plate.

Why are the rheological properties of the lithosphere and asthenosphere so different? As in the case of ice and water, the difference is a matter of temperature and not composition. However, the analogy is not a perfect one. In the case of ice and water, the layering reflects a simple phase change from solid to liquid. In the case of the lithosphere and asthenosphere the situation is more complicated. Temperature increases with depth in the earth until, at depths of about 80 km in the mantle, the temperature reaches about 1400°C, which is close to the melting temperature of mantle rocks under the pressures at that depth. Materials made of a single mineral melt and lose their strength when heated through a temperature range of only a degree or two. Materials made of several minerals soften over a broader range of temperature. The mantle, composed of an assemblage of minerals with different melting temperatures, is not completely molten at any depth. However, at the depth of the asthenosphere, the temperature is very close to the melting temperature of the lowest melting mantle minerals, with the result that these crystals either melt or become soft, rendering the asthenosphere easily deformable. The sharpness of the boundary between the lithosphere and asthenosphere reflects the fact that minerals at temperatures close to melting lose their strength when heated over a very small range of temperature.

The shape of the isothermal surface that defines the boundary between lithosphere and asthenosphere is an important element of plate tectonics. In a static earth it would be horizontal. However, the earth is in motion, and we know from physics that when pieces of matter move rapidly from one thermal environment to another, they tend to carry their isotherms with them. Recall what happens when you put cold feet into warm water—they do not warm instantly. Similarly, when lithosphere plunges into the asthenosphere, it carries its cold isotherms with it, producing a downward deflection of the isothermal boundary between the lithosphere and asthenosphere. The opposite situation occurs in places like the marginal basins between the Japan Arc and the Asian continent, where the lithosphere is getting thinner as plates on either side pull apart. The asthenosphere moves upward in this case, carrying its hot isotherms with it and producing an upward deflection of the lithosphere-asthenosphere boundary. In Figure 1-3 the heavy line defining the boundary between the lithosphere and the asthenosphere is essentially the 1400°C isotherm. If you knew the shape of this isotherm all over the world and nothing else, then as a plate tectonicist you could make a pretty good guess about what is happening tectonically.

Although the idea that a strong lithosphere overlies a weak asthenosphere is strongly associated with the rise of plate tectonics in the early 1960s, the idea and the very words "lithosphere" and "asthenosphere" were advanced several decades before plate tectonics. This idea originated with geophysicists studying gravity. Analysis of the gravity field over mountain ranges requires that part of the mountains' weight floats upon low-density roots extending down into a weak layer (isostatic compensation) and that part is held up by a strong, outer layer (regional compensation). The word "asthenosphere" was introduced in 1914 by Joseph Barrell in a paper analyzing gravity, and the concept of the lithosphere and asthenosphere was a major element in a 1940 textbook by Reginald Daly. Oddly enough, however, by the time plate tectonics appeared on the scene in the early 1960s, the idea of the asthenosphere had pretty well faded from the geologic literature. That this concept had not become part of the geologic mainstream is shown by the fact that the terms lithosphere and asthenosphere were not used in popular U.S. geology texts of the 1950s. It was only after new insight had been gained into the geometry and timing of tectonic processes that these terms were revived to explain the rheological basis of plate tectonics.

Plate Geometry

The intellectual breakthrough that established plate tectonics was based on a simple new geometrical insight. In a five-page article published in *Nature* in 1965, J. Tuzo Wilson noted that movements of the earth's crust are concentrated in narrow mobile belts. Some mobile belts are mountain ranges. Some are deep-sea trenches. Some are mid-oceanic ridges. Some are major faults. Earlier geologic maps of these long, linear features showed many of them coming to dead ends. Wilson postulated that the dead ends are an illusion: the mobile belts are not isolated lineations but rather are all interconnected in a global network. This network of faults, ridges, and trenches outlines about a dozen large plates and numerous smaller ones, each comprising a rigid segment of lithosphere. The geometrical relationships along the boundaries of these moving plates lie at the heart of plate tectonics and of this book.

We saw earlier that a sheet of rigid ice resting on the water of a pond is an interesting thermal and rheological analog of the rigid lithosphere. A sheet of ice also provides a good introduction to plate geometry. Picture a sheet of ice on a pond. During the winter the ice remains frozen and still. With the spring thaw, as the ice begins to break up, the action starts. Cracks develop that divide the ice sheet into a number of plates. At first these plates remain interlocked; then one plate begins to move. Along its trailing edge a crack opens up and fills with water. Along its leading edge the moving plate of ice overrides or plunges beneath another plate. The plate tectonic process has begun on the surface of the pond. As plates of lithosphere move over the asthenosphere, the geometry of their movement is essentially the same. Much of the beauty of plate tectonics lies in the geometric exactness and simplicity of this geometry of movement. Just as Euclidean geometry provides the mathematical framework for much of science and engineering, the geometry of plate motions provides a mathematical and logical framework within which to describe the motion of the earth's tectonic engine.

In this chapter we begin by assuming (as our ancestors did) that the earth is a flat plane. We do this, first, because geometry is a little simpler on a plane than it is on a sphere, and second, because, as every surveyor knows, in a local area one may for practical purposes regard the earth's surface as planar. After we've learned the elements of two-dimensional plate tectonics on a plane, we'll wrap the plane around the sphere.

Let us start as Wilson did while writing his 1965 *Nature* article by cutting out some pieces of paper and moving them around on a table top. Glance at Box 1-1 and imagine that the page is a slab of rock 80 km thick. Cut out the triangle labeled B from your paper lithosphere. You've just made your first pair of plates. Now move plate B to the right in a straight line. Along the left side of the triangle a crack is opened up to disclose the asthenosphere beneath. This part of the boundary between the two plates is termed a **ridge** or **rise.** You might well ask whether it wouldn't make more sense to call the crack between the two plates a valley or trench. We will see that in the real world, although a narrow valley sometimes exists between two diverging plates, the regional topography is invariably that of a broad ridge. The reasons for this are discussed later.

Along the leading edges **bc** and **ca** of the paper triangle the two plates are converging. A boundary where two plates move together this way is termed a **trench.** Again, the choice of this name may appear odd—where plates collide one would intuitively expect a pile of extra-thick lithosphere rather than a trench. Again we find that the earth doesn't always behave like our intuition says it should, for reasons that are discussed below.

Trenches, unlike ridges, are asymmetrical in the following sense: either plate B may be thrust under plate A, or plate A may be thrust under plate B. All trenches have this prop-

Box 1-1. A Triangular Plate.

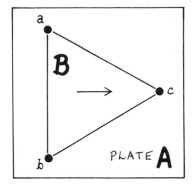

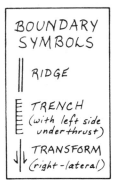

1. Copy figure and cut out triangle.

2. Move triangle a small distance to right, slipping the point of plate B beneath plate A.

3. Along what part of the perimeter does a gap open? Plot this as a **ridge.**

4. Along what part of the perimeter is there underthrusting? Plot this as a **trench.**

5. Your results should look like this.

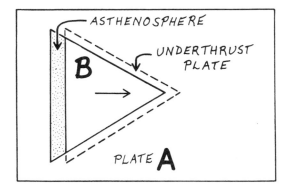

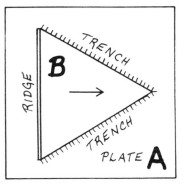

erty, which is termed **polarity.** Symbolically the polarity is specified with a "D" on the "down" or underthrust plate and a "U" on the "up" or overthrust plate (Box 1-1). The polarity also may be indicated with short hachure lines or sawteeth on the overthrust plate. (You'll notice when you start reading articles about plate tectonics that the polarity of a trench is represented in several different ways; plate tectonics is still too young to have developed strict conventions.)

Moving the triangular plate has produced ridges and trenches. Is it possible for any other types of boundaries to exist? Try changing the direction of motion of your triangular plate and see what happens. You'll find that two ridges and one trench are possible, as in Figure 1-4. But this does not exhaust the possibilities. Move plate B parallel to one of the sides, so that along this boundary the two plates neither converge nor diverge (Figure 1-5). This is the third and last type of plate boundary, the **transform fault,** or transform for short.

Because boundaries can move only away from each other (ridge), toward each other (trench), or parallel to each other (transform), there are exactly three possible types of boundary between plates. As was true in our example of a triangular plate, several boundaries of different types are commonly found around the periphery of a given plate, as in Figure 1-5.

As was true of trenches, there are two types of transform faults. Consider two points **a** and **b** which are initially opposite each other across a transform. If you stand at point **a** on plate A (Figure 1-6), you will note that point **b** is moving to your right. If you stand at point **b** you will note that **a** is also moving to the right. This type of transform is termed **right-lateral** or **dextral.** Note that the transform is right-lateral no matter which plate you stand on and regard as your fixed reference frame. The other type of transform is termed **left-lateral** or **sinistral.**

Transforms are very common, and many of them have existed for long periods of time. Those bounding large plates are especially long-lived features. What does this observation tell us about earth processes? It tells us that large plates tend to keep moving in the same direction for long periods of time. Recall the special circumstances required for the existence of a transform boundary: the motion between two plates must be exactly parallel to the boundary between them. If the movement of either plate shifts irregularly while the motion of the other plate remains steady (Figure 1-7), a transform will exist only during those moments when the relative motion of the two plates happens to be parallel to a boundary.

Two alternative interpretations can be made from the observations that transforms are common in nature and that they exist for long periods of time. The first is that once transforms are formed, they control the direction of plate motions. An analogy of a transform by this interpretation

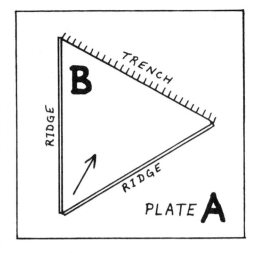

Figure 1-4.
Motion of plate B is oblique to all boundaries, which are either trenches or ridges.

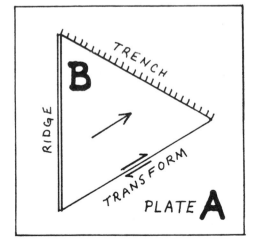

Figure 1-5.
Motion of plate B is parallel to one boundary which is a transform.

Figure 1-6.
Transform fault of the right-lateral or dextral type. If you stand at **a,** you will note that **b** is moving to your right. If you stand at **b,** you will note that **a** is also moving to your right.

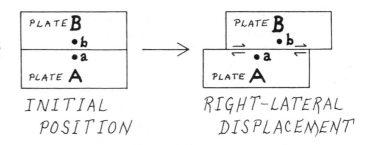

INITIAL POSITION

RIGHT-LATERAL DISPLACEMENT

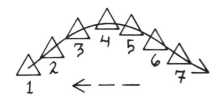

Figure 1-7.
Triangular plate moving along an irregular path. Dashed arrow is direction of relative motion of an adjacent plate. Only at position 4 does the direction of relative motion of the two plates become parallel to a boundary to produce a momentary transform.

would be a long, straight cut made in a sheet of ice on a pond: when the ice starts to move during spring break-up, plates of ice on opposite sides of the cut will tend to move parallel to the cut and to each other. The rationale for this interpretation is that since transforms are cracks or zones of weakness that cut through the lithosphere, they might be expected to guide the direction of plate motion.

An alternative interpretation is that plates are driven by forces unrelated to transforms. Transforms exert no control on the direction of plate motion, but simply align themselves parallel to the direction of motion between the two plates.

Opinion among plate tectonicists concerning these interpretations is divided. Most favor the view that the stresses generated along transforms are not the main driving forces that determine the direction of plate motion. In other words, the earlier analogy between a transform and a cut in a sheet of ice is inappropriate. Regardless of whether transforms control plate motions or simply record them, they provide our primary source of information about the direction of motion between pairs of plates. Moreover, from their persistence, it is safe to conclude that the process or condition responsible for plate motions, once started, continues in the same mode of operation for a long time.

Euler Poles

Defining Euler Poles

Euler poles play a central role in the geometry of plate tectonics. They are named for the 18th-century mathematician Leonhard Euler, pronounced "oiler." We introduce this important concept with a riddle: the entire boundary of a plate is a transform; what is the shape of the plate? The answer: a circle.

Box 1-2. A Circular Plate.

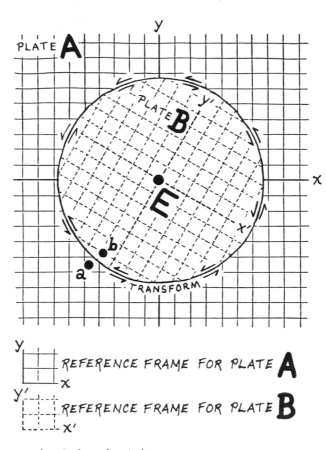

1. Copy figure and cut out plate B along the circle.

2. Place a pin through **E** to form an Euler pole.

3. Rotate plate B clockwise, keeping plate A fixed.

4. Note that the entire boundary of plate B is a left-lateral transform fault.

5. Note that the Euler pole **E** is the only point that keeps the same coordinates in both reference frames as plate B rotates.

To see how a circular boundary can be a transform, cut the round plate B from Box 1-2. Now place the point of your drafting compass through the center **E** of the circle so that plate B can pivot like a wheel. The pivot point **E** is the Euler pole. Hold plate A stationary so that its reference frame (the grid of solid lines) remains fixed and rotate plate B clockwise. Point **b** is moving past point **a** in a left-lateral sense. Now start again, holding plate B stationary so that its reference frame remains fixed, and rotate plate A counterclockwise. Point **a** is now moving past point **b** in a left-lateral sense. From the viewpoint of plate tectonics these two rotations are equivalent. The choice of the "A" fixed reference system (solid grid lines) or "B" fixed reference system (dashed grid lines) is merely a matter of convenience.

The Euler pole **E** is the pivot point for the motion of the two plates relative to each other. **E** has another interesting property. Keeping the reference frame of plate A fixed and rotating plate B, it is obvious that the coordinates of point **b** in A's grid are constantly changing. This is true for all points on plate B. The only exception is the Euler pole **E,** which keeps the coordinates (x_E, y_E) in the "A" coordinate system. Similarly, if plate B remains fixed while A rotates and if **E** is now viewed as a point on plate A, it is the only point on A that remains fixed in the "B" reference frame. The Euler pole is exactly like the hinge point of a pair of scissors, which is the only point that doesn't move relative to either blade of the scissors. The Euler pole of two plates is the only point that remains stationary relative to both plates.

Euler poles can be used to describe the motions of plates with shapes other than round. Box 1-3 demonstrates an Euler pole for a plate with all three types of boundaries. The transforms are segments of circles centered on the Euler pole. In this example, the Euler pole lies outside the boundary of plate B but is still the pivot point for the motion of the two plates. As before, the Euler pole **E** is the only point that remains stationary relative to both plates.

In two-dimensional plate tectonics most transforms are straight lines. Therefore Euler poles are not very useful because they describe the motion of plates bounded by segments of circles. However, we will see that on a sphere, *all* transforms are segments of circles. Therefore all plate motions on a sphere can be described efficiently and compactly using Euler poles. In plate tectonics, the end result of analyzing thousands of observations made over the globe is a table of Euler poles.

Finding Euler Poles

Imagine for a moment that you are an oceanographer surveying a new ocean basin where you have reason to suspect the presence of two plates. How would you go about locating

Box 1-3. Plate with Arcs as Transforms.

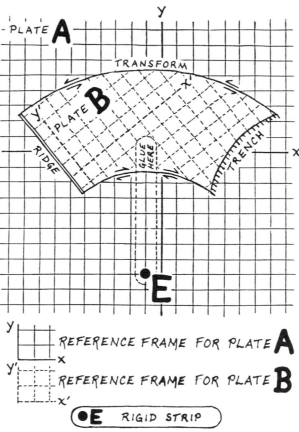

1. Copy figure and cut along boundary of plate B.

2. Cut out lower RIGID STRIP and glue to plate B as shown.

3. Place pin through **E** to form Euler pole.

4. Rotate plate B clockwise, slipping leading edge beneath plate A.

5. Note which parts of the boundary are ridges, which parts are trenches, and which parts are transforms.

6. Note that as plate B rotates, the coordinates of **E** remain the same in both the coordinate system (or reference frame) of plate A and also in the coordinate system of plate B.

the Euler pole that describes their past motion? You could start by looking for transforms in the form of circles or arcs, but at sea you will quickly learn that most transforms aren't perfect arcs, especially when viewed at close range. However, you may notice some long, narrow mountain ranges

Box 1-4.   Finding Euler Pole from Fracture Zones.

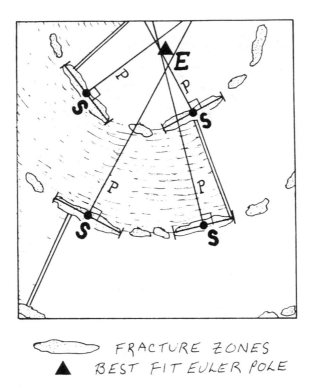

FRACTURE ZONES
BEST FIT EULER POLE

1. Identify local linear geologic features.

2. Draw a straight line segment S through each feature.

3. Construct the perpendicular bisector P of each segment.

4. Eyeball the point **E** that is as close as possible to all bisectors P. This is the Euler pole.

across which the ocean floor steps down from a shallower to a greater depth. These mountain ranges are important. Oceanographers recognized these impressive geographic features decades before plate tectonics came along and termed them **fracture zones.** Locally they appear to be linear whereas over great distances they are segments of circles. In surveying your new ocean basin, you will want to plot all such fracture zones carefully on your navigation charts.

We will see later that many fracture zones mark the trace of present or ancient transforms. Having found some fracture zones in your survey, how do you deduce from them the location of the corresponding Euler pole? Box 1-4 shows how to do this, based on the simple idea that lines drawn perpendicular to the arc of a circle all intersect at the center of the circle. You first draw a best-fitting straight-line segment along the trend of each of your fracture zones (Box 1-4). Use a protractor to read the azimuth of these lines, in degrees clockwise (or east) of north, and record these azimuths, together with the coordinates of the midpoints of the line segments. (You'll want to publish these numbers in a table because plate tectonicists will be very interested in your basic data.) Your next step is to construct the perpendicular bisectors of the line segments, repeating this at different localities spaced as far apart as possible. You'll usually find that the perpendicular bisectors don't all intersect, but most of them nearly do. Usually you can eyeball a Euler pole that is fairly close to most of the perpendicular bisectors. In Chapter 4, you will learn how to find Euler poles mathematically on a sphere.

Isochrons and Velocities

Magnetic Stripes

Trees grow by generating annual layers or rings just beneath the bark. A geologist would call a tree ring an **isochron,** that is, a surface or line that marks the location of material which formed at some specific time in the past. Imagine that 20 years ago you were a forester investigating tree growth and that you had injected a tree with black dye to mark the ring which formed that year. Then, 10 years ago, you repeated the experiment. A cross section of the tree today would display the two isochrons shown in Figure 1-8 for 20 ybp (years before present) and 10 ybp separated by 10 annual rings. Obviously trees in the forest don't have artificial isochrons, but they do display variations in the thickness of their growth rings which are characteristic of past climatic fluctuations. An expert can readily determine the age of a specimen of wood from an archeological site by comparing its ring widths with those of trees of known age from the same region.

In plate tectonics, the chronometer used to determine isochrons on the seafloor is provided by the earth's magnetic field. The heart of the timing system is located in the earth's liquid core, where the geomagnetic field is generated by electrical currents. This magnetic chronometer is binary in the sense that it has two stable states: a **normal** state in which the magnetic field is directed toward the north, and

Figure 1-8.
A tree ring is an example of an isochron, a surface or line which marks the location of material which all formed at the same time. If the distance Δx between the 20 yr and 10 yr tree rings is 5 cm, then the bark of the tree is moving away from the center at a velocity $V = \Delta x/\Delta t = 5$ cm/10 yrs = 0.5 cm/yr. Although this is a fast-growing tree, some plates are moving apart 20 times faster.

a **reversed** state in which the field is directed toward the south. For at least two billion years the field has switched back and forth between these two states at irregular intervals that may be as short as 20 thousand years or as long as several tens of millions of years or more. The geomagnetic field aligns the ferromagnetic domains in rocks on the seafloor as they cool from a molten state at a ridge. From sensitive magnetometer readings made at the sea surface, it is possible to "read" the magnetic memory of the rocks on the seafloor.

The seafloor is generally found to be magnetized in stripes of alternating polarity. Like tree rings, the stripes are of varying widths, and ages can be determined by comparison with a standard pattern of known age as determined by isotopic dating. Using this approach, which is described in more detail in Chapter 8, isochrons have now been determined for almost all of the seafloor. Magnetically determined isochrons near an active ridge are almost always parallel to the ridge and usually have mirror symmetry across it. The reason for the symmetry is discussed in Box 1-5. The explanation offered is not completely realistic for several reasons: the crack between the plates is not bounded by vertical planes as shown but rather is narrow at the top and wider at the bottom; moreover, the plates do not move apart in a series of equal, finite steps but rather by a more irregular process. However, the explanation in Box 1-5 starts with the right assumptions, is qualitatively correct, and ends with the right results.

Rates of Spreading

Imagine that as an oceanographer you've learned from your fathometer readings the location of a ridge (Figure 1-9c). From your magnetometer readings you've determined the isochrons as shown. Can you determine from these data the velocity of seafloor spreading? Note first that the 10 my (million year) isochrons are 1000 km apart. Ten million years ago, both of these isochrons were together at the ridge. Thus, during the time span $\Delta t = 10$ my a total width of $\Delta x = 1000$ km of new oceanic lithosphere formed as the two plates moved this distance apart. The velocity $_A V_B$ of plate B relative to plate A is thus

$$_A V_B = \frac{\Delta x}{\Delta t} = 100 \text{ km/my} = 100 \text{ mm/yr} \qquad (1.1)$$

In addition to giving us spreading velocities, isochrons also help us roll back the process of seafloor spreading for the

Box 1-5. Symmetrical Ridge Formation.

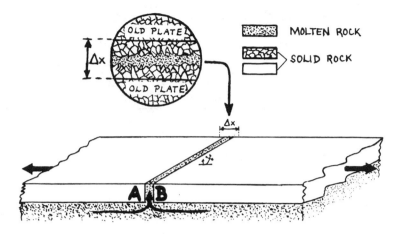

1. Plates A and B crack apart and separate a distance Δx.

2. Magma rises to fill the opening and begins to solidify inward from the old walls of the crack (inset).

3. Material at the center of the filled crack is the last to crystallize and therefore is the weakest.

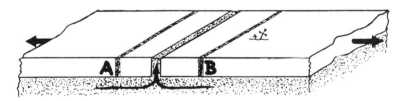

4. After time Δt the plates separate another distance Δx.

5. A crack reopens at the weak center of the old opening.

6. Magma again rises from the asthenosphere to fill the opening.

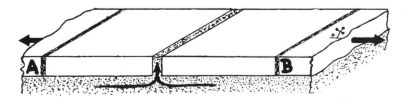

7. The process continues, building both plates.

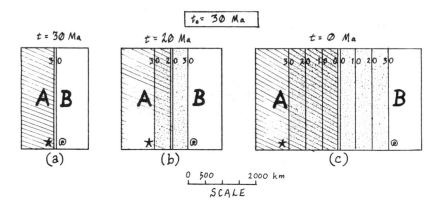

Figure 1-9.
Symmetrical growth of plates A and B by accretion at a ridge. t is the time of the plate reconstruction, $t_0 = 30$ Ma is the time when the plates begin to diverge, and $t = 0$ is the present. The starfish and snail remain firmly attached to the moving seafloor. During the past 30 my each plate has grown by a total width of 1500 km. Both strips of new lithosphere (stippled pattern) were added on the side of the plate adjacent to the ridge.

purpose of finding out where plates were at various times in the past. For example, researchers have been able to accurately determine the ancient position of North America next to Europe by matching up corresponding isochrons. To see how this works, let's start with the simple set of present-day isochrons shown in Figure 1-9c. We know that 20 million years ago (20 Ma) the 20 my isochrons were superimposed at a ridge, as is shown in Figure 1-9b. In showing isochrons as they looked at 20 Ma, the question arises of how to label the isochron that is forming at the ancient ridge, "0 Ma" or "20 Ma." We'll try to avoid confusion by (1) always keeping present-day ages attached to the isochrons when we make reconstructions for earlier times; and (2) labeling as t_0 the time when the ridge first formed and as t the time when a snapshot was taken of the ridge. Figure 1-9b is a picture of the isochrons as they looked at time $t = 20$ Ma, assuming that the ridge started spreading at time $t_0 = 30$ Ma. The isochron just forming at the ridge is labeled "20 my." Figure 1-9a is a picture of the ridge at $t_0 = 30$ Ma.

Velocities determined from isochrons in this way are the true velocities between plates only if the plates are spreading in a direction perpendicular to the ridge. If they spread at some other angle, the true velocity will be greater. In Figure 1-10 the isochrons for plates A and B are spaced the same distance apart as those for plates C and D. The 5 my isochrons are spaced 400 km apart, so the velocity is given by

$$_cV_D = \frac{\Delta x}{\Delta T} \tag{1.2}$$

$$= 400 \text{ km}/5 \text{ my}$$
$$= 80 \text{ km}/\text{my}$$
$$= 80 \text{ mm}/\text{yr}$$

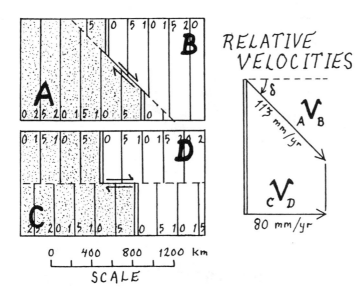

Figure 1-10.
Isochrons spaced the same distance apart produced by different relative plate velocities. In both cases the components of velocity perpendicular to the isochron are the same.

This is the true relative velocity for plates C and D. However, the true velocity for plates A and B is greater than this and is given by

$$_AV_B = \frac{V}{\cos \delta} \qquad (1.3)$$

where δ is the angle between the perpendicular to the isochrons and the direction of the transform (Figure 1-10). In other words, the isochrons give us the velocity in the direction perpendicular to the ridge. This apparent velocity is the true plate velocity only if the ridge is perpendicular to the transforms, which is usually (but not always) a good approximation.

Rises

Up to this point we have used the names rise, trench, and transform to describe the geometry of the boundaries of plates cut from pieces of paper several centimeters long. We turn now to the real earth, where rises, trenches, and transforms are major geologic features, some of them thousands of kilometers long. In discussing each of these three features in turn, we will begin with a geologic description and then

Figure 1-11.
Cross sections of the Mid-Atlantic Ridge, which has a rift valley, and the East Pacific Rise, which has no rift valley. Note the difference in horizontal scales—the true slope of the East Pacific Rise is much less than that of the Mid-Atlantic Ridge.

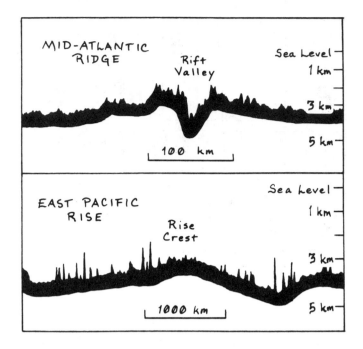

compare the plate tectonic explanation of their origin with earlier theories.

Discovery and Description

The ocean floor is our planet's last frontier. When oceanographers set out to explore it in the 19th century, they soon discovered that the seafloor is dominated by two physiographic provinces. The first is the flat areas at depths of 4 to 6 km called **abyssal plains.** The second is a set of mountain ranges called **rises** or **ridges** that rise several kilometers above the abyssal plains. They typically are several thousand kilometers wide at their bases, and their slopes are so gentle that if you were standing on the Mid-Atlantic Ridge, you would probably not realize you were on a mountain range. Yet it is one of the largest ranges on earth, one that runs down the middle of the Atlantic Ocean almost from pole to pole.

At the summit of the Mid-Atlantic Ridge is a feature not found at the summit of any continental mountain range: a deep gash as large as the Grand Canyon (Figure 1-11). This feature is called a median valley or rift. It is found at or near the top of the Mid-Atlantic Ridge along most of its length.

Let's take an undersea stroll, starting in the valley at the bottom of the rift. We look around at a flat valley about 10 km wide covered with volcanic lava flows and low volcanic hills cut by tension cracks running down the length of the valley. Steep scarps rise to heights of a kilometer on both sides of the valley. Rocks at the center of the valley are bare of sediment. As we climb out of the valley and hike down the flank of the rise, we notice first a fine sprinkling of light-colored sediment on the volcanic rock. Many fault scarps cut the surface, most of them striking parallel to the trend of the rise. Farther down the flank, the sediments thicken to form a thick blanket that completely masks the volcanics. An occasional volcano protrudes above the general level of the slope. The scene we are viewing is the commonest landscape on earth.

If we were to take a similar hike in the Pacific basin, we would find similarities and we would find differences. We would find a rise with dimensions similar to the Mid-Atlantic Ridge called the East Pacific Rise (Figure 1-11). Unlike the Mid-Atlantic Ridge, this rise is not located in the center of the basin but off on the eastern side. The rocks are volcanic at the crest and sedimentary on the flanks, as in the Atlantic, but there is no rift at the top. Different names, ridge and rise, were given to these features in the Atlantic and Pacific oceans before plate tectonics, because marine geologists were uncertain whether they were completely different species or merely varieties of the same animal.

Now let's look at mid-oceanic ridges through the eyes of a geophysicist. First we note that small to moderate sized earthquakes occur near the crests of all rises. Next we note that heat flow is unusually high near the crests. Finally, we note that the gravitational field above the rises isn't nearly as great as we would expect from the great mass of the rise. Rises are not only distinctive topographic features but they also have distinctive geophysical signatures.

These observations, all known by the 1960s, raised many questions. What is the origin of these enormous mountain ranges? Why do rifts run down the centers of some of the rises but not others? Why are they covered with bare volcanic rocks near their crests and with sediments on their flanks? Why do they have earthquakes and high heat flow along their crests? Why are they in the center of the Atlantic basin but off to the sides of the Indian and Pacific Oceans? How old are they?

Theories Before Seafloor Spreading and Plate Tectonics

Before the concepts of continental drift, mantle convection, and plate tectonics were accepted, there were no viable theories to explain the rises. If you believed that continental drift had not occurred, then a corollary was that the ocean basins must be very old, probably Precambrian. Being parts of the ocean basins, the rises were thus presumably also old. But this left unanswered most of the obvious questions about the origin of rises. The intellectual state of affairs in the decade before plate tectonics is well represented in the most influential U.S. textbook of that time, which described rises as among the greatest mountain ranges on earth, yet offered no explanation at all for their origin.

A novel explanation for the rifts along the crests of many rises was offered in the late 1950s by the theory that the earth was expanding and splitting apart at the seams. But this theory raised so many other questions that it never gained much credence. The idea that came closest to explaining the rises was mantle convection as put forward by Arthur Holmes and David Griggs, but their versions of mantle convection did not explain some important features of rises. In short, when plate tectonics appeared on the scene in the early 1960s, rises had been intensively studied by marine geologists and geophysicists but their origin was still not well understood.

Plate Tectonic Explanation of Rises

Understanding rises and ridges was one of the key breakthroughs in plate tectonics. The first conceptual step was to recognize that new ocean floor forms at the rises as molten rock comes from depth, solidifies near the ocean bottom, and spreads bilaterally in either direction away from the rise, like a double conveyor belt. The closely related idea of mantle convection was presented (although not widely accepted) several decades before plate tectonics. In the older convection theories, the rising material was thought to be the upwelling limb of a large mantle convection cell. The zone near the top of the rise where new seafloor formed was thought to be hundreds of kilometers wide.

In plate tectonics, the rift at the crest of a rise is simply a crack between two plates that are moving apart. New seafloor forms in a zone that is only a few kilometers wide as magma flows into the space between the two diverging plates and solidifies. The magma is drawn from adjacent parts of the asthenosphere rather than from a convection cell rising from deep in the mantle, as postulated by the older convection theories.

It is hard to overstate the importance of the discovery that new seafloor forms in a zone only about 10 kilometers wide. On the scale of a map of an ocean basin, this zone can be plotted as a thin line between the two diverging plates. The narrowness of this zone plays an important role in the plate tectonic model for explaining the geology and geophysics of rises. It explains why the trailing edge of a plate is a sharp boundary rather than a blurry gradational zone. It explains the absence of sedimentary rock at the rise crests—the rock is too young for sediment to have accumulated. It explains why fresh eruptions of lava and vents of hot, mineralized water produced by the circulation of seawater through very hot rocks are both concentrated in a very narrow zone at the crest of the rises. It explains why the earthquakes associated with the formation of new seafloor almost all fall within a zone approximately 10 km wide at the crest of the rise. It is the sharpness of this boundary that makes the paleomagnetic tape recorder possible and provides the basis of the simple geometrical relationships that make plate tectonics so useful.

The narrowness of the zone where new seafloor forms reflects the simple structure of the ocean floor, which near the rise crests consists of thin, rigid lithosphere overlying soft asthenosphere. If the rheology of our planet were uniform, as it was thought to be in older theories of convection, new seafloor would form over a much broader zone, earthquakes would be distributed over a correspondingly broad region, the paleomagnetic tape recorder would not work, there would be no magnetic stripes to reveal the simple geometric patterns of plates, and we would probably still be arguing about continental drift.

Explanation of High Topography

Studies of the gravitational field over mid-oceanic rises shows that these enormous submarine mountain ranges are similar to their continental counterparts in being **isostatically compensated.** This refers to the fact that mountain ranges and other unusually high regions are underlain by rocks that have unusually low densities. In effect, the mountain ranges

Figure 1-12.
Lithosphere thickening and sinking as it cools. Lines are isotherms within the lithosphere.

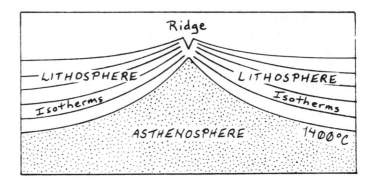

appear to be floating objects that are riding high because of their low density.

In plate tectonics, the depth below sea level of the top of the oceanic lithosphere varies because the density of the lithosphere varies. At the rise crest, the lithosphere is hot and light and floats high. As it moves away from the ridge the lithosphere cools, becomes thicker and denser, and subsides (Figure 1-12). According to the theory of the physics of a cooling plate, the subsidence of ocean floor as it cools while moving away from the ridge is expressed by the mathematical form

$$Z = C\sqrt{T} \qquad (1.4)$$

If Z, the distance that the ocean floor has subsided in moving away from the rise crest, is in meters and the age T is in millions of years, then the value of the constant C that best fits observations is $C = 300$. The flanks of rises are observed to have the shape predicted by this equation, but the equation breaks down at ridge crests and on very old ocean floor. If you know the depth of the seafloor, you can use this equation to estimate the age at which the seafloor formed (Figure 1-13). If the depth of an active ridge crest is 3 km, then the depth of the ocean floor at an age of 100 my is 6 km. Note that the depth of seafloor depends upon its age and not its distance from the ridge. At a given distance from ridges, the ocean will be deeper for slowly spreading ridges than for fast spreaders. The slopes of fast-spreading ridges are less steep than are the slopes of slow-spreading ridges. Lithosphere older than about 100 my has essentially cooled to an equilibrium state at a depth of roughly 6 km and doesn't continue to sink.

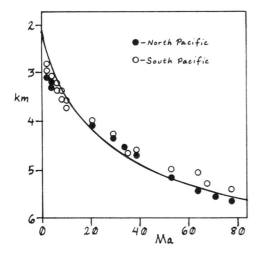

Figure 1-13.
Average depth to ocean floor plotted against the age of the ocean floor. Data from fast-spreading centers and from slow-spreading centers all fall close to this curve.

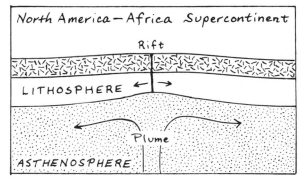 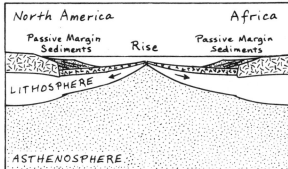

Initiation of Rises

Although rises are very mobile in the sense that they migrate rapidly away from the interiors of plates, they also appear to be long-lived features. How do they originate? In ocean basins it is not unusual for a ridge to die at one location and jump to a new one, suggesting that new rifts may form rather easily in the oceanic lithosphere. Presumably the site of the new rift is a zone of weakness conducive to the pulling apart of the plate. Fracture zones are possible sites of such zones of weakness. Others are places where the oceanic crust has been thermally thinned by rising plumes of hot magma. This appears to account for small-scale ridge jumping that has been observed near the plume of magma that produced the volcanoes of the Galapagos Islands. A third mechanism is related to changes in plate motion: when a change occurs, rises sometimes propagate like wedges into old seafloor in a direction that makes them more nearly perpendicular to the new direction of motion of the two flanking plates.

Rifts forming within continental plates appear to be somewhat less common than those within oceanic plates but are of even greater long-term geologic consequence. One intriguing hypothesis for their origin is that plumes of hot magma rising from deep in the mantle happen fortuitously to occur beneath a continent in a more or less linear array. This zone of thinned, weakened crust then coalesces to form a new spreading center (Figure 1-14). The Atlantic basin may have formed this way. During the earliest stages of continental breakup, volcanism was common on both sides of the incipient rift. During the early stage of rifting when the

Figure 1-14.
Formation of Atlantic ocean basin by rifting initiated by a hot mantle plume rising beneath the Africa–North America supercontinent. As the rift widens, sediments eroded from the continents are deposited upon the trailing edge of the continents.

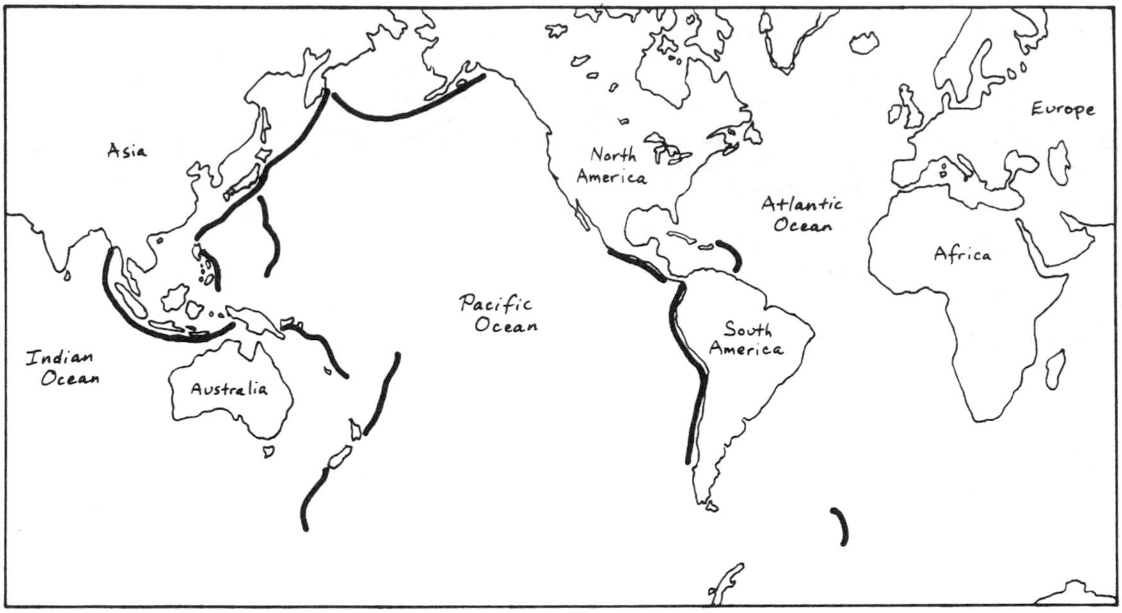

Figure 1-15.
Worldwide pattern of deep oceanic trenches.

ocean basin was shallow and ocean circulation was inter-mittent, salt deposits were common. As rifting continued, coral reefs were formed in shallow water. As the Atlantic continued to widen and deepen, the edges of the rifted continents, termed **passive margins,** subsided and were covered with sediments eroded from the continents. This sequence of events has come to be regarded as a geologic signature characteristic of continental rifting. It can be used to identify rifting in situations where the ancillary geologic evidence for rifting is not as clear cut as it is in the Atlantic.

Trenches and Island Arcs

Discovery and Description

Oceanic trenches are exactly what their name implies: long, narrow depressions of the seafloor. In plan view, trenches generally have the form of arcs. They are found around much of the boundary of the Pacific basin and in other scattered locations around the globe (Figure 1-15). The deepest points on the earth's surface occur in trenches. During the era of ocean exploration, the search for the deepest trench on earth excited public interest almost as much as had the earlier search for the highest mountain peak.

With the discovery of these deep, narrow features came the challenge of understanding their origin. The challenge

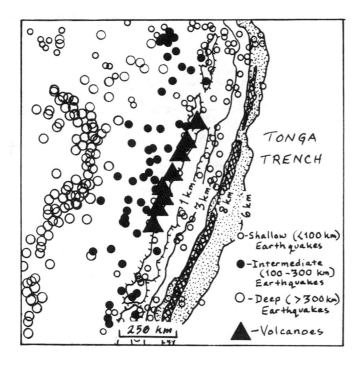

Figure 1-16.
Map of Tonga Trench showing narrow bathymetric depression and belts of shallow earthquakes, active volcanoes, intermediate earthquakes, and deep earthquakes.

increased a notch when seismologists discovered that most of the earth's deep earthquakes originate in belts next to trenches. One example is the Tonga Trench, where a belt of deep earthquakes occurs just west of the trench (Figure 1-16). The mystery deepened when geologists recognized that a belt of active volcanoes is almost always found on the same side of the trench as the belt of deep earthquakes. Moreover, these volcanoes have a mineralogy and chemistry, termed andesitic, that is different from those of volcanoes that form away from trenches. Finally, while measuring gravity from submarines, geophysicists found that above the trenches the gravitational field is smaller than would be expected on the basis of the known distribution of rock and water in and around the trench.

Plate Tectonic Explanation

The plate tectonic model for trenches is elegantly simple (Figure 1-17). A trench is simply a place where oceanic lithosphere bends downward and sinks into the asthenosphere beneath an overriding plate—a process termed **subduction.** The sinking slab of lithosphere tends to carry its isotherms with it. As a result, the slab remains colder than the asthenosphere around it and brittle enough to generate

Figure 1-17.
Subduction as it occurs on planet Earth.

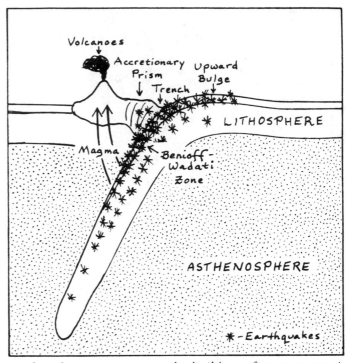

earthquakes in response to the build-up of stresses associated with subduction. These earthquakes are in effect acoustic pingers that delineate the location of the subducting slab. They generally lie along planes called **Benioff-Wadati planes** or **zones,** which dip below the horizontal at various angles from nearly 0° (horizontal) to nearly 90° (vertical). The belt of low gravity along the trench reflects the fact that at the trench, the seafloor is being actively dragged down by the dynamics of subduction. Where the upper surface of the subducting slab has sunk to a depth of about 100 km, volcanoes form above the slab, partly as a result of the upward migration of fluids and volatiles from the oceanic crust carried down with the subducting slab and partly as a result of thermal and mechanical perturbation of the asthenosphere above the subducting slab. Volcanic belts are close to trenches if the subduction zone dips steeply and far from the trench if it dips gently. Andesitic magma is uniquely associated with oceanic trenches, so much so that in older rocks it has become recognized as a tectonic signature rock: where andesite is found, the geologist knows that a subduction zone once existed. This plate tectonic model explains most of the geologic and geophysical features of trenches and island arcs with elegant simplicity.

Where plates are converging slowly, some of the parallel belts usually associated with trenches may be absent. An example discussed in Chapter 2 occurs along the coast of Oregon and Washington, where the small Juan de Fuca and Gorda plates are slowly converging with North America. No bathymetric trench is present, nor is there a seismically recognizable Benioff-Wadati zone, yet the active Cascade volcanoes are a characteristic andesitic mountain chain. Convergence velocities in excess of about 30 mm/yr appear to be required to produce a full-blown trench and Benioff-Wadati zone.

Many elements of this subduction model were anticipated decades ago by proponents of mantle convection. As early as 1940, Reginald Daly realized that since the asthenosphere is too hot and too ductile to build up the elastic strain required for large earthquakes, the occurrence of deep earthquakes requires that cold, brittle material must somehow be thrust to great depth. In the 1939 quote from David Griggs given at the beginning of the chapter, deep earthquakes behind trenches were attributed to stress produced along descending convection currents. Starting with these rather loosely described ideas, plate tectonicists defined a geometrically explicit model in which the rate of subduction was specified, permitting mathematical modeling of the thermal and stress regime in the sinking slab. Because the plate tectonic model was so explicit, it could be tested against field observations. The quality of the fit was so good that seismologists quickly became converts to plate tectonics.

An important innovation of plate tectonics is the emphasis it places on the **polarity** of subduction zones. In ordinary fluid convection, fluid moves symmetrically from both sides into the zone of downwelling (Figure 1-18). In plate tectonics, only one plate sinks into the asthenosphere. Since the two plates may be either oceanic or continental, the four obvious combinations are:

Subducting plate	Upper plate	Example
Oceanic	Oceanic	Marianas
Oceanic	Continental	Peru
Continental	Continental	Himalayas
Continental	Oceanic	??

The way in which subduction is expressed geologically is different for each of the above cases.

Subduction of an oceanic plate produces the "standard" sequence of geologic features shown in Figure 1-19, in which

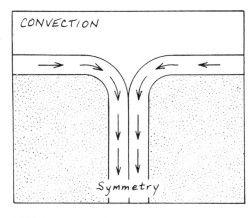

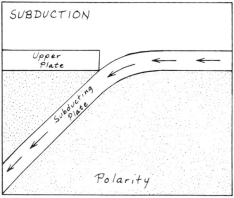

Figure 1-18.
Two theoretical types of convection: symmetrical downwelling and subduction with polarity.

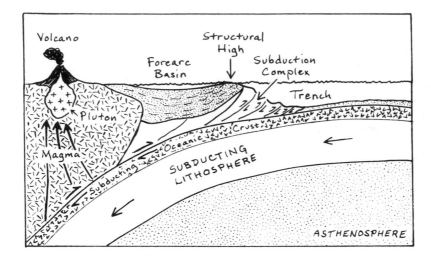

Figure 1-19.
Geologic cross section of the subduction process.

part of the oceanic crust is added to the accretionary prism and part appears to undergo subduction.

Where the subducting plate is carrying a continent, the geologic result is completely different (Figure 1-20). The subducting plate is made up of two layers, a lower layer of

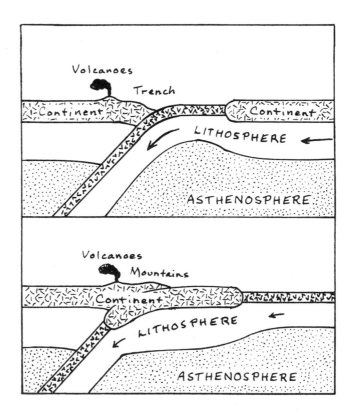

Figure 1-20.
Subduction of continental plate under continental plate.

dense mantle rock and an upper layer of thick continental crust that is much too buoyant to undergo subduction. Therefore as the heavy mantle layer of the lithosphere sinks into the asthenosphere, the lighter crustal layer peels off and becomes interleaved and stacked with the crustal layers of the upper plate. The result is a greatly thickened continental crust characterized by large-scale thrust faulting. Some of the world's great mountain ranges, including the Alps and the Himalayas, were formed in this way.

There is no known example of the subduction of a continent beneath an oceanic plate and for a very good reason. Continental crust is more buoyant than the asthenosphere, whereas oceanic crust is less buoyant; when a continent riding upon a subducting plate reaches the trench, the polarity of the trench reverses to create the stable situation of a "heavy" oceanic plate being subducted beneath the "light" continental plate.

Fracture Zones

Discovery and Description

Fracture zones are long, narrow mountain ranges that were discovered in the early 1950s. They cut across the major features of the ocean floor, including both rises and abyssal plains. A typical fracture zone is about 60 km wide and consists of several irregular ridges and valleys aligned with the overall trend of the fracture zone. The depth of the ocean floor commonly changes across a fracture zone. On a map, fracture zones are arcuate features of great length. Several extend almost completely across the Atlantic basin and one, the Mendocino fracture zone, extends three-quarters of the way across the Pacific basin (Figure 1-21).

Fracture zones have many of the characteristics of faults. Their rough topography is similar to that produced by shearing along strike-slip faults. Moreover, where they cut across mid-oceanic rises or magnetic isochrons, these features are offset, often by extremely large distances. However, other observations suggest that fracture zones are not simple strike-slip faults. For instance, fracture zones offset mid-oceanic rises by large amounts, and since rises are young, active features, then presumably fracture zones are still-active faults undergoing rapid displacement. If fracture zones are simple faults, they should thus be characterized by vigorous seismicity along their entire length. Earthquakes observed along fracture zones, however, are curiously intermittent: they occur

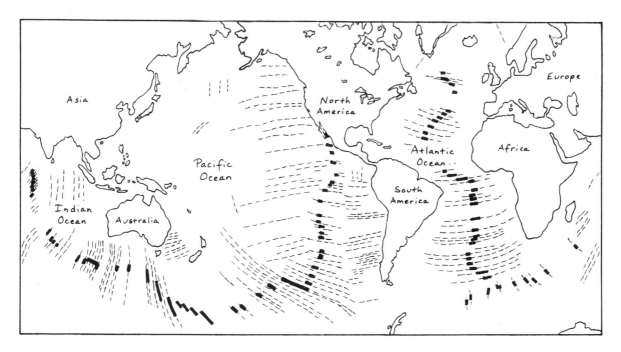

Figure 1-21.
Major fracture zones accompanied by earthquakes (heavy lines) and without earthquakes (dashed lines).

along only those parts of fracture zones that lie between the offset segments of mid-oceanic rise crests (Figure 1-22). Elsewhere, fracture zones are seismically quiet. The longest fracture zones on earth are those in the Pacific basin, and these have almost no seismic activity.

The fracture zones in the eastern Pacific basin pose another dilemma. Several of them appear to have displaced magnetic isochrons by distances of up to 1000 km. These fracture zones continue right up to the edge of the continental margin. Yet the continental margin and the pattern of geology on the adjacent continent are not offset by the fracture zone. This would seem to indicate that the faulting on the fracture zone was older than the continental crust along the continental margin. While this could not be ruled out in the 1950s when various models were being explored, we now know that the ocean floor carrying the fracture zones is much younger than much of the western margin of North America.

Many fundamental questions about fracture zones were crying out for an explanation during the decade prior to plate tectonics. What is their origin? Why does the depth of the ocean change across fracture zones? Is the offset of rises and magnetic isochrons by fracture zones due to strike-slip faulting? If so, what is the age of the faulting? Why is seismicity confined to segments of fracture zones between active mid-oceanic rises?

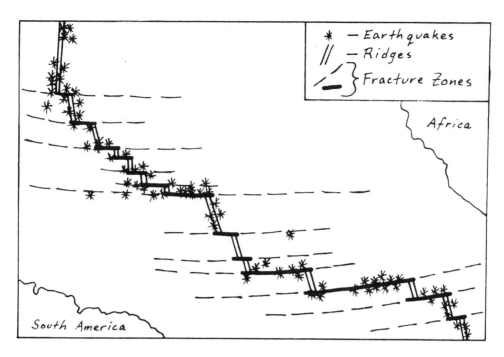

Figure 1-22.
Close-up of Mid-Atlantic Ridge showing association of earthquakes with sections of fracture zones linking ridge segments, and absence of earthquakes along fracture zones away from ridge segments.

Plate Tectonic Explanation

We saw earlier that the plate tectonic explanation of rises and trenches had in part been anticipated by earlier tectonic models based on mantle convection. On the other hand, the plate tectonic model of fracture zones, advanced by Wilson in 1965, was completely new. It astounded everyone because at first it seemed completely counter-intuitive to the usual way of looking at offset on opposite sides of a fault.

The relationship between offset rise crests and fracture zones is the key to understanding fracture zones. In viewing the pattern in Figure 1-23, the intuitive explanation is that the mid-oceanic rise formed straight and was subsequently offset by two fracture zones. The plate tectonic explanation is that the boundary between the two plates initially formed in a stair-step pattern of perpendicular ridges and transforms. The offsets came first, and the rise was never straight. The amount of offset today is the same as at the time of formation.

Between the two rise crests, rocks on opposite sides of a transform fault are moving past each other as fast as 100 mm per year. To a geologist this speed is enormous. It is not surprising that shearing motion of this magnitude produces earthquakes in the lithosphere and generates the rough topography characteristic of fracture zones. These active por-

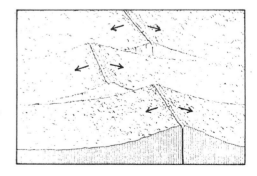

Figure 1-23.
Creation of fracture zones in regions of shear between ridges. Fracture zone is an active fault only between ridge segments.

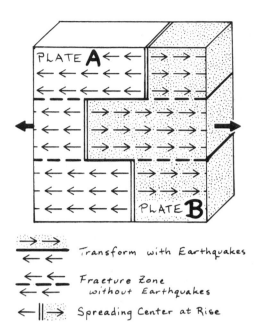

Transform with Earthquakes

Fracture Zone
without Earthquakes

Spreading Center at Rise

Figure 1-24.
Arrows show velocities on plates A and B. Between the rises are transforms (solid lines), where rocks are shearing past each other generating earthquakes. Shearing and associated earthquakes do not occur elsewhere along the fracture zones (dashed lines).

tions of fracture zones are **transforms.** Note that the shearing takes place only on the segment of fracture zone between rise crests (Figure 1-24); at the crest the shear is transformed into spreading of the ridge. Beyond the zone between the two crests, the rocks on opposite sides of the fracture zone are moving together as parts of the same plate. Except between rise crests, fracture zones are not tectonically active features. They are simply the scars left behind by the intense shearing that takes place between rise crests.

The observation that the depth to the ocean floor changes across fracture zones is explained in a straightforward way by the plate tectonic model (Figure 1-23). The rocks on opposite sides have different ages and therefore have subsided by different amounts. The greater the amount of ridge offset across a transform, the greater the difference in ages across the fracture zone and hence the greater the difference in ocean depth on the two sides.

The process of producing transforms and fracture zones is shown in Box 1-6 for a simple pair of plates. Note that the transform fault is active only between the two ridges, and it is only here that the fracture zone is a transform. Transform faults are to fracture zones what volcanic eruptions are to volcanoes: transform faulting and volcanic eruptions are the active geological processes; fracture zones and volcanoes are the geologic evidence that the processes once took place.

Velocity Fields

By combining plate geometry with velocities obtained from isochrons, plate tectonicists are able to make special maps that were not possible in earlier tectonics. These maps display arrays of arrows, called velocity fields, which show the velocity of each point on the surface of each of the plates. Here is how velocity maps are constructed.

Consider a train going down a straight track. If the train is undeformable like a plate, the velocity of the engine relative to the earth is the same as that of the caboose. In fact, all points on the train are moving with the same velocity. If the earth is regarded as plate A and the train as plate B, then all points **b** on the train have the same velocity $_AV_b$ relative to plate A. We say that the **velocity field** over plate B is uniform. Similarly, two plates separated by transforms that are straight lines also have uniform velocity fields. The velocity vectors are all parallel and have the same length.

Box 1-6. Hypothetical Experiment to Distinguish Between Inactive Fracture Zones and Transforms.

1. Plates A and B begin spreading apart 50 Ma. As usual, a fracture zone develops along an offset in the ridge.

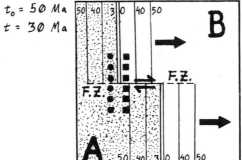

2. Astrogeophysicists arrive 30 Ma and set up a double row of survey markers on the seafloor, as shown, and go home to Krypton, never to return.

3. Seafloor spreading grinds on.

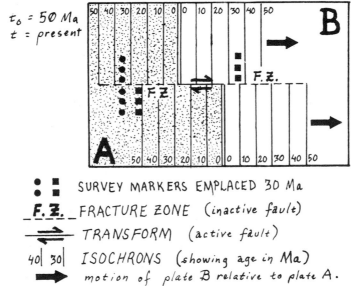

4. While scuba diving, you discover the markers on the seafloor. What do you make of the pattern? Note that when the circles and squares were attached to the seafloor at 30 Ma, plate B had already moved eastward and both plates had grown symmetrically, as may be seen from the symmetry of the 50 and 40 my isochrons. Note also that the transform fault (double arrows depicting shear) is active only between the two ridges, and it is only here that the fracture zone is a transform. To the east of the transform, the rocks on both sides of the fracture zone labeled F.Z. are part of plate B and both sides are moving together. To the west, the rocks on both sides of the fracture zone (F.Z.) are part of plate A.

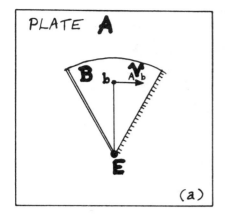

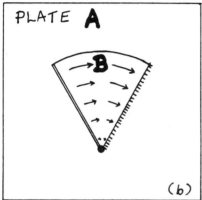

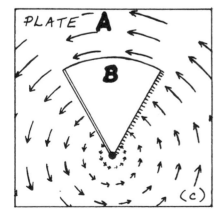

Figure 1-25.
Velocity fields for a simple plate. The linear velocity of point **b** in (*a*) is given by 1.5. The velocity fields in (*b*) and (*c*) are both consistent with the assumed relative motion of plates A and B about Euler pole **E**.

However, velocity fields are not uniform if there is an element of rotation in the relative motion of two plates. On a phonograph turntable, for example, the instantaneous velocity increases from zero at the center to a maximum at the rim. To see how this works for plates, cut a pie-shaped wedge (plate B) from a piece of paper (plate A) and place the Euler pole **E** at the apex (Figure 1-25). Give the wedge B an angular velocity $_A\omega_B$ relative to plate A. A typical angular velocity would be 10^{-8} radian per year. (Recall that a radian is a sector of a circle whose arc length equals its radius and that one radian equals 57.3 degrees.) What is the instantaneous linear velocity $_AV_b$ relative to plate A of the point **b** on plate B, where **b** is located a distance *r* from the Euler pole **E**? The point **b** moves along a circle centered on **E**. At any instant its velocity will be directed tangentially to the circle and its magnitude will be given by

$$_AV_b = r \times {}_A\omega_B \text{ (radians)} \qquad (1.5)$$
$$= r \times {}_A\omega_B \text{ (degrees/57.3)}$$

For example, at a distance of 1000 km = 10^9 mm, the instantaneous velocity of **b** is 10 mm/yr. The velocity field over the pie-shaped plate is shown in Figure 1-25b.

Although geologic observations are usually able to tell us the velocity of two plates relative to each other, they cannot tell us the true velocity of either plate in absolute terms. In other words, the plate tectonics game is played using rules that are relativistic in the Einsteinian sense: a turntable rotating on a fixed earth is observationally indistinguishable from an earth rotating about a fixed turntable. The Earth rotating about the Sun is observationally indistinguishable from the Sun moving around the Earth. Thus, for the pie-shaped plate of our previous example, an alternative interpretation would be the one shown in Figure 1-25c in which $_B\omega_A = -_A\omega_B$ and plate B is stationary. In both cases we would observe the same geologic features: the same trench, the same transform fault, and the same wedge of new lithosphere.

The plate shown in Figure 1-26 is much more interesting. Zero velocity is assumed for plate A. Note that the segments of boundaries shown as transforms (**af**, **bc**, **de**) are arcs concentric about **E** and are exactly parallel to the velocity field $_AV_B$ (small arrows). If this weren't the case, the model would be inconsistent with a basic axiom of plate tectonics: plate motions are exactly parallel to transforms. The velocity vectors near the trench make various angles with the local trend of the curving trench. At points **g** and **h** the relative velocity $_AV_B$ is locally perpendicular to the trench, but elsewhere movement is oblique. Along real trenches which curve, such as the Aleutian trench, the direction of relative motion changes along the trench in this manner. Ridges show a tendency to be nearly linear and to be nearly perpendicular to velocity fields (segment **cd**). However, this relationship is only approximate—ridges sometimes deviate from the direction perpendicular to plate motion by as much as 15 degrees. So it is the orientation of transforms rather than that of ridges that indicates the direction of relative motion between two plates.

Putting Plate Tectonics to Work

You now have the basic intellectual equipment needed to solve plate tectonic problems on a plane. Definitions of the main concepts are summarized in Box 1-7. Let's see if you can put it all together and use it. As chief scientist on a cruise to the south Atlantic, you have gathered data to produce the map shown in Figure 1-27. The tectonic pattern consists of three east-west fracture zones and a set of isochrons. All of these features are located on one plate that we'll call plate A. Since a plate can't interact with itself, you know at the outset that one or more additional plates must have been on the scene when the pattern was produced. Your challenge is to figure out the starting configuration of plates and several subsequent configurations.

Here are some hints of how to get started. (Don't read them if you have some ideas of your own.) (1) The fracture zones obviously mark a former transform trending east-west. So this is the direction of relative plate motion. (2) The isochrons (which are part of plate A) get younger to the east. Therefore they were formed by a ridge migrating eastward relative to A. (3) The spreading velocity may be found from the spacing Δx between isochrons measured in an east-west direction. This is 500 km in a time $\Delta t = 10$ my, which corresponds to a velocity of 50 mm/yr. Note that this is a half-spreading velocity $_AV_R$, the rate at which plate A grew toward the east. (4) The abrupt termination of fracture zones and isochrons toward the east tells you that plate A terminates here, cut off by a hungry trench that once occupied this now-quiet part of the seafloor. With these hints in mind, try to roll your mental movie camera backwards to 70 Ma, the age

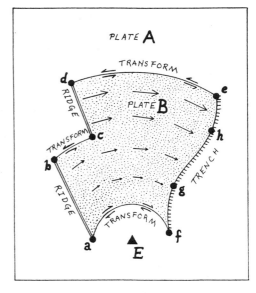

Figure 1-26.
Velocity field for a geologically more realistic plate.

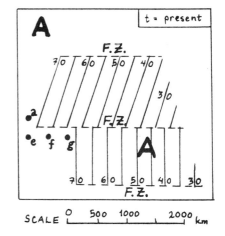

Figure 1-27.
A pattern of isochrons (solid lines), inactive fracture zones (dotted lines), and submarine volcanoes (dots) on the seafloor. Dashed line on right is a depression. Try to roll back time to discover the plate that generated this pattern.

Box 1-7. Definitions and Symbols.

Lithosphere	Rigid outer layer of the earth.
Asthenosphere	Fluid layer beneath the lithosphere.
Plate	Non-deformable block of lithosphere with a perimeter consisting of boundaries of the following three types.

Ridge

Boundary where two plates are diverging. Along the opening crack, magma rises from the asthenosphere and solidifies on both diverging plates. Ridges are symmetrical in the sense that the two plates usually grow at the same rate. Relative plate motion across a ridge is not necessarily perpendicular to the ridge.

Trench

Boundary where two plates are converging. One plate moves beneath the other, eventually to be absorbed into the mantle. Trenches are always asymmetrical in the sense that one plate is underthrust, and its leading edge is "destroyed," whereas the other plate is not shortened. Relative motion across a trench is generally not perpendicular to the trench.

Transform

Boundary along which plate motion is exactly parallel to the boundary. Lithosphere is neither created nor destroyed along a transform. Geometrically, transforms are always circles concentric about the Euler pole for the two plates. In the limiting case of a pole at infinite distance on a plane, the transform is a straight line.

Euler Pole

The pivot point about which two plates rotate relative to each other. The Euler pole is the only point that does not move relative to either plate. The Euler pole for two plates may be found by constructing perpendiculars to local segments of their transform faults.

Fracture Zone

Narrow submarine mountain range marking the location of a present or past transform. Both active and inactive fracture zones are concentric about the Euler pole.

of the oldest isochron. When you've got it worked out, compare with Box 1-8.

Having had this brief glimpse of how plate geometry works, a number of questions will occur to the thoughtful reader, just as they did to the geophysicists who discovered plate tectonics. What drives the plates? Are they being pushed from behind or pulled from in front or dragged along from below? Are rift zones, trenches, and plates themselves under compression or tension? Is there a typical life cycle describing the birth, life, and death of plates? When in the history

Box 1-8. History of a Plate Named B.

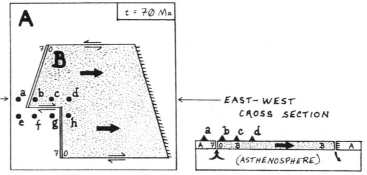

70 Ma A former plate B is born and begins to move toward the east. Luckily, a double row of submarine volcanoes labeled a-h are strategically placed to help us follow the action. Because of the polarity of the trench to the east, we know that plate B is doomed to die.

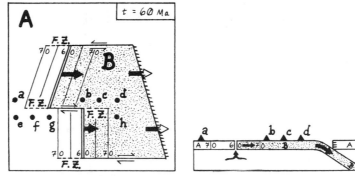

60 Ma In 10 my the ridge has migrated 500 km eastward from plate A and plate A has grown the same amount. Plate B has moved 1000 km east of plate A. To the east, 1000 km of plate B has been subducted into the trench. To the west, plate B has grown by 500 km, like plate A, so the net length of plate B has decreased by 500 km.

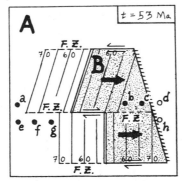

53 Ma Most of the original lithosphere of plate B ($t > 70$ Ma) has now gone down the trench, and some of the new ($t < 70$ Ma) lithosphere is starting down in the northeast.

(continued)

Box 1-8. *(continued)*

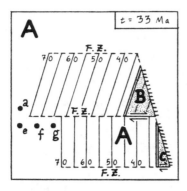

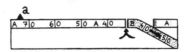

33 Ma   Because the ridge is offset by a transform, when the ridge meets the trench, plate B is split into two separate plates (B and C). Plate C goes under at 26 Ma and subduction stops along the southern part of the trench. The last piece of plate B goes under at 19 Ma. At this time plate B has been destroyed, all subduction stops, and plate A is declared the winner.

$t = 19$ Ma
to present

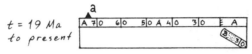

of the earth did plate tectonics start? What is the source of energy that drives plate movements?

Almost everyone agrees that plates are driven by some sort of thermal convection. Beyond this, opinion is divided into two schools of thought. One school holds that the plates ride as passive passengers on large-scale convection cells in the mantle. The other holds that the plates are not passive passengers but are themselves an active part of the convection process. In Chapter 10 we discuss these ideas and give the rationale that leads us to favor the view that plates are active parts of the convection process.

Problems

1-1.   In the following sketches of a two-piece jigsaw puzzle, plate A is stationary and $_A V_B$, the linear velocity of plate B, is given by the arrow. The units of velocity are mm/yr = km/my. In (*l*) and (*m*) the motion is not linear and plate B has an angular velocity of 5°/my about the Euler pole **E.** In all cases plate B is being thrust under plate A, except where the polarity is indicated with "U" and "D".

(a) Label the boundaries in the first column as ridge, trench, or transform with appropriate symbols.

(b) Sketch in the second column what the plates would look like today ($t = 0$) if they first separated at time $t_0 = 10$ Ma. Show isochrons at intervals of 4 my and color or stipple plate B to distinguish it from plate A. The anchor, starfish, and shell are resting firmly attached to the sea-floor, which otherwise is rather monotonous. Keep track of where they move.

(c) Now sketch in the third column what the plates would look like today if they first separated at time $t_0 = 20$ Ma. Don't be reluctant to use a pair of scissors—the experts do! (You have our permission to photocopy this page if you don't want to damage the book.)

$$t = t_0$$

$$t = 0$$
$$t_0 = 10 \ Ma$$

$$t = 0$$
$$t_0 = 20 \ Ma$$

1-1a

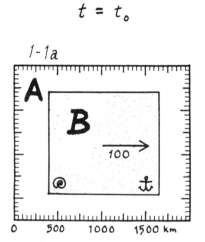

0   500   1000   1500 km

1-1a-SOLUTION

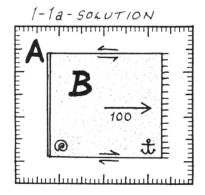

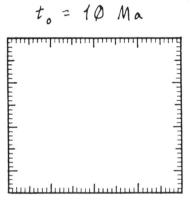

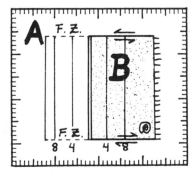

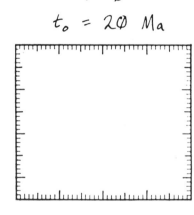

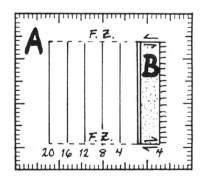

1-1b

1-1c

1-1d

1-1e

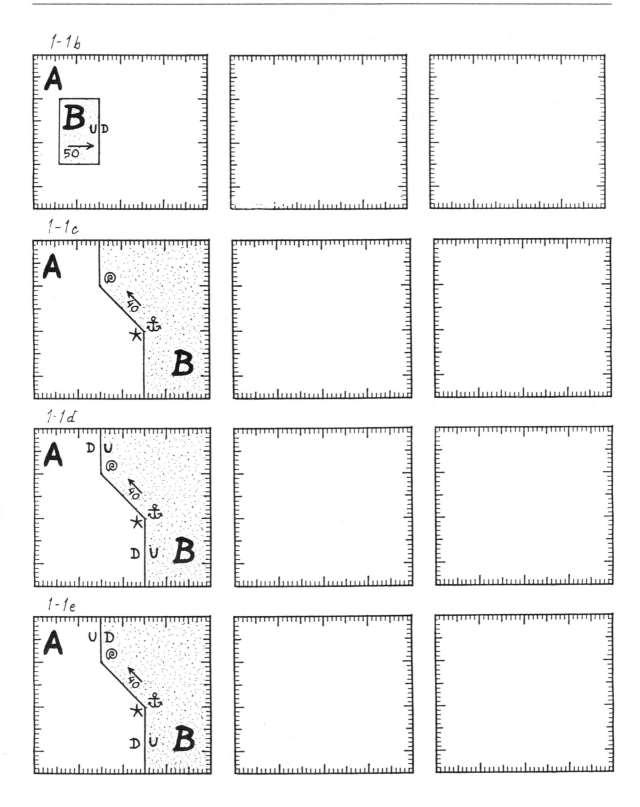

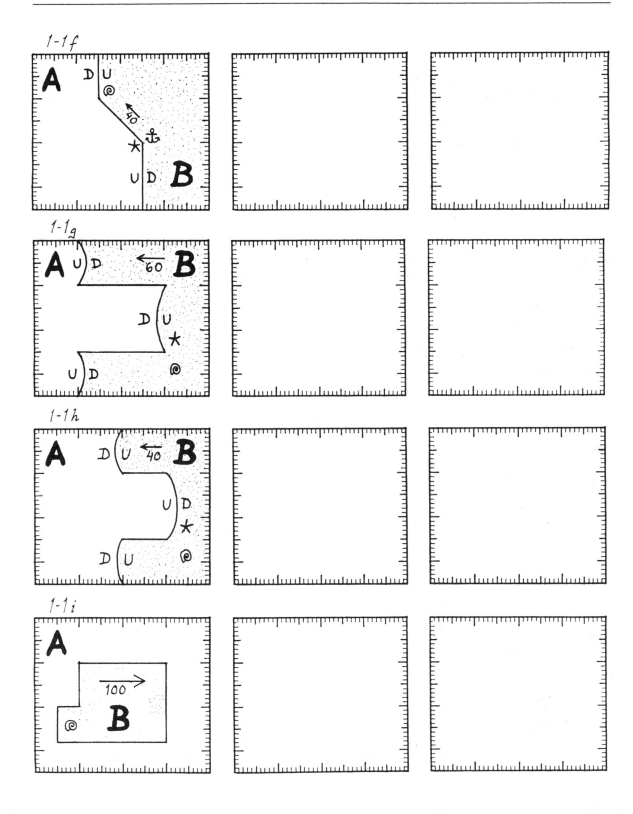

1-1f

A   D U
@
←40
⚓
★
U D B

1-1g

A U)D   ←60 B
D U
★
@
U)D

1-1h

A D(U ←40 B
U D
★
D(U @

1-1i

A
→100
@ B

1-1 j

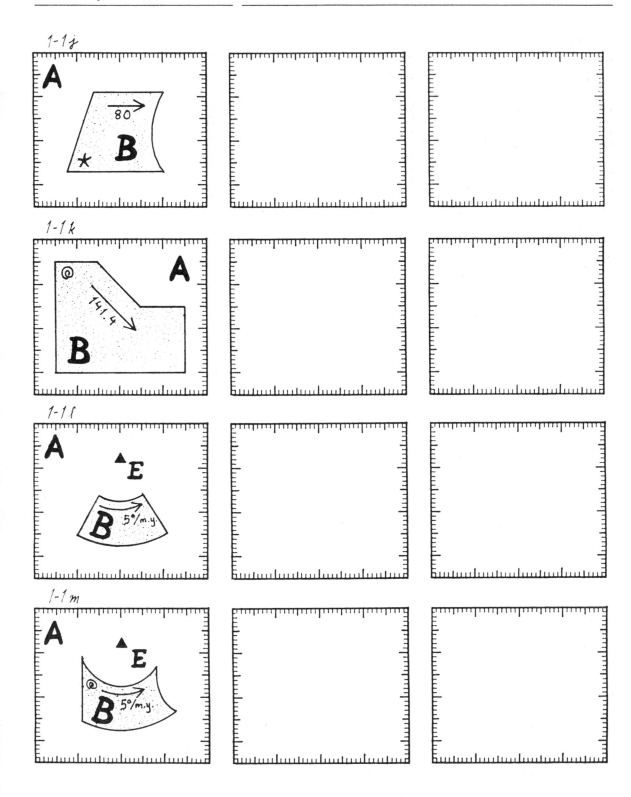

1-1 k

1-1 l

1-1 m

1-2. Some mini-jigsaw puzzles are shown below, without velocity arrows this time but with the boundaries identified as being ridge, trench, or transform. Transforms are shown as straight lines with a "T", but without a pair of small arrows (because we don't want to give the answer away). Some of the pairs of plates are impossible in the sense that there is no velocity field that is consistent with the boundaries.

(a) First try to find a velocity field consistent with all of the boundaries. Regarding plate A as fixed, sketch this velocity field over the other plate or plates.

(b) If you can't find a consistent velocity field, mark the diagram "impossible." But remember that to a scientist, "impossible" is a very strong word.

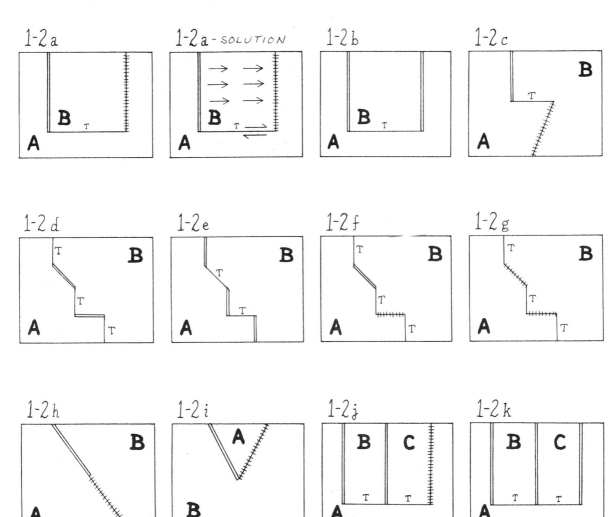

1-3. The north-south magnetic lineations over the Pacific Ocean off the coast of California, Oregon, and Washington give excellent isochrons which are offset along east-west fracture zones as shown. Oddly enough, these isochrons get **younger** toward the present North American coastline. This is quite different from the pattern in the Atlantic Ocean, where the isochrons are youngest in the center and get **older** toward the coasts. To explain these observations, an ancient trench which is no longer active (at least as a trench) has been postulated in the position shown. See if you can work out a plausible plate tectonic history to account for this peculiar pattern. To make the problem tractable, make the simplifying assumption that the Pacific plate and the North America plate have not moved relative to each other. (This assumption is wrong, but we'll wait until we've learned a few more tricks before we introduce this complication.) Also assume that a new crust formed symmetrically about a ridge.

(a) Draw a series of maps showing the pattern of plates at times $t = 10$ Ma; 20 Ma; 30 Ma; and 50 Ma.

(b) Show all ridges, active transforms, inactive fracture zones, and active trenches with appropriate symbols.

(c) Color each plate separately.

(d) Make a table listing the rates of relative motion between all plates and between the ridges and plates.

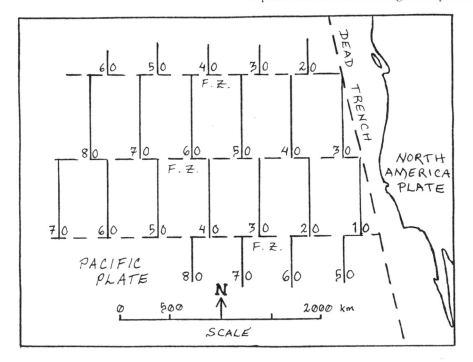

Suggested Readings

Texts

The New View of the Earth, Seiya Uyeda, W. H. Freeman, San Francisco, 217 pp., 1978. *Non-mathematical introduction to plate tectonics for the general reader. Key ideas of plate tectonics are presented with simple elegance. Ideal companion volume to present technique-oriented book.*

Plate Tectonics, Xavier Le Pichon, Jean Francheteau, and Jean Bonin, Elsevier, New York, 302 pp., 1973. *Intermediate level text for students with some mathematical background.*

Principles of Geology, James Gilluly, Aaron Waters, and A. O. Woodford, W. H. Freeman, San Francisco, 631 pp., 1951. *"Influential textbook of the 1950s" referred to in the text.*

The Rotation of the Earth, Walter Munk and Gordon Mac-Donald, Cambridge Univ. Press, Cambridge, 323 pp., 1960. *Lucid and witty review of the physics of the earth's rotation (referred to in this chapter) which conveys the intellectual landscape in geophysics on the eve of plate tectonics. Discusses the importance of finite strength; reviews (and rejects) paleomagnetic evidence in support of polar wander.*

Classic Papers

The Strength of the Earth's Crust, Joseph Barrell, The Journal of Geology, v. 22, p. 425–433, 1914. *Classic paper introducing the terms "lithosphere" and "asthenosphere."*

Radioactivity and Earth Movements, Arthur Holmes, Transactions Geological Society of Glasgow, v. 18, p. 559–606, 1929. *Seminal paper developing mantle convection as the cause of continental drift and continental rifting.*

A Theory of Mountain Building, David Griggs, American Journal of Science, v. 237, p. 611–650, 1939. *Discussion of mantle convection that anticipates many ideas of plate tectonics.*

Strength and Structure of the Earth, Reginald Daly, Prentice Hall, New York, 434 pp., 1940. *Text built around the concepts of the lithosphere and asthenosphere.*

History of Ocean Basins, Harry Hess, *in* Petrological Studies: A Volume in Honor of A. F. Buddington, ed. by A. E. J. Engel, H. L. James, and B. F. Leonard, Geological Society of America, p. 599–620, 1962. *Classic "geopoetry" paper tracing the geologic and geophysical consequences of seafloor spreading published on the eve of plate tectonics.*

Plate Tectonics and Geomagnetic Reversals, Allan Cox, W. H. Freeman, San Francisco, 702 pp., 1973. *Forty key papers that established plate tectonics, printed in their entirety along with author's account of the discovery of plate tectonics.*

Plate Tectonics, John M. Bird and Bryan Isacks, American Geophysical Union, Washington, D. C., 1021 pp., 1980. *Articles on plate tectonics originally published in the Journal of Geophysical Research.*

Plate Tectonics on a Plane

A New Class of Faults and their Bearing on Continental Drift, J. Tuzo Wilson, Nature, v. 207, p. 343–347, 1965. *Classic discovery article in which the key geometrical ideas of plate tectonics are first presented.*

Rises, Trenches, Great Faults, and Crustal Blocks, W. Jason Morgan, Journal of Geophysical Research, v. 73, p. 1959–1982, 1968. *Classic article in which Wilson's ideas are developed and used to calculate plate velocities.*

Geology of Rises and Trenches

Mountain Belts and the New Global Tectonics, John F. Dewey and John M. Bird, Journal of Geophysical Research, v. 75, p. 2625–2647, 1970. *Classic article linking geologic observations to the geometry of rises and trenches.*

2

Plates in Velocity Space

One of the keys to the precision and power of plate tectonics is its ability to find out how fast plates are moving relative to each other. Before learning how this is done, you will need to develop some mental tools for handling relative velocities.

The Velocity Line

The velocities we deduce in plate tectonics are generally **relative velocities** rather than absolute velocities for the following reason. The plates of lithosphere floating on the asthenosphere resemble a group of ships floating on the ocean. Sailors on the ships can, with careful triangulation, work out their velocities relative to each other. However, unless they can see the ocean bottom or a land mass, they have no way of knowing the absolute velocity of the ships, that is, the velocity relative to the seafloor. In the case of lithospheric plates, the isochrons on either side of a ridge, together with a transform offsetting the ridge, tell us the velocity of two plates relative to each other, but they tell us nothing about the "absolute" velocity of the plates relative to a reference frame fixed in the deeper parts of the earth beneath the asthenosphere. Analyzing plate motions is like traveling on a train at night: you can see the light of another train and estimate its velocity relative to your own, but you cannot see the ground and estimate your velocity relative to it.

We'll begin by reviewing the familiar problem of two trains traveling down a track at different velocities. The track runs east-west and is straight. We'll regard west not as a separate direction but rather as the negative of east. Thus we can represent the velocity of any train by a positive or negative number. For example, $V = -30$ kph is the velocity of a train traveling west at 30 kilometers per hour. We can represent the velocities of many different trains by plotting the velocities as points on a **velocity line** along which each number represents a velocity (Box 2-1). "One-dimensional velocity space" is another way to describe the velocity line. The velocity of any train on a straight track can be represented by a point in this space.

In using the velocity line diagram, you don't need to know the absolute velocity of any of the moving objects in order to determine all of their relative velocities. You can simply assume arbitrarily that one of the velocities is zero and plot the other velocities relative to it. To get the relative velocity of any two trains or plates, say that of train A relative to train B, the first step is to plot the velocities (not the positions!) of the two trains as points **A** and **B** on a velocity axis or line. Next, draw an arrow with its tail end at **B** and its head at **A** (Box 2-1). The length and direction of this arrow specifies the velocity $_B\mathbf{V}_A$, which is the velocity of train A relative to train B. The arrow will remain the same if you later decide to change your fixed reference frame (Box 2-2). Because the velocity line is one-dimensional, the vector $_B\mathbf{V}_A$ can be expressed in an alternative scalar form $_BV_A$, positive numbers corresponding to vectors pointing east and negative numbers to vectors pointing west. When we move to the plane in the next section, we will no longer be able to represent the vectors by single scalar numbers.

The same velocity line used to describe the trains in Box 2-2 can also be used to describe a set of plates named A, B, and G that are moving eastward or westward relative to each other. (However, we should give plate A a realistic velocity of 60 mm/yr and not 60 kph.) It's a simple matter to discover a set of plate boundaries that are consistent with the velocity line diagram of Box 2-1. Just cut a piece of paper into three pieces and label them B, G, and A. Hold G steady and give A and B the appropriate velocities to the right (east) and left (west). If the plates are arranged as shown in Figure 2-1, the BG and GA plate boundaries are ridges across which the spreading velocities are $_GV_B = -20$ mm/yr (B moving westward relative to G) and $_GV_A = 60$ mm/yr (A moving eastward relative to G).

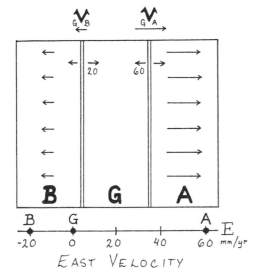

Figure 2-1.
Plates with velocities proportional to those of the trains of Boxes 2-1 and 2-2.

Box 2-1. The Velocity Line.

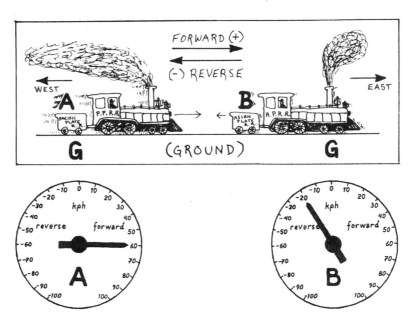

Train A is going east at 60 kph and train B is going west at 20 kph, both relative to the ground G. To keep your signs straight, observe that the speedometers read positive when the trains are going forward, which in this problem is toward the east. Eastward velocities will be regarded as positive and westward velocities as negative. The velocities of trains A and B and the ground G relative to each other may be represented by plotting three points **A, B,** and **G** on a **velocity line** as follows.

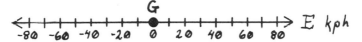

1. Label a number axis *E* for eastward velocity. Positive numbers represent eastward velocities and negative numbers represent westward velocities. We can assume arbitrarily that either train A, train B, or the ground G isn't moving. Let's be reasonable and assume that the ground is standing still. Plot a point **G** corresponding to its zero velocity at the origin.

2. The 60 kph velocity of train A relative to the ground G can be represented in several different ways. One is by an arrow with length proportional to 60 kph, as shown above. When describing the velocity of train A relative to the ground G, use the convention of labeling the head of the arrow at **A** and the tail at **G.** Place this arrow along the velocity line with its tail at the origin (**G**). The arrow head then is at 60 kph. Mark **A** at this point. Because train A is moving eastward relative to the ground G, the arrow points in the positive direction of the velocity axis. This arrow is labeled "$_G\mathbf{V}_A$." The subscripts before

(continued)

Box 2-1. *(continued)*

and after the **V** refer to the moving object and the reference object, respectively; therefore the expression "$_GV_A$" is read "the motion of A relative to G." Equivalent representations of the velocity of train A relative to the ground G are thus (1) point **A** on the velocity line, (2) the arrow from **G** to **A,** and (3) the symbol $_GV_A$. Because this is a one-dimensional problem, we can represent a velocity **V** by a simple scalar number V, positive values of V corresponding to vectors pointing to the right and negative values to vectors pointing left. So we can represent the velocity of train A relative to the ground G by $_AV_G = V_A - V_G = 60 - 0 = 60$ kph.

3. To find the velocity $_AV_G$ of the ground G relative to train A, start with the velocity line and draw an arrow with its tail at **A** and its head at **G.** Since this is pointed toward the negative number axis, it represents a westward velocity of 60 kph. Taking advantage of the fact that a one-dimensional vector can be expressed as a scalar number, we may write $_AV_G = V_G - V_A = 0 - 60 = -60$ kph, the minus sign indicating a "−east" or westward velocity. This says that if you're sitting in train A, the ground appears to you to be moving westward at 60 kph.

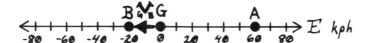

4. The velocity $_GV_B$ of train B relative to the ground is 20 kph toward the west, which is represented by an arrow 20 kph units long pointing toward the negative end. Place this arrow on the velocity line with its tail at the origin (**G**). The arrowhead is then at −20 kph. Place **B** at this point.

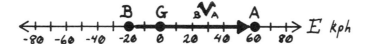

5. To find the velocity $_BV_A$ of train A relative to train B, draw an arrow with its tail at **B** and its head at **A.** Its length is 80 kph and it points toward the positive end, which means that train A is traveling eastward at 80 kph relative to train B. Using the scalar form, $_BV_A = V_A - V_B = 60 - (-20) = 80$ kph.

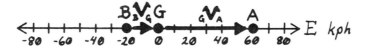

6. The arrow $_BV_A$ is composed of two arrows: $_BV_A = {_BV_G} + {_GV_A}$. (Memory aid: during vector addition, identical subscripts cancel when next to each other.)

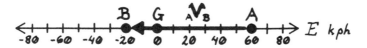

(continued)

Box 2-1. *(continued)*

7. To find the velocity $_AV_B$ of train B relative to train A, switch the head with the tail on the $_BV_A$ arrow. Using the scalar form, $_AV_B = -_BV_A = -80$ kph. The physical interpretation of "$_AV_B = -80$ kph" is the following. If you were sitting in train A and regarding it as your frame of reference, train B would appear to you to be traveling at -80 kph (westward at 80 kilometers per hour). If train B is 20 meters to the east of train A, you have 0.9 second to meditate.

Box 2-2. Change of Reference Frame on Velocity Line.

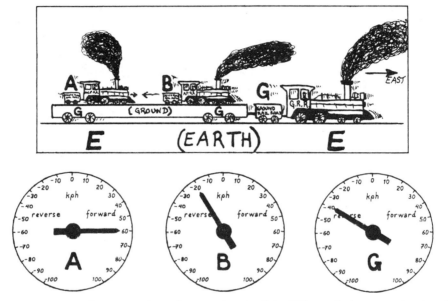

Naturally you jump out of train A, only to discover that the "big G" beneath the two trains A and B is not the ground; it's the bed of a flat car being pushed by a train named G which is going west (backward) at 40 kph relative to the earth E, that is $_EV_G = -40$ kph. On the velocity line, **G** must therefore be replotted at -40 kph. Velocities **A** and **B** are replotted to have the same distances from **G** as before.

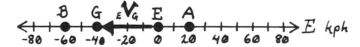

The train G and the flat car G are moving at the same velocity, so both of their velocities are represented by the same point **G**. Note that the relative velocities of A, B, and G are the same as in Box 2-1. Relative velocities don't change when a group of velocity vectors is moved. As seen from G, train A is still moving east at 60 kph; as seen from train A, train B is still moving west at 80 kph; and trains A and B will still have the same collision in 0.9 second.

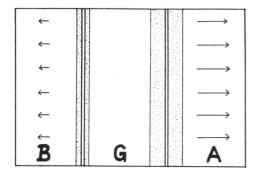

Figure 2-2.
Growth of plate G with ridges on both sides. Ridges move away from G at half the rates of plate A and plate B.

Figure 2-3.
Same relative velocities as in Figure 2-2, but zero has been shifted on the velocity line. Does Figure 2-2 still represent the growth of plate G?

Is it possible for a plate like G to have ridges on both sides? As new crust is accreted to G, it moves laterally away from the ridges. Wouldn't this tend to compress G and make it buckle? If so, this would violate a major element of plate tectonics theory, the assumption that plates are rigid and do not undergo deformation. This question arose early in the development of plate tectonic theory when it was realized that Africa is surrounded mainly by ridges but is tectonically stable. However, this apparent dilemma vanishes when you realize that the two ridges adjacent to plate G (or Africa) are not stationary but are moving away from the plate. Imagine three rigid rafts floating on water without quite touching. As the two outer rafts move away from the stationary center raft G, water rises gently to fill the cracks between the rafts. Similarly, molten magma from the asthenosphere rises in the "cracks" (i.e. ridges) between the diverging plates and crystallizes onto G and the adjacent plates (Figure 2-2). Plate G is therefore not being shortened by compression but instead is growing by accretion.

The situation remains pretty much the same if we assume that plate B rather than G has zero velocity (Figure 2-3). This is analogous to using the South America plate as our fixed frame of reference. If we now start traveling eastward, each time we cross one of the ridges there's a step increase in eastward velocity. An analogous situation exists if we assign zero velocity to plate A. Whichever plate is assumed to have zero velocity, the relative plate velocities and the spreading velocities at the ridges remain the same. Changing the location of the zero on the velocity line is simply the equivalent

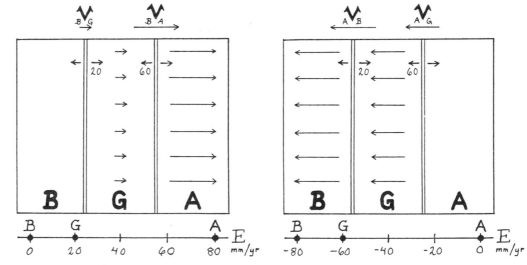

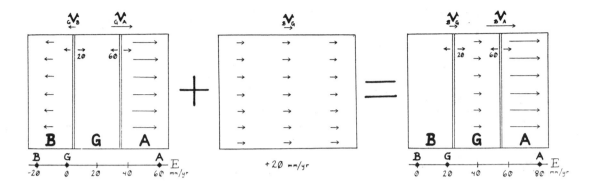

of adding the same uniform velocity field to all plates (Figure 2-4). We can do this at any time in order to gain a different perspective on the problem.

Now shuffle your paper plates and lay them out in a different sequence, keeping their velocities the same as before. You'll find that the plate boundaries are no longer two ridges but are either two trenches or a ridge and a trench. The example shown in Figure 2-5 would correspond roughly to the Nazca plate (A), the South America plate (B), and the Africa plate (G).

With a little more shuffling, many other geometrical arrangements of the three plates can be devised which are consistent with the original set of velocities. For example, transforms can be generated by placing plate boundaries in an east-west direction. In Figure 2-6 the result is that plate G is bounded by both a left-lateral and a right-lateral transform (remember that these refer to relative and not absolute motion). If you view G as fixed, then B is moving westward at 20 mm/yr and A is moving eastward at 60 mm/yr. The ridge between them is moving eastward at 20 mm/yr relative to G and at 40 mm/yr relative to B.

Velocity diagrams for plates on the real earth are more complicated than those of trains on a straight track for several reasons. One is that the total number of plates is quite large—more than a dozen have now been recognized. Another is that plate motions are not restricted to one direction.

The Velocity Plane

To handle plates moving on a plane we need to expand the preceding analysis from one to two dimensions. To express the relative velocity of two plates, say F and B, moving on a

Figure 2-4.
Shifting the origin of the velocity line is equivalent to adding a constant velocity field to all plates.

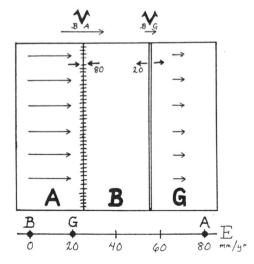

Figure 2-5.
Changing the sequence of plates changes the pattern of ridges and trenches, even though relative velocities are the same.

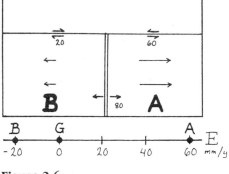

Figure 2-6.
The same velocities with a different plate geometry give transforms along east-west boundaries.

plane, a **velocity vector** is needed. A velocity vector can be thought of as a velocity arrow with both a length and a direction. We will use the italic symbol $_BV_F$ to denote the length of the vector, which is a simple number, and the boldface symbol $_B\mathbf{V}_F$ to denote the complete vector.

Box 2-3 gives examples of such vectors. A ferry F moving due northeast from a buoy B at a velocity $_B\mathbf{V}_F$ = 4 km/hr has two components of velocity: a **north component** $_BN_F$ = 2.82 km/hr and an **east component** $_BE_F$ = 2.82 km/hr. We can show both components by plotting the vector $_B\mathbf{V}_F$ on a **velocity plane** in which the vertical axis represents northward velocity and the horizontal axis represents eastward velocity (Boxes 2-4 and 2-5).

In Boxes 2-4 and 2-5 we assume arbitrarily that the buoy has zero velocity, although this was not stated as one of conditions of the model. Would our conclusions about the velocity of the ferry F, the truck T, and the turtle A, relative to the buoy, still be valid if we discover that this assumption is not true? Suppose, for example, that ocean currents are carrying the buoy over the ocean floor toward the northeast at a velocity of 2.82 kph (Figure 2-7). It's reasonable to now assume that the velocity of the ocean floor ("O" in Figure 2-7) is at rest and to plot this point at the origin in the velocity plane. This shifts the entire pattern of relative velocity vectors in Box 2-5 in velocity space by 2.82 kph; however, the lengths and directions of all the other vectors remain the same as before, so the relative velocities are still the same.

The compactness of plotting points rather than arrows in velocity space is demonstrated in Figure 2-8. To completely describe all of the velocity relationships among the four points **B, F, T,** and **A** one could draw the 12 relative velocity vectors shown. Note that these may be drawn in arbitrary positions on the velocity plane because relative velocity vectors may be moved horizontally and vertically without changing their direction D and length V. These two parameters, D and V, completely specify the relative velocity vector regardless of the position of the vector in the velocity plane. The information provided by the 12 velocity vectors in Figure 2-8 is expressed more compactly in Figure 2-7 by plotting the four points in velocity space. Moreover, as long as the points retain the same distances and directions relative to each other, the group of points may be plotted anywhere in the plane. We might want to place **A** at the origin, for example, to describe a flea's-eye view as seen by a passenger on Alice the turtle.

Box 2-3. Relative Velocity Vectors in Two Dimensions.

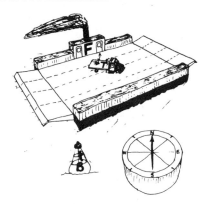

Let's free our moving objects from the constraint of moving along a straight track or road and allow them to move in any direction on a plane. We now find that we must use a vector rather than a simple number to describe their velocities. The relative velocity vectors may be represented by arrows, as shown in the following example about a turtle named Alice.

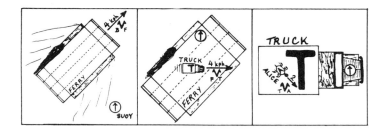

Inventory of Relative Velocity Vectors

$_BV_F$ VELOCITY OF FERRY F RELATIVE TO BUOY B.

To a boy with a compass sitting on the buoy B, the ferry F appears to be moving northeast at a velocity of 4 kph. To a sailor on the ferry, the boy on the buoy appears to be moving southwest at the same velocity.

$_FV_T$ VELOCITY OF TRUCK T RELATIVE TO FERRY F.

The truck is moving diagonally across the deck of the ferry. The ferry appears to be stationary to a sailor on deck who has a compass. Luckily the motion of the ferry doesn't affect the reading of the compass, which tells the sailor that the truck is going due east relative to the ferry. The truck speedometer shows 4 kph, so $_FV_T$ is 4 kph in an easterly direction.

$_TV_A$ VELOCITY OF TURTLE A RELATIVE TO TRUCK T.

A turtle named Alice is crawling diagonally across the bed of the truck, which naturally is the frame of reference for Alice and the truck driver. Alice is fast. Glancing in the rearview mirror and noting the compass attached to the cab of the truck, the driver estimates that Alice is crawling southeast at 2 kph.

Box 2-4. Adding Two-dimensional Relative Velocity Vectors.

Polar coordinates *V* and *D*

1. The **length** $_BV_F$ of the vector describing the velocity of the ferry relative to the buoy is 4 kph.

2. The **direction** $_BD_F$ in which the ferry appears to the boy to be moving is exactly northeast; we will always measure degrees clockwise from north, so $_BD_F = 45°$.

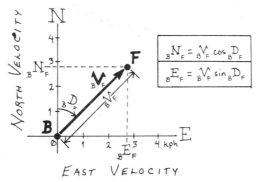

$$_BN_F = {}_BV_F \cos {}_BD_F$$
$$_BE_F = {}_BV_F \sin {}_BD_F$$

Cartesian coordinates *N* and *E*

1. The north component *N* of the relative velocity vector $_B\mathbf{V}_F$ is $_BN_F = 4 \cos 45° = 2.82$ kph.

2. The east component *E* of relative velocity is $_BE_F = 4 \sin 45° = 2.82$ kph.

3. Negative values of *N* correspond to southward velocities.

4. Negative values of *E* correspond to westward velocities.

5. Cartesian components can be converted to polar components by using the equations

$$_BV_F = \sqrt{{}_BN_F{}^2 + {}_BE_F{}^2}$$
$$_BD_F = \tan^{-1}\left({}_BE_F/{}_BN_F\right)$$

or by pressing a button on a modern calculator.

Adding Vectors

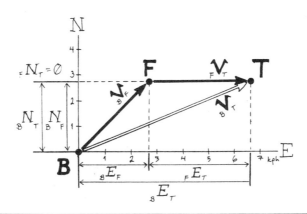

(continued)

Box 2-4. *(continued)*

Vectors are added by adding the Cartesian components separately:

$$_BN_T = {}_BN_F + {}_FN_T$$
$$_BE_T = {}_BE_F + {}_FE_T$$

This pair of equations is usually expressed by the more compact notation

$$_B\mathbf{V}_T = {}_B\mathbf{V}_F + {}_F\mathbf{V}_T$$

Box 2-5. The Velocity Plane.

The relative velocities of Alice, the truck, the ferry, and the buoy can be described compactly by plotting them in two-dimensional velocity space. We'll call this kind of plot the **velocity plane.** Here's how to use it.

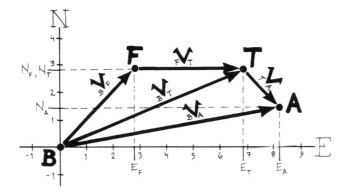

1. Label a set of coordinates to show north (N) and east (E) components of velocity, as in Box 2-4. Assume arbitrarily that the velocity of one of the objects is zero and plot this at the origin. Let's pick the buoy— it seems to be anchored. Plot **B** at the origin.

2. Plot **F** at coordinates (N_F, E_F) to show that F is moving in a northeastward direction from B at 4 kph.

$$N_F = {}_BN_F = 4 \cos 45° = 2.82 \text{ kph}$$

$$E_F = {}_BE_F = 4 \sin 45° = 2.82 \text{ kph}$$

3. Plot **T** 4 velocity units to the east of **F** to show that as viewed from the ferry the truck is moving eastward at 4 kph.

$$N_T = N_F + {}_FN_T = 2.82 + 0.00 = 2.82 \text{ kph}$$

$$E_T = E_F + {}_FE_T = 2.82 + 4.00 = 6.82 \text{ kph}$$

(continued)

Box 2-5.   *(continued)*

4. Plot **A** 2 velocity units southeastward from **T**.

$$N_A = N_T + {}_TN_A = 2.82 - 1.41 = 1.41 \text{ kph}$$

$$E_A = E_T + {}_TE_A = 6.82 + 1.41 = 8.23 \text{ kph}$$

5. To find the velocity ${}_F\mathbf{V}_A$ of Alice relative to the ferry, first read the velocity coordinates of Alice, (N_A, E_A) = (1.41, 8.23); second read the velocity coordinates of the ferry, (N_F, E_F) = (2.82, 2.82). Then

$$_FN_A = N_A - N_F = 1.41 - 2.82 = -1.41 \text{ kph}$$

$$_FE_A = E_A - E_F = 8.23 - 2.82 = 5.41 \text{ kph}$$

$$_F\mathbf{V}_A = \sqrt{{}_FN_A{}^2 + {}_FE_A{}^2} = \sqrt{(-1.41)^2 + 5.41^2} = 5.6 \text{ kph}$$

$$_FD_A = \tan^{-1}({}_FE_A/{}_FN_A) = \tan^{-1}(5.41/-1.41) = 104.6°$$

Note that where adjacent subscripts are the same they cancel, as is true for "F" on the right. Similarly

$$_B\mathbf{V}_A = {}_B\mathbf{V}_T + {}_T\mathbf{V}_A$$

where ${}_T\mathbf{V}_A$ is Alice's velocity relative to the truck T. Alice's velocity relative to the buoy is therefore found as follows.

$$_B\mathbf{V}_A = {}_B\mathbf{V}_F + {}_F\mathbf{V}_T + {}_T\mathbf{V}_A$$

Vector Description	North Velocity	East Velocity
${}_B\mathbf{V}_F$	2.82 kph	2.82 kph
${}_F\mathbf{V}_T$	0.00 kph	4.00 kph
${}_T\mathbf{V}_A$	−1.41 kph	1.41 kph
${}_B\mathbf{V}_A$	1.41 kph	8.23 kph

$$_B\mathbf{V}_A = \sqrt{{}_BN_A{}^2 + {}_BE_A{}^2} = \sqrt{1.41^2 + 8.23^2} = 8.35 \text{ kph}$$

$$_BD_A = \tan^{-1}({}_BE_A/{}_BN_A) = \tan^{-1}(8.23/1.41) = 80.3°$$

The fact that the objects in the preceding example were riding piggy-back on each other is not crucial to the solution. The same analysis applies to many other problems. For example, F, T, and A might be ships on the ocean, each ship using radar to determine the positions of the other two ships relative to it (Figure 2-9). By taking two observations at different times, each ship can determine the velocity of the other ships relative to itself. At 08:15 hours ship F logs the position of ship T at 6 km to the west and at 09:15 hours the

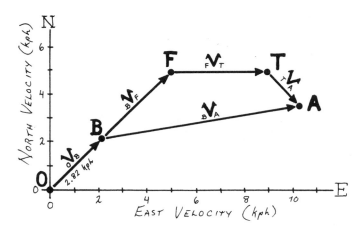

Figure 2-7.
The assumption made in Box 2-5 that the buoy is not moving is now changed to reflect the fact that the buoy is drifting northeastward at a velocity of 2.82 kph. Points **F**, **T**, and **A** are all shifted parallel to **B**, so the directions and lengths of the relative velocity vectors don't change.

position of T is logged at 2 km to the west, so that the velocity $_FV_T = 4$ km and the direction $_FD_T = 90°$. Note that $_FD_T$ does not refer to the position of F, which is to the west, but to the velocity of F, which is eastward. Because the values of D and V are the same as in the previous example, points **F** and **T** plot in the same positions on the velocity plane.

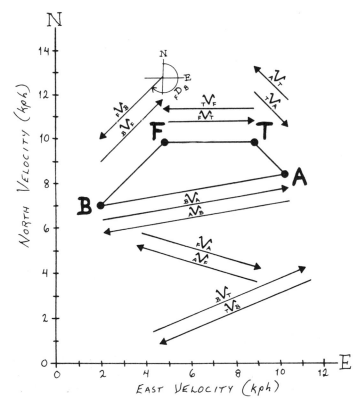

Figure 2-8.
Points **B**, **F**, **T**, and **A** are plotted in velocity space, together with arrows showing the 12 possible velocity vectors. Note that the relative velocity between any two objects, say B and F, depends only on the length of the vector, in this case $_FV_B$, and its direction, $_FD_B$. It is independent of the position of the arrow in velocity space.

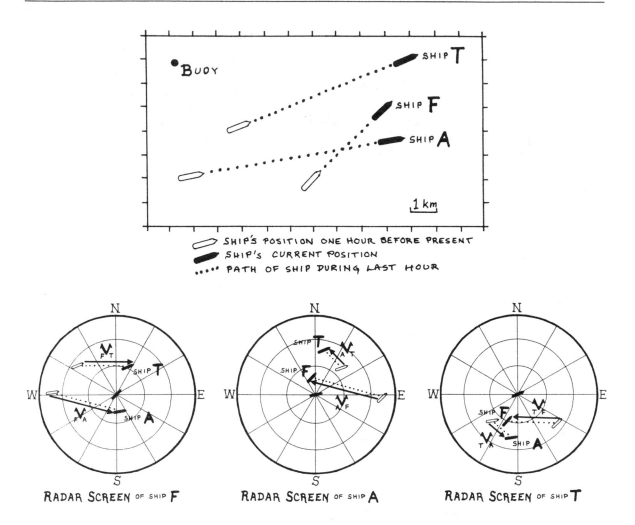

Figure 2-9.
Map shows the change in the positions of three ships during a 1-hour interval. Radar images show the three ships at the beginning (open symbols) and at the end (solid symbols) of the 1-hour interval. Each radar screen can be considered a map in which the ship carrying the radar remains at the center of the map. Relative velocity vectors can be read directly from this map. The scale of the screens is 2 km between concentric circles.

Plates in Velocity Space

Cut a piece of paper into three pieces and label the pieces F, T, and A. Can you devise a set of plate motions that is consistent with the relative velocities shown in Figure 2-7?

First, let's again change to velocity units suitable for describing plates: millimeters per year rather than kilometers per hour. We start by assuming that one of the plates, let's say F, has zero velocity. The next step is to start one of the plates, say T, moving at the velocity specified in Figure 2-7. We can always do this. The problem therefore is solved, except for the most interesting part: determining the nature of the boundary between F and T. Figure 2-10 explores the range of possibilities. The next step (not shown) is to repeat

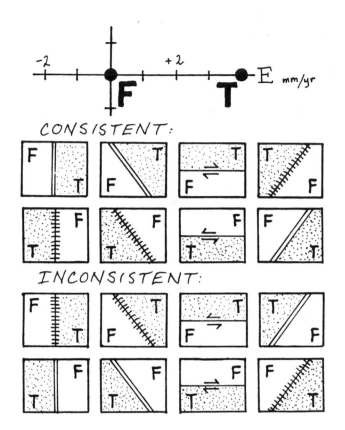

Figure 2-10.
The velocity space diagram shows plate T
moving eastward relative to plate F. All of the
plate boundaries in the upper group are
consistent with this motion. The boundaries
in the lower group are inconsistent with this
motion because plate T must move westward
with respect to plate F to form these
boundaries.

the process for F and A. Some plate motions that are consistent with the relative velocity diagram of ships A, F, and
T are given in Box 2-6. As was true for the velocities along
a line, the nature of the boundary depends upon the initial
geometry of the plates.

Usually we're interested in working the problem the other
way around. From geologic observations we know where
the boundaries are for a set of three plates, and we know
some of their velocities. Using this information, we want to
construct a triangle in velocity space to help us determine
the unknown velocities and to check the internal consistency
of our geologic observations. Velocity space diagrams are
one of the most powerful tools in the kit of the plate tectonicist. This chapter contains a few examples (Box 2-7) and
lots of problems.

The following example, which is based on the first use of
a velocity diagram to solve a real-life geologic problem, shows
the power of analyzing plate motions in velocity space. For
many years, earth scientists puzzled over certain geologic
relationships near the northern end of the San Andreas fault.

Box 2-6. Plate Motions Consistent with Velocity Space Diagrams.

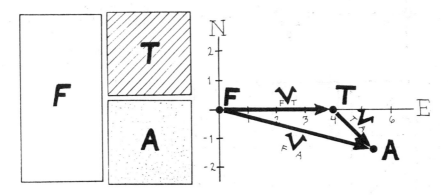

The Problem

1. Cut a piece of paper into three paper plates and label them F, T, and A, as shown.

2. Hold plate F fixed and move the other plates as indicated by the velocity space diagram.

3. Determine whether the plate boundaries are ridges, trenches, or transforms.

4. Repeat the problem after rotating the entire set of plates 90° clockwise.

5. Repeat until the plates are back in the original positions.

Solutions

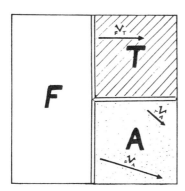 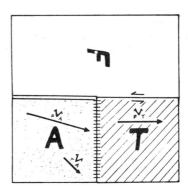

$_F\mathbf{V}_T$ is perpendicular to the FT plate boundary, showing that T is moving away from F; this divergence produces a ridge. Similarly $_F\mathbf{V}_A$ shows that A is diverging away from F to form a ridge and $_T\mathbf{V}_A$ shows that A has a component of motion away from T, forming a third ridge.

$_F\mathbf{V}_T$ is parallel to the FT plate boundary, so this is a transform (left-lateral). $_F\mathbf{V}_A$ has a component of motion away from the FA boundary, so this is a ridge. $_T\mathbf{V}_A$ has a component directed toward the TA boundary, so this is a trench.

(continued)

Box 2-6. *(continued)*

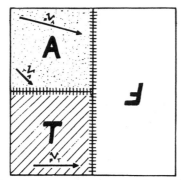 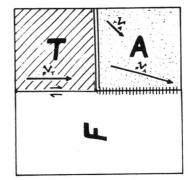

$_F\mathbf{V}_A$ and $_F\mathbf{V}_T$ have components directed toward the boundary of F, so this is a trench. $_T\mathbf{V}_A$ has a component directed toward the TA boundary, so this is a trench.

You should have the idea by now.

Box 2-7. Making and Interpreting a Velocity Triangle.

The Problem

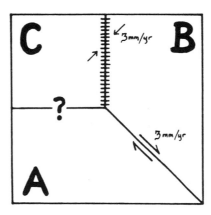

The AB boundary is a transform with right-lateral movement of 3 mm/yr. The BC boundary is a trench along which convergence is occurring obliquely at an angle 45° east of north.

1. Plot the motion of A, B, and C in velocity space.

2. Holding plate A fixed, plot the vectors on plates B and C showing their velocities relative to plate A.

3. Decide whether the boundary between plates A and C is a ridge, trench, or transform.

4. Find $_A\mathbf{V}_C$.

(continued)

Box 2-7. *(continued)*

The Solution

Assume that one of the plates, say A, is not moving and plot this point at the origin.

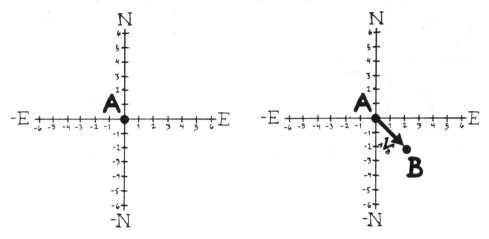

The boundary between plates A and B is a right-lateral transform with a direction $D = 135°$. The velocity $_AV_B$ of plate B relative to plate A is 3 mm/yr, so $_AV_B$ is an arrow 3 units long headed southeast with its tail at the origin **A** and its head at **B.**

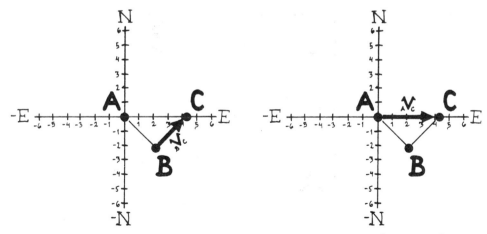

The BC boundary is given as a trench with oblique convergence at a rate of 3 mm/yr, that is $_BV_C = 3$ mm/yr. Plate C is moving in a northeast direction relative to plate B, so $_BD_C = 45°$. $_BV_C$ is therefore an arrow 3 units long pointing northeast with **B** at its tail and **C** at its head, as shown. This completes the plot of the motion of A, B, and C in velocity space. To find the velocity $_AV_C$ of plate C relative to plate A, draw the arrow with its tail at **A** and its head at **C.** We find that $_AD_C = 90°$ and $_AV_C = 4.2$ mm/yr. Plate C is headed eastward at 4.2 mm/yr relative to plate A.

Regarding plate A as stationary, the velocity of plate B is $_AV_B$ and that of plate C is $_AV_C$.

(continued)

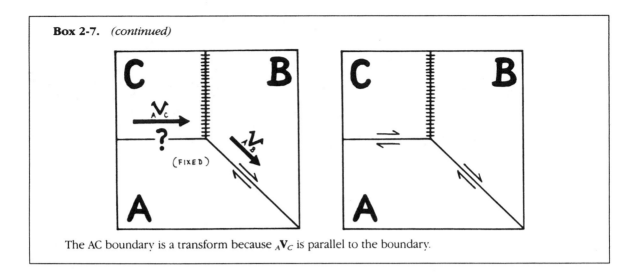

Box 2-7. *(continued)*

The AC boundary is a transform because $_A\mathbf{V}_C$ is parallel to the boundary.

This fault is characterized by violent earthquakes and enormous right-lateral displacements of geologic formations along its entire length up to latitude 40° N. The northward extension of the fault lay offshore and was inaccessible to the geologist, so there wasn't a clear-cut answer to the question of whether the fault continued northwestward from latitude 40° N. Oddly enough, at exactly this same locality the Mendocino fracture zone terminates abruptly against the coast of California. Moreover, many earthquakes occur along this eastern end of the Mendocino fracture zone.

These facts suggested the possibility that at 40° N, the San Andreas fault changed direction from southeast-northwest to east-west (Figure 2-11). However, this would require an enormous deformation of the earth's crust near the kink in the fault. You can appreciate just how enormous if on Figure 2-11 you cut out a piece of Pacific plate along the Mendocino F.Z. and along the San Andreas transform and try to slide this piece northward and westward around the corner without opening up a crack. The piece of paper will end up folded and torn. The earth's crust is not disrupted in this way at the junction of the Mendocino fracture zone and San Andreas fault. In particular, there is not the compression one might expect south of the junction. Geologists knew that there had to be some other explanation, but it eluded them.

The first breakthrough in understanding this intriguing problem came in the mid-1960s as geophysicists began to suspect that seafloor spreading was taking place only a few

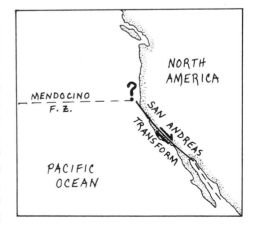

Figure 2-11.
Is the Mendocino fracture zone really a continuation of the San Andreas right-lateral transform fault?

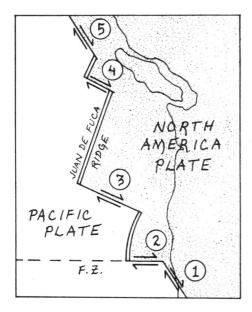

Figure 2-12.
The Juan de Fuca Ridge and some offsetting transforms (2, 3, and 4) link the San Andreas fault (1) with a major transform (5) west of Canada. Note, however, that the transforms are not all parallel.

hundred kilometers west of the coasts of Oregon and Washington along the Juan de Fuca Ridge. At first it seemed that this ridge might simply mark the boundary between the North America plate and the Pacific plate (Figure 2-12). However there are lots of problems with this model. First, the supposed boundary between North America and the Pacific plate incorporates some transforms that are not parallel. We now know that the trend of the Mendocino fracture zone (transform segment 2 in Figure 2-12) is not parallel to segment 3 because the lithosphere between these transforms is, in fact, breaking up to form some complicated microplates. However, even if we ignore transform segment 2, this still leaves transform segments 3 and 4, which are parallel to each other, and segments 1 and 5, which are parallel to each other but not to 3 and 4. As we saw in Chapter 1, nonparallel transforms like this mean either that plate tectonics doesn't work or that we've made a mistake. So for a while it looked as if the discovery of the Juan de Fuca Ridge didn't help much.

The second breakthrough came when geophysicists posed the following question: Is it possible that the small piece of ocean floor between the Juan de Fuca Ridge and the coast of North America is a separate plate? Let's call this hypothetical plate F (Figure 2-13). If the boundary between F and North America (N) is drawn as shown in Figure 2-13, then all of the transform segments except for segment 2 are consistent: Segments 1 and 5 are parallel to the motion between the North America and Pacific plates whereas segments 3 and 4 are parallel to the motion between the Pacific plate and plate F.

Assuming for the moment that F really does exist as a separate plate, is the hypothetical boundary between F and

Figure 2-13.
Introducing a hypothetical plate F eliminates most of the inconsistency in the directions of transforms. The velocity diagram tells us the velocity $_P\mathbf{V}_F$ of plate F relative to plate P.

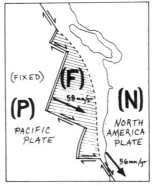

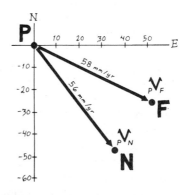

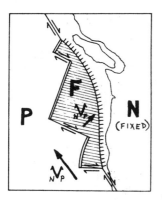

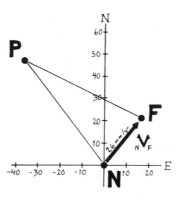

Figure 2-14.
Regarding the North America plate N as fixed, the relative velocity $_N\mathbf{V}_F$ of plate F is toward the northeast. Therefore the NF boundary is a trench.

N a trench, a transform, or a ridge? It was at this point in their analysis of the San Andreas-Mendocino-Juan de Fuca problem that geophysicists were first inspired to analyze plate motions by plotting relative velocity vectors on the velocity plane. Let's try to follow their reasoning step by step. We begin by assuming arbitrarily that the Pacific plate (P) is fixed and we therefore plot **P** at the origin to show that it has zero velocity. The average trend of the San Andreas fault gives us the direction $_P D_N$ of the movement of the North America plate (N) relative to the Pacific plate (P). This is the direction of the relative velocity vector $_P\mathbf{V}_N$. The magnitude of the velocity is known, from direct geodetic measurements of displacements across the fault, to be 56 mm/yr. This is the length $_P V_N$ of the vector. With this information we can plot the relative velocity vector $_P\mathbf{V}_N$ on the velocity plane. We then plot point **N** for "North America" at the head end of the arrow.

The direction of the vector $_P\mathbf{V}_F$ for the motion of F relative to P is determined by transforms 3 and 4. The length of the vector is 58 mm/yr, as determined from the spacing between magnetic isochrons adjacent to the Juan de Fuca Ridge. With this information we can plot the vector $_P\mathbf{V}_F$ and then place point **F** at the head of the arrow. Finding the relative velocity vector $_N\mathbf{V}_F$ for F relative to N is now a simple matter: We draw an arrow with its tail at point **N** and its head at point **F.** This vector has a length of 26 mm/yr and points northeasterly.

We're now in a position to determine the nature of the boundary between N and F. First, for convenience let's change our frame of reference and regard N as fixed, which we do by shifting point **N** to the origin (Figure 2-14). The relative velocity vectors of P and F relative to N can be read easily from Figure 2-14 (or from Figure 2-13, for that matter) and transferred to a map of the plates. When we do this, we see that F is converging toward the boundary with N. Therefore the boundary is a trench.

When this interpretation was first proposed, it was received with considerable skepticism. First of all, there's the serious objection that seafloor bathymetry shows that no trench is present. Second, although a few earthquakes occur along the coasts of Oregon and Washington, there is not a well-defined inclined zone of intense earthquake activity as there is beneath South America and other areas where subduction is taking place. Moreover, preliminary surveys of shallow layers of sediments offshore did not show the folding and disruption to be expected if the oceanic lithosphere were being shoved beneath the continent. The only glimmer of hope for the proponents of plate F was the observation that there is a line of volcanoes including Mounts Rainier, St. Helens, Baker, and Shasta, located above the supposed subduction zone. Moreover, the lavas from these volcanoes are andesite, a rock type that is generally found only in volcanoes above subduction zones.

With this encouragement, the plate tectonicists proposed the following explanation for the embarrassing shortage of earthquakes in Oregon and Washington. Except along subduction zones, earthquakes rarely occur at depths greater than 30 km. This is not surprising. Above 30 km the lithosphere is sufficiently cool and brittle to store up elastic energy and therefore it breaks with a snap when it's deformed beyond its strength. However, the earth's temperature increases with depth at a rate of $30°/km$. At a depth of 30 km the temperature is $900°C$, which is red hot. At this temperature the lithosphere is ductile like tar and deforms smoothly by plastic flow. It doesn't snap. Earthquakes are therefore confined to depths of less than 30 km in the lithosphere with one important exception: fast subduction zones. There the lithosphere descends so fast that part of it stays cool enough and brittle enough to generate earthquakes. Where a slab of lithosphere is being subducted at fast rates of around 100 mm/yr, the interior of the slab remains cool enough to generate earthquakes to a depth of 700 km. However, where subduction is slow (less than 30 mm/yr), as beneath Oregon and Washington, the lithosphere heats up before it has descended very far. Therefore earthquakes are fewer in number and occur at shallower depths. This complicated explanation began as an ingenious way to account for observations which didn't fit a theory. The idea of quiet slow subduction seemed pretty far-fetched at first, as new ideas often do in science, but it fits observations from many parts of the world and has ended up being generally accepted.

The clincher in favor of plate F came when oceanographers decided to have a closer look at the boundary between the North America plate and the hypothetical plate F. Using echo-sounding equipment with better resolution and greater power of penetration than that used in the original surveys, they found that along the boundary the sediments on the seafloor were deformed exactly as one might expect them to be in a zone where the lithosphere is being underthrust. Some of the sediments are being scraped from the lithosphere, compressed, and added to the edge of the continent. An incipient trench along the subduction zone is filled by this accumulation of folded sediments, including sediments shed from the flanks of the continent.

A massive amount of data now substantiates the reality of this subduction zone. In view of this, it is interesting to recall that this major geologic feature was discovered by geophysicists who needed plate F to complete triangles they were drawing in velocity space.

Triple Junctions

Points where three plates meet, which are called **triple junctions,** are especially important tectonically. An example of tectonic action near a triple junction is shown in cartoon form in Figure 2-15, where the triple junction J is the point where the Pacific (P), Juan de Fuca (F), and North America (N) plates meet. If the triple junction J moves up along the coast, point **e** will find itself in a transform environment; if J is stationary or moves southward, **e** will remain in a subduction environment. In this section we will show how to calculate the velocities of triple junctions.

First, however, let's ask ourselves an interesting topological question. What is the maximum number of plates that can meet at a point? If the earth were cut like a pie, a large number of plates could touch where the cuts all intersect; however, plate boundaries on the earth look much more like random slices than pie cuts. Try creating some plates by drawing random lines on a piece of paper. How many plates come into contact where two lines cross? Obviously the answer is four. Your experiment of randomly cutting a plate has created not triple but rather quadruple junctions.

On the real earth, four or more plates almost never come together at a point. Virtually all multiple-plate junctions are

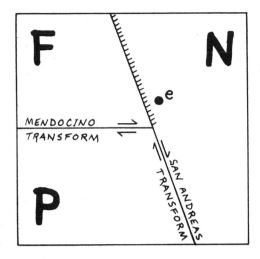

Figure 2-15.
A triple junction marks the juncture of the Pacific plate (P), the Juan de Fuca plate (F), and the North America plate (N). Two transforms and a trench meet at this triple junction. As the triple junction moves northwestward, the tectonic environment at **e** will change from subduction to transform in character.

triple junctions. To gain some insight into why this should be the case, regard one of the random lines you drew as a transform, cut the paper along this line, and slide the two sides past each other. The result will be to change all of the quadruple junctions along the line into triple junctions. You have just shown that although quadruple junctions are easy to conceptualize in a static mode, they are dynamically unstable. How about triple junctions—are they dynamically stable? We will find that some are and some are not.

Triple junctions migrate along plate boundaries. Because they have velocities, much can be learned about them by plotting their motions in velocity space. A useful analogy is the following. Consider marbles rolling at different velocities along a boundary between two plates. In velocity space what would be the locus of these velocities? It turns out that they fall on straight lines (Box 2-8). The relation of these velocity lines to the velocity of the two plates depends upon whether the boundary is a trench, transform, or ridge. Marbles rolling along trenches plot in velocity space on a line passing through the velocity of the upper plate and trending in the same direction as the trench. Marbles rolling along transforms plot in velocity space on a line passing through the velocities of both plates and trending in the same direction as the transform. Marbles rolling along the ridge between two diverging plates plot in velocity space on the perpendicular bisector of the velocity vector between the two plates (Box 2-8). (This assumes that spreading is symmetrical and perpendicular to the ridge.)

The trick in finding the velocity of a triple junction in velocity space is to recognize that the triple junction remains on all three of the boundaries radiating from the junction. In effect it is one marble rolling simultaneously along all three boundaries. Therefore the triple junction lies at the intersection of three velocity lines in velocity space, each line describing the velocity of a marble rolling along one boundary.

Velocity space diagrams also indicate whether a triple junction is stable: if the three velocity lines do not intersect, then the location of the triple junction is not defined in velocity space, and the triple junction is unstable. If you project the plate geometry of an unstable triple junction forward in time, you will find that it evolves into a new plate geometry. In this regard, unstable triple junctions are like quadruple junctions: they may be hypothesized (and may, in fact, exist) at an instant in time, but they are dynamically unstable. An example of an unstable triple junction is shown

Box 2-8. Velocities of Marbles Rolling along Boundaries.

Triple junctions migrate along the boundaries between pairs of plates as if they are marbles rolling parallel to the boundaries. A marble (or triple junction) will remain on a plate boundary if it has a velocity corresponding to any point on the dashed velocity line *ab*.

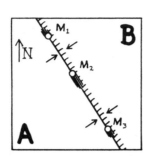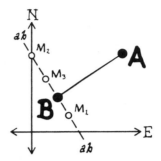

The velocity line *ab* for a trench is parallel to the trench and, because the trench moves with the overthrust plate, it passes through the point in velocity space representing the overthrust plate. This relationship does not require that the direction of convergence be perpendicular to the trench.

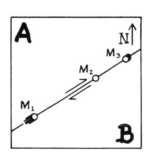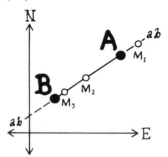

The velocity line *ab* for a transform is parallel to the transform and, because the transform doesn't move with respect to either plate, it lies along the line through both **A** and **B** showing the relative velocity of plates A and B.

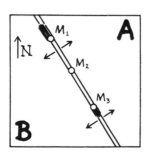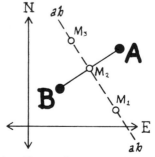

The velocity line *ab* for the ridge is parallel to the ridge. If spreading is symmetrical and perpendicular to the trend of the ridge (as shown in this example), then *ab* is the perpendicular bisector of the line segment $\overline{\textbf{AB}}$ showing the relative velocity of plates A and B.

Figure 2-16.
The triple junction at t_0 (above) is shown by velocity analysis to be unstable because velocity lines *ac* and *ab* do not intersect. At time $t_0 + \Delta t$ (below), the triple junction TJ has evolved from an unstable ridge-transform-ridge geometry to a stable transform-ridge-transform geometry.

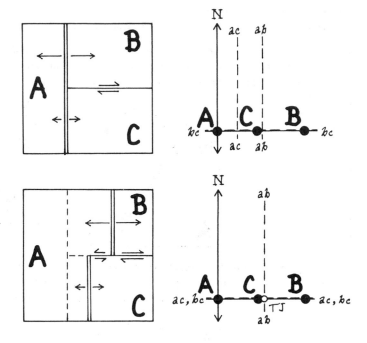

in Figure 2-16. In the initial configuration at time t_0, the triple junction geometry is ridge-transform-ridge. Analysis in velocity space shows that this is unstable. After a short time Δt, a new triple junction will have evolved with the geometry transform-ridge-transform. Analysis (Figure 2-16) and intuition indicate that this junction is dynamically stable. Some more examples of stable and unstable triple junctions are given in Box 2-9.

The growth of the San Andreas fault along the coast of California illustrates how important the role of triple junctions can be. Initially the Farallon plate lay between the Pacific (P) and North America (N) plates, and the San Andreas transform did not exist (time t_1 in Figure 2-17). By 25 Ma (time t_2) the San Andreas had formed, bounded on both its north and south ends by triple junctions. The geometry of the northern junction was transform-transform-trench, that of the southern junction was ridge-transform-trench. Because of this geometry, the northern junction migrated northwest along the edge of the continent and the southern junction migrated to the southeast. The growth of the San Andreas transform was a direct consequence of the migration of these two triple junctions. Eventually (time t_3) the geometry of the

Box 2-9. Migration of Triple Junctions.

A triple junction is a point where the three plates, A, B, and C, meet. It is also the intersection of the boundaries between the three pairs AB, BC, and AC. The velocity of any point moving along one of these boundaries will lie on a line in velocity space (Box 2-8). Three such lines (*ab, bc, ac*) describe the velocities of marbles moving with all possible velocities along the three boundaries intersecting in a triple junction. Since the triple junction is like a single marble rolling simultaneously along the three boundaries, it lies at the intersection of *ab, bc,* and *ac.* If these lines intersect in a single point, the triple junction is stable. This means that as time progresses, ridge, transform, and trench boundaries will remain the same and the angles between them will not change. If *ab, bc,* and *ac* do not intersect at a single point, the triple junction is unstable and will exist only for a moment, after which a different plate geometry will evolve (Figure 2-16). The following examples and analyses should help.

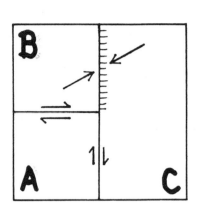 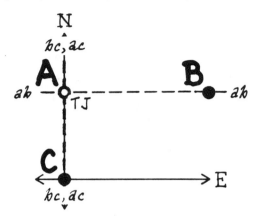

1. Because *ab* and *ac* must both pass through **A,** this **trench-transform-transform** triple junction is **stable** only because *bc,* which must pass through **C,** also passes through **A.** This means that for a trench-transform-transform triple junction, the trench **must have the same trend** as one of the transforms.

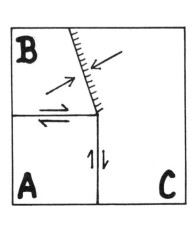 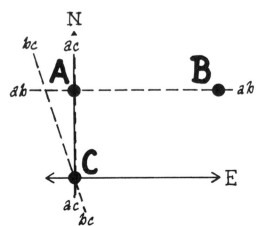

2. If the trench of a **trench-transform-transform** triple junction **does not have the same trend** as one of the transforms, the triple junction is always **unstable.**

(continued)

Box 2-9. *(continued)*

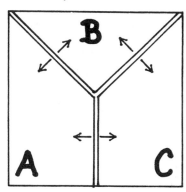

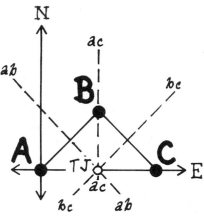

3. Because the perpendicular bisectors of the sides of a triangle always intersect in a single point, **ridge-ridge-ridge** triple junctions are **ideally stable.**

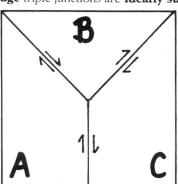

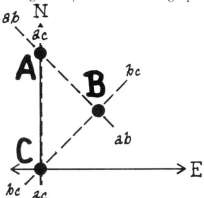

4. Because the sides of a triangle never intersect in a single point, **transform-transform-transform** triple junctions are **always unstable.**

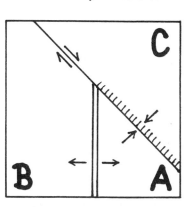

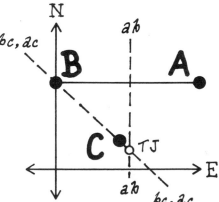

5. Because both *bc* and *ac* must pass through **C,** this **ridge-trench-transform** triple junction is **stable only if** *ab* also passes through **C** or *ac* is equal to *bc* (trench and transform have the same trend), as is shown here.

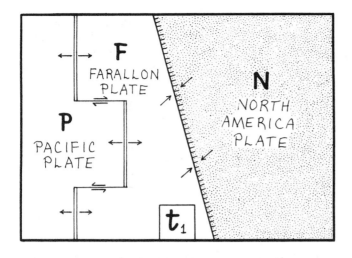

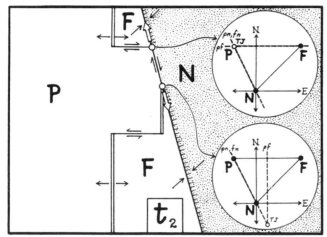

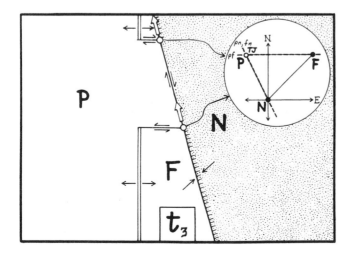

Figure 2-17.
After the Pacific plate (P) has made contact with the North America plate (N), the San Andreas transform begins to grow (t_2). The triple junction (open circle) at the northwest end of the transform migrates to the northwest relative to North America (open arrow) and the southern triple junction migrates to the southeast. When the geometry of the southern triple junction changes (t_3), the southern junction begins to migrate toward the northwest.

southern junction changed to transform-transform-trench and the junction began to move toward the north. Although these cartoons depart somewhat from historical reality, they serve to show how important triple junctions can be in tectonic processes.

Problems

2-1. A piece of paper lithosphere has been cut into three pieces A, B, and C in different ways. In all cases the paper plates have the velocity diagram shown at the top. All of the angles shown are multiples of 45°.

(a) Show whether the plate boundaries are ridges, trenches, or transforms.

(b) Draw a pair of arrows on opposite sides of each boundary showing the direction of relative motion (note that these arrows need not be perpendicular to ridges and trenches).

(c) Show the magnitude of the relative velocity across each boundary in mm/yr.

Hint: If you have trouble visualizing the motion, hold plate B fixed and draw the velocity fields over plates A and C.

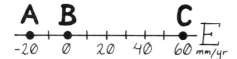

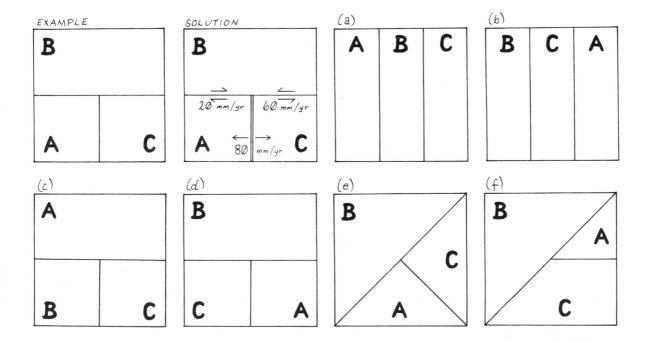

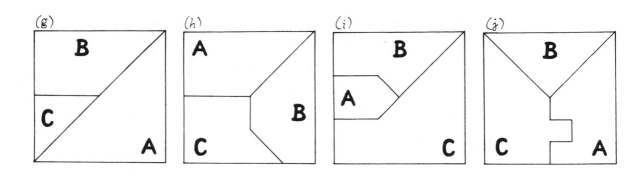

2-2. The plates are now moving as shown below. Proceed as
before.

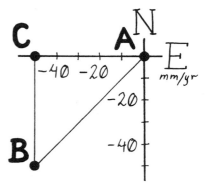

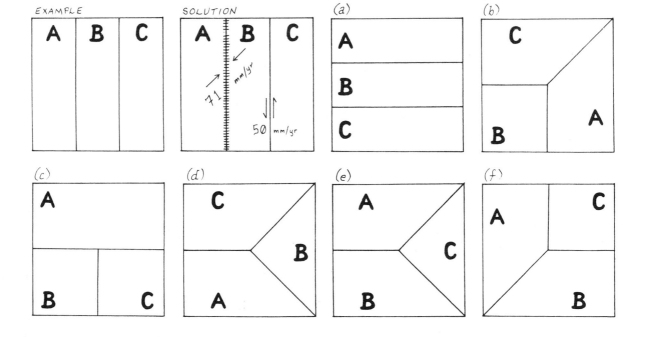

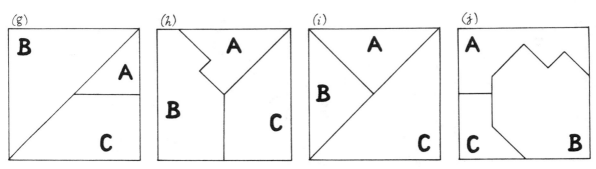

2-3. Make a velocity diagram for each set of plates and determine whether the dashed boundary is a ridge, transform, or trench; then show the relative velocity across the boundary, giving both the direction (small arrows) and the magnitude (in units of mm/yr). In (*g*) and (*h*) double-headed arrows are used to show that the boundary is a transform, but it's up to you to determine whether it's left-lateral or right-lateral.

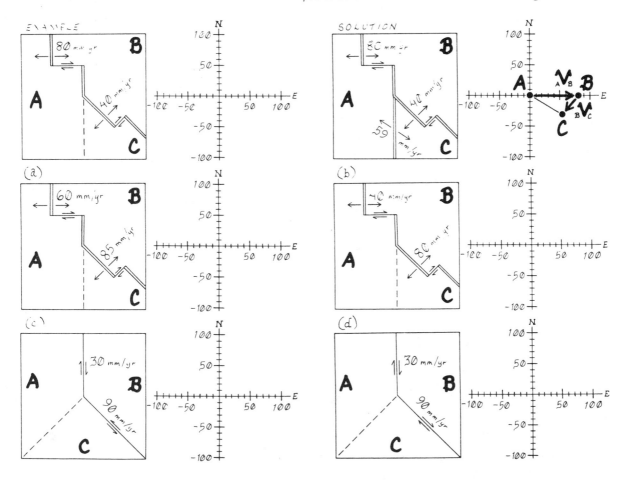

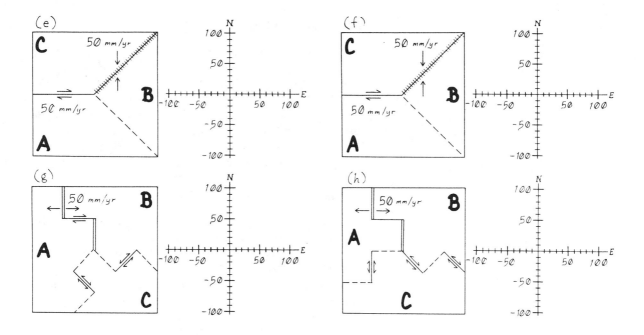

2-4. Devise a model to explain why, in Figure 2-12, transform segment 2 is not parallel to transform 3.

2-5. In Problems 2-1, 2-2, and 2-3, analyze all of the triple junctions in velocity space to determine if the triple junctions are stable. For the stable triple junctions, determine the velocity vector of the triple junction relative to plate C.

Suggested Readings

Plate Tectonics on a Plane

Rises, Trenches, Great Faults, and Crustal Blocks, W. Jason Morgan, Journal of Geophysical Research, v. 73, p. 1959–1982, 1968. *Classic article presenting elements of plate tectonics on a plane.*

Velocity Space

The North Pacific: An Example of Tectonics on a Sphere, Dan McKenzie and Robert Parker, Nature, v. 216, p. 1276–1280, 1967. *Classic paper introducing the use of vector diagrams for plotting plate velocities; first vector analysis of the Mendocino triple junction.*

Triple Junctions

Evolution of Triple Junctions, Dan McKenzie and W. Jason Morgan, Nature, v. 224, p. 125–133, 1969. *Exposition of theory of triple junctions; gives vector diagrams for all possible triple junctions and discusses their stability.*

On the Stability of Triple Junctions and Its Relationship to Episodicity in Spreading, P. Patriat and Vincent Courtillot, Tectonics, v. 3, p. 317–332, 1984. *Extension of previously listed paper; thorough discussion of ridge-ridge-ridge triple junctions.*

Mendocino Triple Junction

Implications of Plate Tectonics for the Cenozoic Evolution of Western North America, Tanya Atwater, Bulletin Geological Society of America, v. 81, p. 3513–3536, 1970. *Classic paper linking plate tectonics and continental geology.*

Geometry of Triple Junctions Related to the San Andreas Transform, William Dickinson and Walter Snyder, Journal of Geophysical Research, v. 84, p. 561–572, 1979. *Classic paper linking California geology to the migration of the Mendocino triple junction.*

Lithospheric Behavior with Triple Junction Migration: An Example Based on the Mendocino Triple Junction, K. P. Furlong, Physics of Earth and Planetary Interiors, v. 36, p. 213–223, 1984. *Develops idea of "no slab window" in the wake of a migrating triple junction.*

Juan de Fuca Plate

Recent Movements of the Juan de Fuca Plate System, Robin Riddihough, Journal of Geophysical Research, v. 89, p. 6980–6994, 1984. *More complete account of evolution of Juan de Fuca plate than given above.*

Propagation As a Mechanism of Reorientation of the Juan de Fuca Ridge, D. Wilson, R. Hey, and C. Nishimura, Journal of Geophysical Research, v. 89, p. 9215–9255, 1984. *Convincing case made for the role of propagating rifts in the recent evolution of the Juan de Fuca plate.*

3

Getting Around on a Sphere

Until now we've treated plates as if they were flat rafts float-ing on a planar asthenosphere. Now we want to mold these two-dimensional objects onto a spherical earth. Before we can get started, however, we need some tools for describing the locations and motions of plates on a globe. The main tool we will use will be a projection, which is a represen-tation on a sheet of paper of the latitude and longitude lines of the earth. Road maps are everyday examples of projections.

On a plane, straight lines were generally used to describe plate boundaries and directions of plate movement. On a sphere, straight lines are not very useful: they may be tan-gent to a sphere but they do not exist on the surface of a sphere. Geometrically the closest analog of a line on a sphere is a circle. It turns out that most of the elements of plate tectonics described by lines on the plane are described by circles on the globe. The situation is analogous to the way in which circles of latitude and circles of longitude on the globe are mapped into two sets of perpendicular lines on a Mercator projection. Plate tectonic geometry on a sphere turns out to be mainly a matter of relationships between circles.

Circles on a Sphere

The simplest way to draw a circle on a sphere is with an ordinary drafting compass. The circle's location is specified

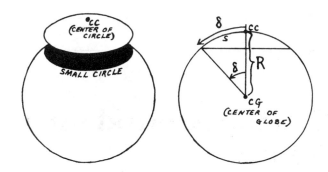

Figure 3-1.
Point **CC** at the center of the circle is called the pole. The size of the circle is given by the angle δ measured at the center of the globe. The great circle distance from **CC** to a point on the circle is, by definition, also equal to δ. The linear distance *s* measured in kilometers over the curved surface of the globe is given by $s = R\,\theta$, where $R = 6{,}380$ km is the radius of the globe and θ is measured in radians (1 radian = 57.3°).

by giving the coordinates of the center of the circle (**CC** in Figure 3-1), which is called the **pole** of the circle. The circle's size is specified by its **angular radius**, defined as follows: draw two radii from the center of the sphere to the surface, one ending at the center of the circle and the other ending at some point on the circle; the angular radius is the angle δ between these two radii (Figure 3-1). This same angle δ is, by definition, also the **angular length** of an arc drawn on the surface of the sphere from the center to some point on the circle. Less commonly, the radius is specified by the linear distance *s* (Figure 3-1) measured over the curving surface of the sphere. The equator, for instance, is a circle centered at the North Pole with an angular radius of δ = 90°. Note that it could be described equally well as a 90° circle centered on the South Pole.

Circles can also be made on a sphere by cutting the sphere with a plane. If the plane passes through the center of the sphere, the intersection of the plane with the surface of the sphere will be a **great circle** (δ = 90°). Otherwise the intersection will be a **small circle** (δ < 90°). Meridians of longitude are simply great circles formed by planes passing through the earth's rotation axis (Figure 3-2). Circles of latitude are small circles formed by planes perpendicular to the earth's rotation axis and therefore parallel to the equator, the only circle of latitude that is a great circle.

Because spheres and planes are topologically different, the relationship between transforms and their Euler poles is different on spheres and planes. On a plane there's no limit to the maximum size of a circle, so there's no limit to the maximum distance between a transform and its Euler pole. In fact, most of the transforms in Chapters 1 and 2 were straight lines, which correspond to Euler poles at infinite distance. On a sphere, however, there's an upper limit of 90° to the size of a circle. It's impossible to have a bigger one.

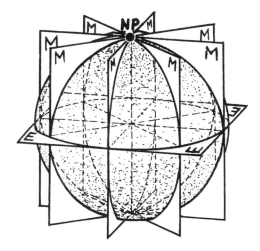

Figure 3-2.
Meridians of longitude are the intersections with the surface of the globe of a set of planes, *M*, passing through the globe's rotation axis. The equatorial plane *E* passes through the center of the globe and is perpendicular to all of the planes *M* and to their line of intersection along the rotation axis.

It might seem that a circle centered on the North Pole with an angular radius of 100° would be bigger than one with a radius of 90°. However, the 100° circle centered on the North Pole could also be described as an 80° circle centered on the South Pole, and the convention in spherical geometry is to specify the size of the circle by the smaller angle.

In one sense, great circles are the spherical equivalent of lines. For example, the shortest path between two points on a globe, as found by stretching a string tight between the two points, lies along a great circle between the points. (This is why ships and airplanes travel along great circle routes wherever possible.) On a sphere, great circles rather than small circles are always used to define and measure angular distance: the **great circle distance** between the end points of a great circle route is the angle between two radii originating at the center of the spheres and terminating at the end points (Figure 3-1).

In plate tectonics, small circles on the globe play the role of lines on the plane in two important relationships. The first is that transforms which would have been straight lines on the plane are small circles on the globe. The second is that the analog of motion in a straight line on a plane is motion along a great circle on a sphere.

Spherical Coordinates

Let's see how we use latitude and longitude to specify a location. San Francisco is located at about 38° North latitude, which means that it is 38° north of the equator on a small circle of radius (90° − 38°) = 52° centered on the North Pole. To simplify calculations, we will regard latitudes north of the equator as positive and latitudes south of the equator as negative. San Francisco is located at about 122° West longitude, which means that it's on a meridian great circle 122° west of the Greenwich meridian, the **global zero reference meridian**. The longitude of San Francisco can be described as 122° West, − 122° East, or 238° East. Note that if you add ±360° to a longitude, you're back where you started, so 10° East, 370° East, and − 350° East are all the same longitude. The following are some of the ways, all equivalent, in which to describe the coordinates of San Francisco: (38° North latitude, 122° West longitude); (38° North latitude, − 122° East longitude); (38° N, − 122° E); (38° N, 238° E); (38°, 238°). The last is a shorthand expression that

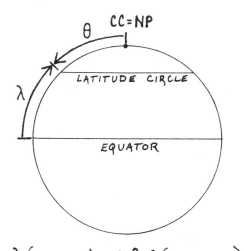

$$\lambda \; (\text{LATITUDE}) = 90° - \theta \; (\text{COLATITUDE})$$

Figure 3-3.
Latitude λ and colatitude θ.

we'll generally use with the understanding that latitude is always listed first and is negative if south of the equator; longitude is measured east from Greenwich. We will list the coordinates of Santiago, Chile as (−33°, 289°) to describe 33° South latitude and 289° East longitude, which is the same as 71° West longitude.

If we think of circles of latitude as small circles that are concentric about the North Pole, then the angular radius of the small circle at a given latitude λ is called the colatitude θ = 90° − λ (Figure 3-3). The range of θ is from 0° at the North Pole (λ = 90°) to 180° at the South Pole (λ = −90°). The colatitude is simply the angle between a point on the globe and the North Pole.

Fixed Reference Frame

An ordinary reference globe standing in the corner of a library consists of two parts. The first part is a sphere on which is printed a set of latitude and longitude circles and the outlines of continents. The second part is a rigid wooden or metal framework that stands on the floor and supports the globe at the poles by two pivot points, the globe being free to rotate within the rigid framework. We'll refer to the rigid framework as the **fixed reference frame** and we will think about it in two ways: as a physical three-dimensional object like the fixed framework around the geographic globe (Figure 3-4) and as a fixed set of points and curves on a

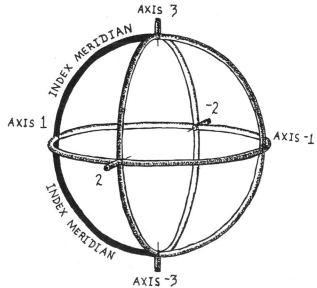

Figure 3-4.
Fixed reference frame. Globe within the frame can rotate about either the ± **2** axis or the ± **3** axis, as shown by pins.

piece of paper (Figure 3-5). The fixed reference frame consists of three perpendicular great circles drawn on a sphere. The circles intersect at six points. Imagine six vectors originating at the center of the sphere and ending at the six points of intersection of the circles. These compose a set of mutually perpendicular axes or unit vectors that we'll refer to as ±**1**, ±**2**, and ±**3**. We will refer to the great circle that contains vectors **3** and **1** by the symbol <**3, 1**>. The three axes and the three reference circles <**1, 2**>, <**2, 3**>, and <**3, 1**> compose the principal elements of the fixed reference frame shown in Figure 3-4.

The type of projection we will use is called a Lambert equal area projection, in which points on the globe are represented by points on or within a circle drawn in the plane of the projection. The <**3, 1**> reference circle is represented as the outer circle of the projection. The other two reference circles, <**1, 2**> and <**2, 3**>, are represented on the equal area projection by perpendicular straight lines through the center of the circle. The three axes of the fixed reference frame, which are points on the globe, also appear on the projection as points. In addition to these principal elements, the fixed reference frame also has secondary elements in the form of a set of great circles and a set of small circles. As meridians of the fixed reference frame, the great circles intersect at the ±**3** axis; in Figure 3-5, they are represented on the projection as meridians converging on ±**3**. The small circles are latitude circles of the fixed reference frame centered on ±**3**; in Figure 3-5, they are coincident with latitude lines on the underlying globe. An important property of projections is that points on the globe are represented by points on the projection whereas circles on the globe are represented by lines or other continuous curves on the projection.

This type of projection with grid lines like the longitude and latitude grid of a globe viewed from a distant point lying in the plane of the equator is called an **equatorial projection**. The half-circle to the left of the projection through axes **3, 1,** and −**3** is especially important because it will serve as a zero reference analogous to the Greenwich meridian. We will refer to this half-circle as the **index meridian**.

The reason for introducing the idea of a fixed reference frame is to provide a coordinate system that will remain fixed when we rotate the globe. To visualize the relationship between the fixed and rotating spherical coordinate system, take the globe with its outlines of continents and grid of

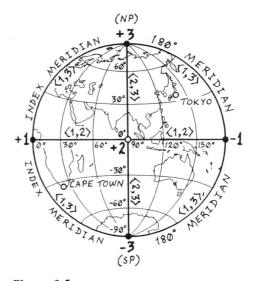

Figure 3-5.
Globe representing the earth placed inside fixed reference frame, shown in an equal area projection. With the globe in the orientation shown, the elements of the fixed reference frame are coincident with the latitude and longitude lines of the globe.

latitude and longitude lines and place it inside our fixed reference frame. Align the globe's geographic North Pole with the **3** axis and the South Pole with −**3**. Note in Figure 3-4 that the axes are fitted with pins stuck into the globe to provide pivot points at +**3** and −**3**. These pivot points permit the globe to rotate around axis **3**. Rotate the globe until the Greenwich meridian (longitude $\phi = 0$) is aligned with the index meridian. Note that the equator of the globe is aligned with the horizontal line on the projection representing the great circle <**1, 2**>. The circle <**3, 2**> on the projection represents the meridian of longitude 90° East on the globe. The longitude meridians on our globe compose a set of great circles all passing through the ±**3** axis; therefore they are coincident with the meridians of the fixed reference frame. The circles of latitude on the globe coincide with the complementary set of grid lines of the fixed reference frame, both being small circles concentric about a common center at ±**3**.

As we look down on our globe we can see points on the side nearest to us and can easily plot these on a projection. How can we represent points on the far side? Let's imagine that the globe is transparent, so that points on both sides can be seen. We will use a commonly accepted convention of representing points on the *near* hemisphere with *open* symbols and points on the *far* hemisphere with *solid* black symbols. If a solid point and an open point are superimposed on the projection, are they **antipodal**, that is, are they located 180° apart on the globe? No, they are not except when they happen to be plotted at the center of the projection. Point **A** (Figure 3-6) on the far hemisphere at (−20°, 350°) is superimposed on point (−20°, 10°) whereas point **A′**, which is antipodal to **A**, is at (20°, 170°). Note that the point antipodal to (λ, ϕ) has coordinates (−λ, $\phi + 180°$), whereas the point coincident on the projection has global coordinates (λ, −ϕ).

On one hemisphere we can plot only half of a great circle. Where does the other half plot on the other hemisphere? We note that the meridian of longitude at 60° East may be generated by the great circle <**3, −3**> passing through **B** = (0°, 60°) on the equator on the near hemisphere as shown in Figure 3-6. The same great circle passes through **B′** = (0°, 240°) on the far hemisphere. Note that the two halves of this great circle do not coincide but are symmetrical on either side of the great circle <**2, 3**>. It's important to remember this when drawing a great circle between two points on different hemispheres.

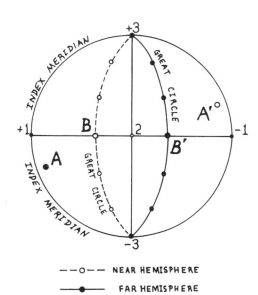

--−o−− NEAR HEMISPHERE

───•─── FAR HEMISPHERE

Figure 3-6.
Points and circles are plotted on both hemispheres of the globe. Points **A′** and **A** are antipodal (that is, 180° apart), as are points **B** and **B′**.

Rotation about Axis 3

Now let's mentally rotate our globe about the **3** axis in a direction corresponding to the actual rotation of the earth. The cities of Roma and Calcutta on the near hemisphere will be moving from left to right, corresponding to motion toward the east, whereas Santiago on the far hemisphere will appear to be moving from right to left, again corresponding to easterly motion. If the globe is rotating at a rate of one revolution per day, the two cities will increase their longitudes relative to the index meridian by 30 degrees every 2 hours, as shown in Figure 3-7.

If your globe didn't already have latitude circles, it would be easy to draw them with a pencil held against the globe in a fixed position as the globe rotates around axis **3**. Latitude circles map onto the equal area projection as the set of curves beginning at the index meridian and running to the right (Figure 3-5). It's important in using a projection not to confuse these small circles with the set of great circles, all of which run from **−3** to **+3**. Graphically the process of rotating a point on the globe such as the point shown in Figure 3-7 at (0°, 60°) by a known angle, say 30°, about the **3** axis couldn't be simpler.

1. Place a piece of tracing paper over the projection. Points plotted on the tracing paper represent points on the rotating globe whereas grid lines and axes printed on the underlying piece of paper are regarded as elements of the fixed reference frame, which does not rotate.

2. Plot the point (0°, 60°) on the tracing paper.

3. Keeping the paper fixed, replot the point on the same small circle grid line 30° to the right of the original point. Use the set of great circles to mark off angular displacements along the small circles. (In Figure 3-7, the great circle grid lines are drawn at intervals of 30°.)

4. Be careful to change symbols when changing hemispheres.

There are two possible directions of rotation about the **+3** axis, positive and negative. If the near hemisphere moves to the right, the rotation is positive about the **+3** axis. If the near side moves to the left, the rotation is negative. In defining the sign of rotations we will use the following right-hand rule. Point your thumb along the rotation axis in the direction of the North Pole (here **+3**) and wrap your fingers around the globe. Your fingers are pointing in the direction

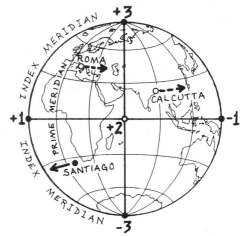

8:00 GREENWICH MEAN TIME

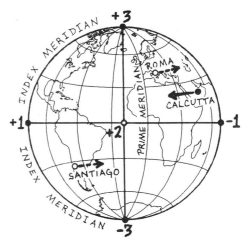

13:20 GREENWICH MEAN TIME

Figure 3-7.
At 8:00 GMT, Roma and Calcutta on the near hemisphere rotate to the right and Santiago on the far hemisphere rotates to the left. At 13:20 GMT, Santiago and Calcutta have both switched hemispheres and therefore their apparent directions of motion have also switched.

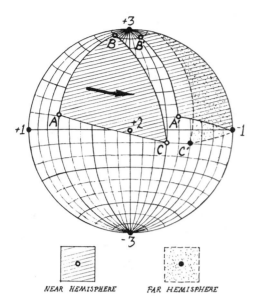

NEAR HEMISPHERE   FAR HEMISPHERE

Figure 3-8.
Triangular plate with vertices at **A**, **B**, and **C**, when rotated 100° about the **3** axis, ends up in the position shown with vertices **A′**, **B′**, and **C′**. Points **A** and **C** move much farther than point **B** because they are farther from the center of rotation, axis **3**.

of positive rotation. If you look down on your thumb, positive rotation is counterclockwise. The earth's actual rotation is positive about the North Pole.

Since we do lots of rotating in plate tectonics, we'll use a shorthand notation. ROT [**3**, 60°] will be used to refer to a rotation of 60° about the **3** axis. This produces exactly the same effect as a rotation of −60° about the −**3** axis or, symbolically, ROT [−**3**, −60°]. The two successive rotations ROT [**3**, 60°] followed by ROT [**3**, −60°] bring the globe back to its starting position, as do ROT [**3**, 60°] followed by ROT [−**3**, 60°]. What rotation is needed to bring the globe back to its starting position after ROT [−**3**, 350°]? Name four equivalent rotations if you can.[1]

The rotation of a triangular plate with vertices **A**, **B**, and **C** by 100° about axis **3** is shown in Figure 3-8. Note that each point starts and ends on the same small circle. Note also that all three rotate through 100° along these small circles. In terms of coordinates this means that each of their longitudes increases by 100°, the longitudes of **A** and **B** increasing from 30° to 130° and the longitude of **C** increasing from 120° on the near hemisphere to 220° on the far hemisphere. Note also, however, that the great circle distance that the vertices move depends on how far away the vertices are from axis **3**, the axis of rotation. The great circle distance is equal to $\Delta\phi\sin\theta$, where $\Delta\phi$ is the change in longitude and θ is the colatitude. **B** is 10° away and moves a great circle distance of only 15°, whereas **A** is 80° away and moves a great circle distance of 98°.

When we rotate Roma by 30 degrees, do its coordinates change? In one sense they do not—after all, Roma has the same longitude at noon as it does at 2:00 P.M. Yet the point representing Roma appears at a different place on our projection, so in another sense the coordinates have changed. Therefore it's useful to distinguish between two coordinate systems. One of them is a set of axes plotted on the globe and rotating with it. This rotating **globe coordinate system** is plotted on tracing paper. The other is attached to the fixed reference frame and doesn't move. The axes and grid lines of our projection are a representation of the fixed reference frame and, being printed on the underlying page, never move. These we will refer to as the **projection coordinate system**. Whenever there's a chance of confusing the two sets of coordinates we will use subscripts to distinguish between them: $(\lambda, \phi)_{PROJ}$ or $(\lambda, \phi)_{GLOBE}$. When the Greenwich meridian of the globe is aligned with the index

---

[1] ROT[**3**, 350°], ROT[**3**, −10°], ROT[−**3**, 10°], ROT[−**3**, −350°].

meridian of the projection, the two coordinates are the same. Starting from this position with London at longitude 0°, if the globe is rotated 96° counterclockwise about the **3** axis, the longitude of London will become 96° in the projection system. Of course it's still 0° in the globe system. In other words, the operation ROT[**3**, 96°] is simply the same as adding 96° to the longitude in projection coordinates. If a point has coordinates $(\lambda, \phi)_{PROJ}$ before the rotation it will have coordinates $(\lambda, \phi + 96°)_{PROJ}$ after the rotation.

It is useful to remember that the index meridian may be set at any longitude, thereby eliminating any initial rotation about the **3** axis.

Rotation about Axis 2

A limitation of rotating about axis **3** is that a given point always stays at the same latitude. In plate tectonics we often need to rotate points to different latitudes. To accomplish this, we will introduce an additional degree of freedom to permit rotating the globe about axis **2**. Luckily our fixed reference frame (Figure 3-4) is fitted with pins along the **2** axis, so all we need to do conceptually is insert the ±**2** pins into the globe, pull out the pins along the ±**3** axis, and rotate. This can be done graphically using the projection in the following way (Box 3-1):

1. Stick a thumbtack through the center of a piece of heavy paper or cardboard measuring roughly 8.5 × 11 inches and cover the head of the tack with tape to hold it there.

2. Copy the page with the equal area projection (Figure 3-20) from the book and stick the tack through the exact center of the projection at axis **2**. Then fasten the page to the sheet of heavy paper using tape around the edges.

3. Put a small piece of transparent tape at the center of a piece of 8.5 × 11 inch transparent drafting or tracing paper.

4. Impale the center of the tracing paper on the thumbtack so that the paper can rotate around the tack.

5. Draw a circle on the tracing paper over the outer circle of the projection.

6. Mark on the tracing paper the location of three new axes, ±**x**, ±**y**, and ±**z** attached to the globe and moving with it. The **z** axis is the North Pole (NP). The **x** and **y** axes lie

Box 3-1. How to Rotate Around Axis **2**.

Point **A** has coordinates (30°, 30°) and point **B** has coordinates (− 60°, 300°). What are their latitude and longitude in projection coordinates after ROT [**2**, 60°], that is, a rotation of 60° about axis **2**?

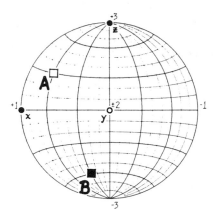

● Plot **A** and **B**

With the tracing paper in the initial position as shown, the PROJ and GLOBE coordinates coincide so that the point **z**, which is the North Pole, plots at ± **3**. Note that **B** is on the far hemisphere.

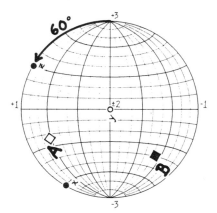

● ROT[**2**, 60°]

Rotate the tracing paper 60°. Since the rotation is positive, the right-hand rule tells us that the plastic globe and our tracing paper rotate counterclockwise. If you look down on a rotation vector from outside the sphere, positive rotation is always counterclockwise. Note that the tracing paper would rotate clockwise in the case of ROT [+ **2**, − 60°] or ROT [− **2**, 60°], which are the same.

● Read **A** = (− 24°, 28°)$_{PROJ}$ and **B** = (− 41°, 215°)$_{PROJ}$

Note that the latitude λ and longitude ϕ of **A** and **B** have changed in the projection but not in the globe coordinates.

in the equatorial plane with **x** through the Greenwich meridian and **y** through the meridian 90°. Initially these axes are coincident with the ±**1**, ±**2**, and ±**3** axes as shown below.

Projection: **1**   **2**   **3** −**1** −**2** −**3**

Globe: **x**   **y**   **z** −**x** −**y** −**z**

After a net rotation of 90° about axis **3**, the projection and globe axes are aligned as follows.

Projection: **1**   **2**   **3** −**1** −**2** −**3**

Globe: −**y**   **x**   **z**   **y** −**x** −**z**

A rotation of −90° about **3** returns **x** and **y** to their original positions.

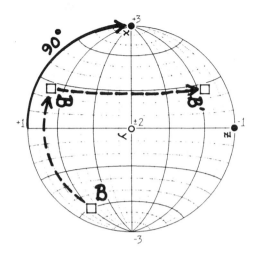

7. Rotate the tracing paper about the thumbtack, remembering that a positive rotation is counterclockwise. You have just rotated about axis **2**. Think of the underlying projection as your fixed reference frame and the overlying sheet of tracing paper as your globe. As the sheet of tracing paper rotates about the thumbtack, try to visualize the globe rotating within the fixed reference frame about the **2** axis.

8. Note that points do not need to be replotted on the tracing paper during rotation about the **2** axis because the tracing paper physically rotates relative to the underlying projection.

*Thou shalt not rotate about axis **1***. At first glance it might seem that this rotation could be accomplished by moving points on the projection along the meridian circles that go from +**3** to −**3**. However, this won't work because the meridian circles are not small circles centered on the **1** axis, as would be required for rotation about **1**. To help us avoid this temptation, we deliberately refrained from installing pivot pins at the +**1** axes on our fixed reference frame (Figure 3-4).

However, not being able to rotate about axis **1** isn't a serious limitation because we can achieve the same effect by successive rotations about axes **3** and **2**, as shown in Figure 3-9.

Distance Between Two Points

There are lots of different ways to get from San Francisco to Tokyo but there's only one shortest route, which is the great

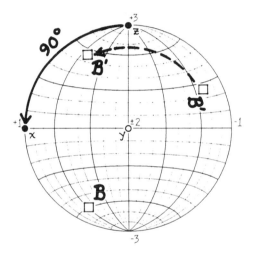

Figure 3-9.
How to rotate about axis **1** in two steps. Suppose you want to rotate point **B** = (−60°, 30°) an angle 140° about axis **x**, which initially coincides with axis **1**. The key is to rotate **x** up to axis **3**, where you already know how to rotate. After you've finished rotating about **x** in this position, you can rotate **x** back to its original position at **1**.

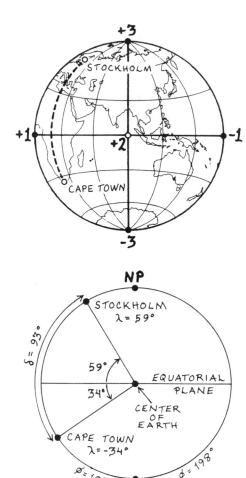

Figure 3-10.
Angular distance δ between two points on the globe, Stockholm and Cape Town. Note that δ is also the angle subtended at the center of the earth by radial lines to the two points.

circle route. As noted earlier, when we speak of the distance between two points on the sphere we mean the distance along a great circle route. To find the distance between two points, therefore, we begin by constructing the great circle which passes through both points.

Let's start with a simple case (Figure 3-10). The coordinates of Cape Town are $(-34°, 18°)$ and those of Stockholm are $(59°, 18°)$. How far apart are they? Since their longitudes are the same, the great circle linking them is the meridian of longitude at 18°. Imagine a triangle with its apex at the center of the earth and with sides ending in Stockholm and Cape Town. This triangle lies in the plane which cuts the surface of the earth along the great circle connecting Stockholm and Cape Town. The angle at the apex of the triangle is simply the difference in latitudes between the two cities: $δ = 59° - (-34°) = 93°$. This is the distance between the two cities in angular measure. Recalling that one degree equals 111 km measured over the surface of the earth, the airline distance for a direct flight between the two cities is $93 × 111 = 10,323$ km. For any two points **A** and **B** located on the same meridian of longitude at latitudes $λ_A$ and $λ_B$, the angular distance between them is simply $|λ_A - λ_B|$, absolute value because we will always want to regard distances as positive. Box 3-2 shows how to find the great circle distance of two points which do not have the same longitude.

From paleomagnetic research we're able to find where the geographic pole was in the past. For purposes of investigating ancient climates we would often like to be able to find the corresponding location of the ancient equator. The inverse problem arises when oceanographers determine from an analysis of sediments on the seafloor the location of the equator in the past. From this datum we would like to be able to calculate the ancient position of the geographical pole. Being able to convert from equatorial planes or other great circles to poles and back is essential in plate tectonics. Boxes 3-3 and 3-4 show how to do this. Box 3-5 shows how to use poles to find the angle between two great circles.

The location of the ancient equator over an oceanic plate is usually marked by the accumulation of an unusually large thickness of fossiliferous sediments, reflecting greater biological activity along the equator. Contour lines showing the thickness of the sediments deposited during some geologic period usually show a definite linear trend, as shown by the heavy line through point **B** in Figure 3-11. This line segment is actually a short segment of the ancient equatorial great circle. Note that the line segment through **B** does not have

Box 3-2. How to Find the Angular Distance Between Two Points on a Sphere.

What is the great circle distance between points **A** and **B**? The key to solving this problem is to rotate the tracing paper until both points lie on the same great circle. If points **A** and **B** are both on the same hemisphere they will lie on the same meridian half-circle from −**3** to **3**, as shown on the left. However, if points **A** and **B** are on opposite hemispheres, they will lie on two complementary half-circle meridians of longitude which together compose a complete great circle that encircles the globe, as shown on the right. Examples of such complementary meridian half-circles are 10° east and 190° east, 60° east and 240° east, φ° east and φ + 180° east. Note that the two half-circles are always symmetrical on either side of the vertical line through the center of the projection corresponding to the great circle <**2, 3**>.

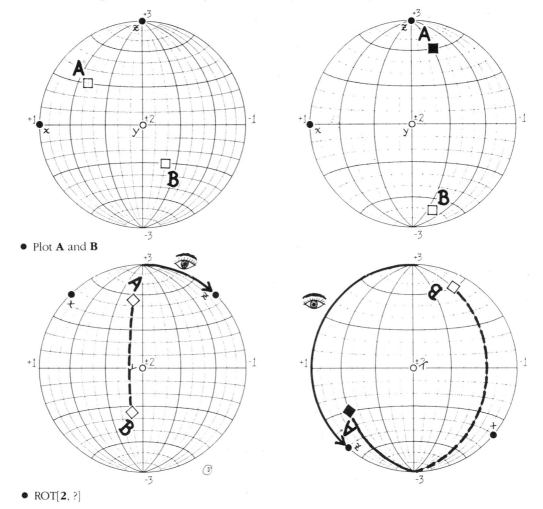

● Plot **A** and **B**

● ROT[**2**, ?]

Rotate until **A** and **B** are along the same great circle. On the left a rotation of − 46° is required and on the right a rotation of 140° is required.

(continued)

Box 3-2 *(continued)*

- Read δ

The distance δ between **A** and **B** along the great circle can be read from the projection. On the left, δ = 90° and on the right, δ = 150°

Box 3-3. How to Plot a Great Circle If You Know the Pole.

The pole **P** of a great circle has coordinates (29°, 48°). The great circle (or equator) corresponding to this pole is plotted in the following steps.

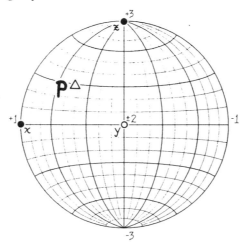

- Plot **P** and **z**

z plots at **3**.

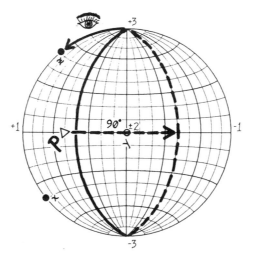

(continued)

Box 3-3 *(continued)*

- ROT[**2**, ?]

Rotate the tracing paper until **P** is on the equator of the projection, which is the horizontal line through the center of the plot. For this problem a rotation of 39.6° counterclockwise is required. The coordinates of **P** are now $(0°, 40.5°)_{PROJ}$.

- Draw great circle

Draw the great circle on the near hemisphere from $-\mathbf{3}$ to $\mathbf{3}$ along the meridian $\phi = 130.5°$, passing through a point on the equator of the projection 90° to the right of **P**. The great circle continues on the far hemisphere (solid line) along the meridian $\phi = 310.5°$ of the projection.

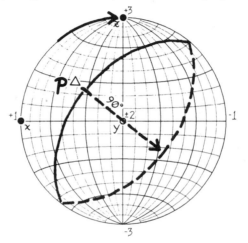

- ROT[**2**, $-$?]

Rotate the North Pole **z** back to **3**. The underlying projection can now be used to read the coordinates of points on the globe because the GLOBE and PROJ coordinates now coincide.

Box 3-4. How to Find a Pole If You Know Two Points on a Great Circle.

Usually, only one great circle can be drawn between two points **A** and **B** on a sphere. This great circle, <**A**, **B**>, is the intersection of the surface of the sphere and a plane passing through **A**, **B**, and the center of the sphere. The line perpendicular to this plane intersects the surface of the sphere at the two poles **P** and **P**′, which are 180° apart and thus antipodal. If the coordinates of **P** are (λ_P, ϕ_P), then the coordinates of **P**′ are $(-\lambda_P, \phi_P + 180°)$. In the special case that **A** and **B** are antipodal, an infinite number of different great circles can be drawn through them, just as an infinite number of planes can be passed through a line.

    The figures on the left show the case of both points on the same projection hemisphere; those on the right show the case of oblique hemispheres. *(continued)*

Box 3-4 *(continued)*

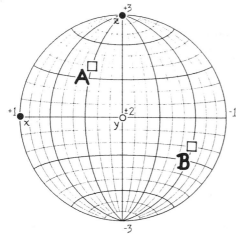 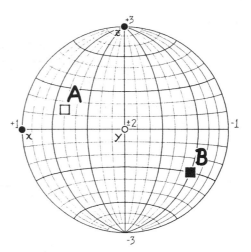

- Plot **A** and **B** and **z**

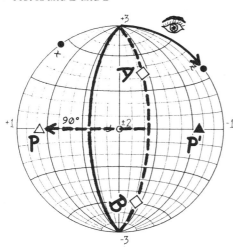 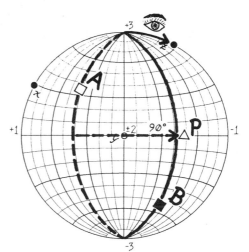

- ROT[**2**, ?]

Rotate the tracing paper until **A** and **B** lie on the same great circle.

- Read ϕ_{PROJ} of **A** and/or **B**

This is the longitude of the great circle <**A**, **B**> in PROJ coordinates.

- Plot **P** = $(0°, \phi + 90°)_{PROJ}$ and **P'** = $(0°, \phi - 90°)_{PROJ}$

The pole **P** and **P'** are plotted on the equator of the projection 90° from the intersection of the great circle <**A**, **B**> with the equator.

(continued)

Box 3-4 *(continued)*

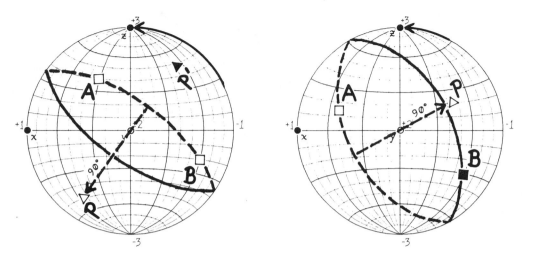

- ROT[**2**, −?]

Rotate **z** back to **3** so that the coordinates of the poles can be read from the grid lines of the projection: **P** = $(\lambda_P, \phi_P)_{GLOBE}$ and **P′** = $(\lambda_{P'}, \phi_{P'})_{GLOBE}$

Box 3-5. How to Find the Angle Between Two Great Circles.

Just as an angle on a plane is defined at the intersection of two lines, an angle on a globe is defined at the intersection of two great circles. Let's say that two great circles intersect at a point **B**. For purposes of identification let's plot two points, **A** and **C**, on the two great circles each at a distance of 90° from **B**. We can now identify the two circles as <**A**, **B**> and <**C**, **B**>. The three points **A**, **B**, and **C**, form a spherical angle with **B** at the apex.

First let's consider the simple case in which the PROJ and GLOBE coordinates coincide and **B** is located at one of the geographic poles on the globe, say −**z** = −**3**. Then **A** and **C** lie on the equator and <**A**, **B**> and <**C**, **B**> are meridians of longitude. The angle W between the great circles is the angular distance between **A** and **C**. This is simply equal to the difference between their longitudes W = $|\phi_A - \phi_B|$.

If the intersection **B** is not initially at −**3**, the secret of finding the angle W quickly is to rotate **B** to −**3** so that **A** and **C** lie on the equator of the projection and the great circles <**A**, **B**> and <**C**, **B**> coincide with meridian great circles on the projection. Here's a simple way to do this. (Note that W is also the angle between the poles of the two great circles.)

(continued)

Box 3-5 *(continued)*

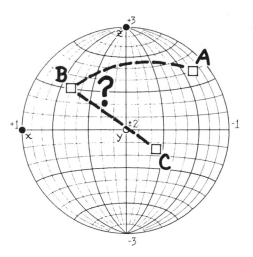

● Plot the great circles and the points **A**, **B**, **C**

Let **B** = $(\lambda_B, \phi_B) = (30°, 40°)_{GLOBE, PROJ}$ be the intersection of the great circles. Let **A** and **C** be points on the two great circles that are 90° away from **B**.

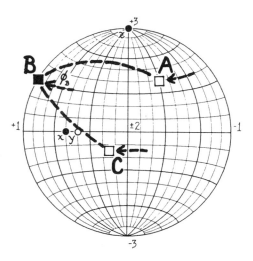

● ROT[**3**, −ϕ_B]

This puts **B** on the index meridian, which is now "$\phi = 40°$". Note that ±**x** and ±**y** have rotated 40° from their former positions at ±**1** and ±**2**.

(continued)

Box 3-5   *(continued)*

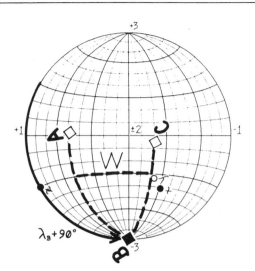

- ROT[**2**, $\lambda_B + 90°$]

This rotates **B** to $-$**3**, and rotates points **A** and **C** to the equator of the projection.

- W = $|\phi_A - \phi_B|$

This is the desired angle between the great circles. The absolute value sign insures that W is always positive. Note that W is measured along the great circle perpendicular to the intersection between the great circles, which is the standard way to measure angles on a sphere. If $W > 180°$, replace W by $360°$ $- W$ so that W will be in the range $0°$ to $180°$ for convenience.

A and **C** do not have to be exactly $90°$ away from **B**, but the closer they are to $90°$ from **B**, the more accurate you will be in finding the angle W.

an east-west azimuth like the present equator but rather makes an angle D with north. This tells us right away that the ancient geographic pole was different from the present one, as does the observation that the ancient equator is north of the present equator. The angle D is called the **local azimuth** of the great circle representing the ancient equator and is measured in degrees clockwise from north. There are many different ways to describe the same azimuth, say one of $120°$: $D = 120°$, North $120°$ East (which means $120°$ east of north), N $120°$ E, South $60°$ East (which means $60°$ east of south), S $60°$ E, etc. Note also that the great circle extends in both directions from **B**, so that the same great circle through **B** could be described by the local azimuth $D = 300°$, $D = -60°$, or N $60°$ W. At a different point **B'** along the same ancient equatorial great circle, the local azimuth D' will be

Figure 3-11.
Contour lines showing the thickness (in tens of meters) of late Cretaceous sediments on the Pacific plate. The Cretaceous equator (heavy line) is along the band of greatest sedimentary thickness. From the trend (or azimuth D) of the equator determined at **B**, it's possible to find the coordinates of the geographic pole in the Cretaceous period (Box 3-8).

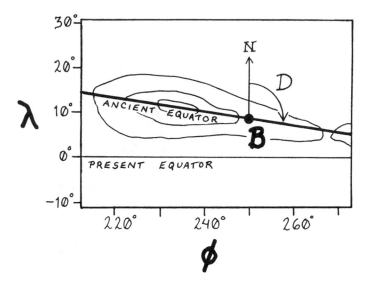

different from D. In fact, as you travel along any great circle route, the local azimuth, which describes the direction you're heading, will change with your position along the route. If you know the local azimuth at any point along a great circle, you can find the pole corresponding to that great circle by the method shown in Box 3-6. Conversely, if you know the pole of a great circle you can find the local azimuth using Box 3-7.

Another problem of considerable interest in plate tectonics is analogous to the following simple problem in navigation. If you fly 1100 km from San Francisco along a great circle route, where will you end up? This depends, of course, on the local azimuth D of the great circle at San Francisco. A geophysical equivalent of this problem is the following. From seismograms recorded at Stanford we know that seismic waves arrived from a direction 30° west of south and that the epicenter was 2500 km away. What are the coordinates of the epicenter? This boils down to the problem of a great circle which starts at point **B**, where its azimuth $D = 210°$. What are the coordinates of a point located 2500 km away on this great circle? Box 3-8 shows how to find the epicenter.

Cartesian Coordinates

Until now we have used the coordinates (λ, ϕ) to describe the latitude and longitude of a point on the globe. If the globe is of unit radius, we can regard the point as being the

Box 3-6. How to Plot a Great Circle and Its Pole If You Know the Local Azimuth of the Great Circle at a Point on the Circle.

Suppose you're flying around the world on a great circle route. When you arrive at point **B** with coordinates $(\lambda_B, \phi_B) = (45°, 120°)$ you notice that you are heading in a direction 70° east of north, so your local azimuth $D = 70°$. Using this information you'd like to plot your entire great circle route on a projection, together with the poles for this great circle. In this operation we begin with **B** at -3, plot **z** and the pole **P** of the great circle in relation to **B**, and rotate the GLOBE and PROJ coordinates together.

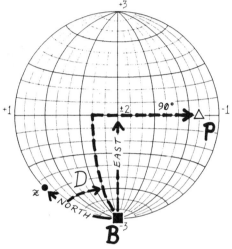

- Plot **B** at -3, **z** at $(-\lambda_B, 0°)_{PROJ}$, and **P** at $(0°, D \pm 90°)_{PROJ}$

With **B** at -3 and **z** on the index meridian, the great circle we're after is the meridian $\phi_{PROJ} = D$. The pole of this great circle is then $\mathbf{P} - (0°, D \pm 90°)_{PROJ}$.

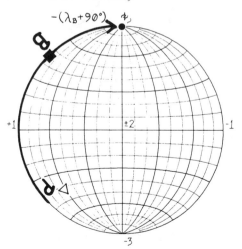

- ROT$[2, -(\lambda_B + 90°)]$

Rotate **B** to $(\lambda_B, 0°)_{PROJ}$

(continued)

Box 3-6.   *(continued)*

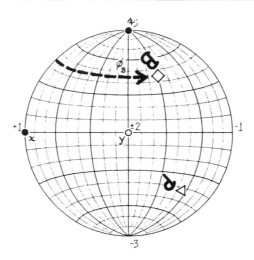

• ROT[**3**, ϕ_B]

This brings the GLOBE and PROJ coordinate systems back into coincidence so we can read the coordinates of **P** = $(-41.6°, 147.2°)_{PROJ, GLOBE}$.

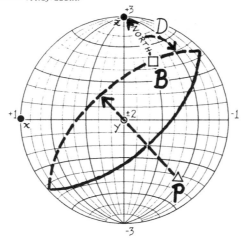

• Plot great circle

See Box 3-3.

If only the pole coordinates are wanted, all you need to do is plot **P** = $(0°, D \pm 90°)$, ROT[**2**, $-(\lambda_B +$ 90°)], and ROT[**3**, ϕ_B]. You can now read the coordinates of **P** from the projection.

Box 3-7.   How to Find the Local Azimuth at a Point on a Great Circle If You Know the Coordinates of the Pole of the Circle.

Suppose you're flying on a great circle route and you know your present position **B** and the coordinates of the pole **P** of the great circle. How can you find what direction you should be heading, that is the local

(continued)

Box 3-7. *(continued)*

azimuth *D* of the great circle? This problem is very similar to the preceding one, and the key, as before, is to rotate **B** to −**3**.

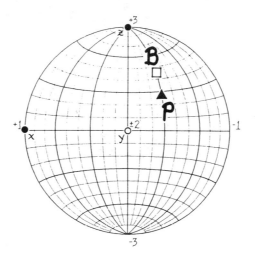

- Plot **B** and **P**

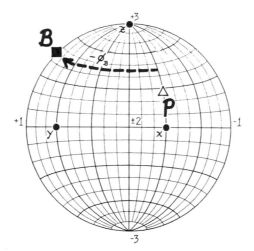

- ROT[**3**, −ϕ_B]

This brings **B** to the index meridian.

(continued)

Box 3-7.   *(continued)*

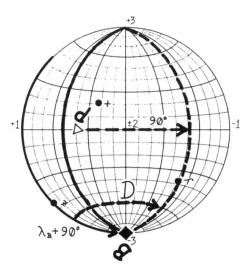

- ROT[**2**, $\lambda_B + 90°$]

The tracing paper is rotated counterclockwise 135°, which rotates **B** to $-\mathbf{3}$ and **P** to $(\lambda_P, \phi_P)_{PROJ}$, which in this example is $(0°, 50°)_{PROJ}$.

- Plot both of the meridians $\phi = \phi_P \pm 90°$

This is the great circle corresponding to pole **P**.

- Read $D = \phi_P \pm 90°$

This is your heading or local azimuth. The two values differ by 180° and correspond to motion forward or backward along the great circle.

Box 3-8.   Locating a Point a Known Distance Away Along a Great Circle of Known Local Azimuth.

A great circle starts at point **B**, where its local azimuth is D degrees east of north. What are the coordinates of point **A** located a distance d degrees away (or $111 \times d$ km away) along this great circle? The key to a quick solution is to plot **B** at $-\mathbf{3}$, which puts the index meridian at $\phi = \phi_B$. We know that the half-circle meridian in the projection along $\phi = D$ is the locus of all points lying along a great circle with local azimuth $D = 150°$. This tells us the longitude of **A** in projection coordinates: $\phi_A = D$. The latitude of **A** in projection coordinates is simply $d - 90°$. To find the GLOBE coordinates of **A**, all we need do is to plot the point $\mathbf{A} = (d - 90°, D)_{PROJ}$ and do the rotations necessary to bring the PROJ and GLOBE coordinates back together.

(continued)

Box 3-8. *(continued)*

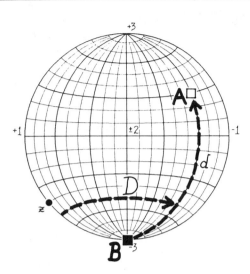

- Plot **z**, **B**, and **A**

Plot **B** at -3 and plot **z** at $(0°, -\lambda_B)_{PROJ}$ as in Box 3-6. Plot $\mathbf{A} = (d\ -90°, D)_{PROJ}$.

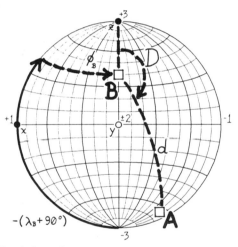

- ROT[**2**, $-\lambda_B -90°$], ROT[**3**, ϕ_B], read **A**

This brings PROJ and GLOBE coordinates together. If d has the maximum value of 180°, **A** will be located at $\mathbf{B}' = (-\lambda_B, \phi_B - 180°)$.

head of an arrow of unit length, the tail of which is at the center of the globe. Given λ and ϕ, the orientation of this unit vector is completely determined. An alternative way to describe the orientation of this unit vector is to give its com-

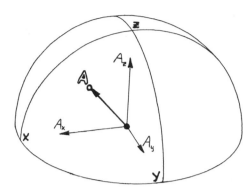

Figure 3-12.
Cartesian components of vector **A**.

ponents in three mutually perpendicular or orthogonal directions using a standard Cartesian coordinate system.

The standard convention for axes is as follows (Figure 3-12):

x: In plane of Equator toward $\phi = 0°$

y: In plane of Equator toward $\phi = 90°$

z: Aligned with rotation axis toward North Pole

Resolving a unit vector $\mathbf{A} = (\lambda, \phi)$ first into polar and equatorial components, we have

$$A_{polar} = \sin \lambda \tag{3.1}$$

$$A_{equatorial} = \cos \lambda \tag{3.2}$$

In Cartesian components we have

$$A_x = \cos \lambda \cos \phi \tag{3.3}$$

$$A_y = \cos \lambda \sin \phi \tag{3.4}$$

$$A_z = \sin \lambda \tag{3.5}$$

When converting back to spherical we use

$$\lambda = \sin^{-1} A_z \tag{3.6}$$

$$\phi = \tan^{-1} (A_y/A_x) \tag{3.7}$$

Since **A** is a unit vector, $A_x^2 + A_y^2 + A_z^2 = 1$, only two of the three Cartesian components are independent. A similar set of equations can be written for the projection coordinate system, substituting the **1**, **2**, and **3** axes for **x**, **y**, and **z**. People who would rather use algebra than projections to solve problems in spherical trigonometry generally work with Cartesian coordinates for reasons that will be apparent below.

Several of the operations of interest in plate tectonics can be done using the scalar or vector products of two unit vectors. Consider unit vectors **A** and **B** radiating from a common origin and subtending an angle δ. As points on the unit sphere, **A** and **B** are separated by the angular distance δ. The **scalar** or **dot product** is by definition scalar quantity

$$\mathbf{A} \cdot \mathbf{B} = \cos \delta = A_x B_x + A_y B_y + A_z B_z \tag{3.8}$$

so the angular distance between **A** and **B** is simply

$$\delta = \cos^{-1}(\mathbf{A} \cdot \mathbf{B}) \qquad (3.9)$$
$$= \cos^{-1}(A_x B_x + A_y B_y + A_z B_z)$$

The **cross product** $\mathbf{C} = \mathbf{A} \times \mathbf{B}$ is defined as a vector perpendicular to the plane containing **A** and **B**. Its sense of direction is given by a right-hand rule, as shown in Figure 3-13. If **A** and **B** are perpendicular to the axis of a right-handed screw and the screw is rotated in the direction going from **A** to **B**, then the screw will advance in the direction of $\mathbf{A} \times \mathbf{B}$. (Another way of looking at the direction is to sweep the palm of your right hand the shortest angle from **A** to **B**. Your thumb will then point in the direction of $\mathbf{A} \times \mathbf{B}$.) The scalar length C of the cross product vector **C** is given by

$$C = \sin \delta \qquad (3.10)$$

Since the distance between two points on a globe is always equal to or less than 180°, C is always positive. The three Cartesian components of **C** are given by

$$C_x = A_y B_z - A_z B_y \qquad (3.11)$$
$$C_y = A_z B_x - A_x B_z \qquad (3.12)$$
$$C_z = A_x B_y - A_y B_x \qquad (3.13)$$

The cross product **C** is not a unit vector unless $\delta = 90°$, but it can be converted to a unit vector by dividing each of the components by their scalar length $C = \sin \delta$. Since the cross product **C** is by definition perpendicular to the plane containing the unit vectors **A** and **B**, the unit vector $\mathbf{C}/C$ is the pole of the great circle through points **A** and **B** (compare with Box 3-4).

Consider now the problem of finding the angle between two great circles intersecting at point **B** (Box 3-5). If $\mathbf{D} = \mathbf{B} \times \mathbf{A}$, then the pole $\mathbf{P}_1$ of the circle through **B** and **A** is $\mathbf{P}_1 = \mathbf{D}/D$. Similarly if $\mathbf{E} = \mathbf{B} \times \mathbf{C}$, then the pole $\mathbf{P}_2$ of the circle through **B** and **C** is $\mathbf{P}_2 = \mathbf{E}/E$. The desired angle Ω between the two circles is equal to the angle between the two poles:

$$\Omega = \cos^{-1}(\mathbf{P}_1 \cdot \mathbf{P}_2) \qquad (3.14)$$

with $0° \leqslant \Omega \leqslant 90°$.

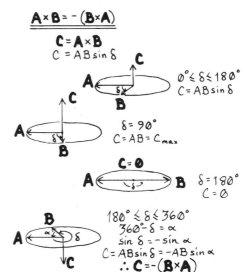

Figure 3-13.
Vector cross products following a right-hand rule.

Box 3-9. Performing Stereonet Operations on a Programmable Pocket Calculator.

Most of the operations you have learned to perform on an azimuthal projection can also be performed on most pocket calculators. The polar-to-rectangular (P-R) and rectangular-to-polar (R-P) conversion functions are the key to efficient use of the calculator. These functions simply convert between two different two-dimensional coordinate systems, one (rectangular coordinates) using displacements, x and y, along two perpendicular axes, and one (polar coordinates) using the distance r from the origin in a direction described by the angle θ from one axis. These correspond to the Cartesian and spherical coordinate systems in three dimensions.

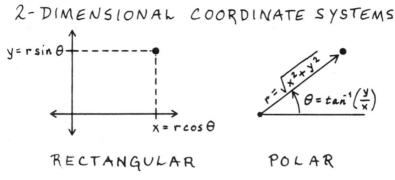

$$\text{2-DIMENSIONAL COORDINATE SYSTEMS}$$

RECTANGULAR — $y = r\sin\theta$, $x = r\cos\theta$

POLAR — $r = \sqrt{x^2 + y^2}$, $\theta = \tan^{-1}\left(\frac{y}{x}\right)$

A rotation about the origin in polar coordinates is simply the addition of the rotation angle to the direction angle.

If we convert to Cartesian (x, y, z) coordinates, we can perform "ROT 2" in polar coordinates in the (z, x) plane perpendicular to the y-axis and "ROT 3" in polar coordinates in the (x, y) plane perpendicular to the z-axis.

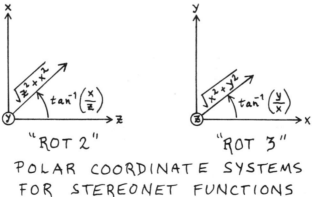

"ROT 2" — $\sqrt{z^2 + x^2}$, $\tan^{-1}\left(\frac{x}{z}\right)$

"ROT 3" — $\sqrt{x^2 + y^2}$, $\tan^{-1}\left(\frac{y}{x}\right)$

POLAR COORDINATE SYSTEMS
FOR STEREONET FUNCTIONS

In the following examples, the language is specific to the Hewlett-Packard HP-41c, but the algorithms can be used on almost any pocket calculator, and, with some translation, the stack manipulation (ENTER ↑ , X<>Y (exchange the contents of the x and y registers), R ↑ (roll up), R ↓ (roll down), etc.) can transfer to any HP calculator. The HP-41c was chosen because of the Hewlett-Packard stack and its alphanumeric labeling capability.

The first step is to write spherical-to-Cartesian (S-C) and Cartesian-to-spherical (C-S) conversions using polar-to-rectangular (P-R) and rectangular-to-polar (R-P) conversions:

(continued)

Box 3-9. *(continued)*

Function	Stack x	y	z	t
LBL "S-C"	r	ϕ	λ	t
X<>Y	ϕ	r	λ	t
R↓	r	λ	t	ϕ
P-R	a	z	t	ϕ
R↑	ϕ	a	z	t
X<>Y	a	ϕ	z	t
P-R	x	y	z	t
RETURN	x	y	z	t

Function	Stack x	y	z	t
LBL "C-S"	x	y	z	t
R-P	a	ϕ	z	t
X<>Y	ϕ	a	z	t
R↓	a	z	t	ϕ
R-P	r	λ	t	ϕ
R↑	ϕ	r	λ	t
X<>Y	r	ϕ	λ	t
RETURN	r	ϕ	λ	t

With these conversions in hand, rotations by the angle Ω about the **2** and **3** axes are simply:

Function	Stack x	y	z	t
LBL "ROT 2"	Ω	r	ϕ	λ
R↓	r	ϕ	λ	Ω
XEQ "S-C"	x	y	z	Ω
X<>Y	y	x	z	Ω
X<> Z	z	x	y	Ω
R-P	a	θ	y	Ω
R↑	Ω	a	θ	y
ST + Z	Ω	a	θ'	y
R↓	a	θ'	y	Ω
P-R	z'	x'	y	Ω
X<> Z	y	x'	z'	Ω
X<>Y	x'	y	z'	Ω
XEQ "C-S"	r	ϕ'	λ'	Ω
RETURN	r	ϕ'	λ'	Ω

Function	Stack x	y	z	t
LBL "ROT 3"	Ω	r	ϕ	λ
ST + Z	Ω	r	ϕ'	λ
R↓	r	ϕ'	λ	Ω
RETURN	r	ϕ'	λ	Ω

Note: "ROT 3" is rarely used—it is faster and takes less memory if you simply add Ω to ϕ.

When operating on the unit sphere, as we do in this chapter, use $r = 1$.

To translate to other HP calculators substitute

X<>Y, R↑, +, LAST X, R↓, X<>Y for R↑, ST + Z, R↓; R↑, R↑, X<>Y, R↓ for X<>Y, X<> Z; and R↑, X<>Y, R↓, R↓ for X<> Z, X<>Y in "ROT 2"; and X<>Y, R↓, +, R↑, LAST X, R↓ for ST + Z, R↓ in "ROT 3".

Constructing Projections

The equal area projections we have been using are but one of many types of projections, each with its advantages. These include the **stereographic projection**, on which small circles on the globe appear as circles on the projection; the **Mercator projection**, on which declination can be measured directly from the map; the **gnomonic projection**, on which all great circles are straight lines; and the **orthographic projection**, which looks as the earth would from outer space. (An orthographic projection was used for Figure 5-2.) In this section we will concentrate on equal area, stereographic, and Mercator projections, these being the projections used most commonly in plate tectonics.

All projections are constructed by a conversion of the spherical coordinates of points on the globe to Cartesian coordinates on a piece of paper. The different projections are the result of different conversions, as we shall see in a moment. In all projections, however, a meridian of longitude can be approximated by converting, plotting and connecting the points $(0°, \phi)$, $(1°, \phi)$, $(2°, \phi)$, etc. Similarly a circle of latitude can be approximated by $(\lambda, 0°)$, $(\lambda, 1°)$, $(\lambda, 2°)$, etc. This is the way in which the projections used in this book were originally generated using a computer. If you have a desktop computer you can also calculate and display similar projections using the equations given in this chapter in a short program.

In the rest of the chapter we will learn the spherical-to-Cartesian conversions for the equal area, stereographic, and Mercator projections. You'll find this especially interesting if you want to program your personal computer to make projections; however, you do not need to know how to construct projections to use them and to appreciate their qualities. So at this point you may want to put on a scientist's rather than an engineer's hat and skip ahead to the problems at the end of the chapter.

Azimuthal Projections

The two most widely used projections in geology are the equal area and stereographic projections. They are called azimuthal projections and can be constructed quite simply using the following analysis (and a digital computer if you have one). Place a flat screen that will become the plane of

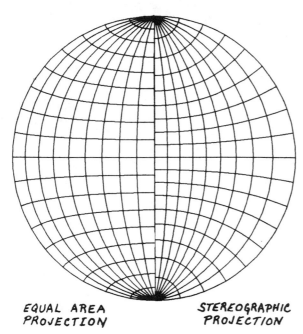

Figure 3-14.
Equal area and stereographic projections.

EQUAL AREA
PROJECTION

STEREOGRAPHIC
PROJECTION

the projection into contact with a transparent globe. Draw a line perpendicular to the flat screen through the point of contact and passing through the center of the globe. Now place a light source at some specified point along the line and use it to project points and lines on the globe onto the screen. This projected image will be an azimuthal map of the surface of the globe. If the point of contact was at the North Pole or South Pole, then all of the meridians of longitude converge on the point of contact. The planes corresponding to the meridians all pass through the axis perpendicular to the point of contact and therefore through the hypothetical light source. The result is a polar projection in which meridians of longitude are straight lines radiating from the center of the projection. Moreover, the spherical angles between the meridian great circles are equal to the plane angles between the corresponding radiating lines.

Equal area and stereographic projections (Figure 3-14) have some properties that are the same and some that are different. They are both convenient to use with a piece of tracing paper pivoted at the center of the projection because a great circle remains a great circle when it is rotated about the origin. It is always possible to graphically find a great circle through any two points in order to find their angular separation. Other projections such as the Mercator do not have this property. On the equal area projection, small circles generally map into ovals on the projection, whereas on the stereographic projection all circles map into circles, although the center of a circle on the globe does not necessarily map into the center of the corresponding circle on the projection. On the equal area projection small circles of given radius located at different latitudes on the globe map into ovals of constant area on the projection, whereas on stereographic projections equal circles on the globe are mapped into circles with varying areas, the circles being smaller near the center of the projection. Equal area projections are commonly used in plate tectonics because data points that are uniformly distributed over the globe appear to be uniformly distributed over the projection.

Let's describe points on the plane of the projection using polar coordinates r (radial distance from center) and D (angle from a reference direction). Then points on the plane lying on a circle of constant radius r correspond to points on the globe lying on a small circle of radius δ centered on the point of contact. To complete the construction of the projection, we need to know how to find r, which is related to the angular distance δ by the function

$$r = c f(\delta) \tag{3.15}$$

where c is a constant that determines the radius r_{max} of the projection

$$c = r_{max}/f(\delta_{max}) \tag{3.16}$$

and $f(\delta)$ is one of the following functions:

$$f(\delta) = \sin(\delta/2) \quad \textit{Lambert equal area} \tag{3.17}$$

$$f(\delta) = \tan(\delta/2) \quad \textit{Wulf stereographic} \tag{3.18}$$

Converting polar coordinates r and D to Cartesian coordinates h (horizontal) and v (vertical), the coordinates of the point on the projection are simply

$$h = r \sin D \tag{3.19}$$

and

$$v = r \cos D \tag{3.20}$$

Polar Projections

Until now we've always kept our fixed reference frame in the same orientation with the **3** axis at the top of the projection and the **2** axis at the center. Sometimes it is more convenient to use a **polar projection** with the -3 axis at the center of the projection and the $+1$ axis at the top. In geographic coordinates this would correspond to the view of the earth with the South Pole at the center of the projection and meridians of longitude radiating out as straight lines from this center (Figure 3-15). If we align the polar projection and globe coordinates so that the -3 and $-\mathbf{z}$ (south polar) axes are at the center of the projection, the Greenwich meridian and index meridian appear as a radial line from the center to the **1** axis at the top of the projection. The direction of increasing east longitude ϕ, as defined for the globe by applying the right-hand rule about the north polar axis and as defined for the fixed reference frame by applying the right-hand rule to the **3** axis, is clockwise for both the globe and projection coordinates. If the globe is given a positive rotation about the -3 axis, a point on the globe moves counterclockwise along a small circle centered on the center of the projection. Note that in Figure 3-15 the axes of the fixed reference frame and the map are both shown on the near hemisphere.

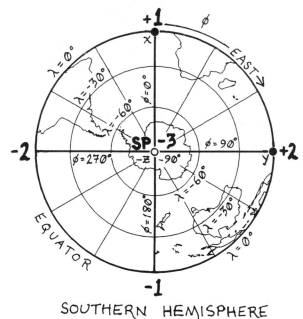

SOUTHERN HEMISPHERE

Figure 3-15.
Polar projection of southern hemisphere.

Figure 3-16.
Polar projection of northern hemisphere.

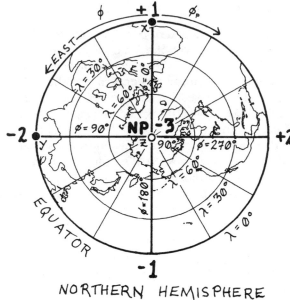

NORTHERN HEMISPHERE

To view the northern hemisphere we could use a projection on the far hemisphere but this would tend to confuse us because it would present a view of the continents as seen from inside the earth—a mirror image of most maps. However if the globe in Figure 3-15 is given a rotation of 180° about the **1** axis of the fixed reference frame, the result is that shown in Figure 3-16: the north pole of the globe now coincides with the −**3** axis of the fixed reference frame. Since the fixed reference frame has not changed, longitude ϕ_P in projection coordinates still increases in the clockwise direction whereas in the globe coordinate system, longitude ϕ_G now increases in the counterclockwise direction.

Polar projections are not as useful as the equatorial projections introduced earlier because the family of great circles of the polar projection are all straight lines that pass through the center of the projection. None of these lines passes through two points at different longitudes whereas using the equatorial projection, it is always possible to find a great circle that passes through any two points simply by rotating the tracing paper representing the globe over the underlying grid representing the fixed reference frame. If the grid lines of a polar and an equatorial projection are superimposed, the great circles of the latter are also great circles on the polar projection and can be used to solve problems in the polar projections.

Constructing Polar Projections

We will first construct a polar projection in which the point of contact and center of the projection is the $-z$ axis or South Pole. We want to plot point **A** (Figure 3-17), which lies on a great circle that makes an angle D with the Greenwich meridian $\phi = 0$, so

$$D = \phi \qquad (3.21)$$

is the longitude of point **A**. The distance δ from the center of the hemisphere at $-z$ to **A** is

$$\delta = 90° + \lambda \qquad (3.22)$$

where λ is the latitude of **A**.

As an example, let's find the coordinates of **A** $= (-40°, 40°)$ on a polar equal area projection that maps a hemisphere $(\delta_{max} \leq 90°)$ onto a circle of radius $r_{max} = 10$ cm. We first solve for the scale constant c (from 3.16):

$$c = 10/\sin 45° = 14.14 \text{ cm}$$

The length of r is then (from 3.15):

$$r = 14.14 \sin 25° = 5.98 \text{ cm}$$

The coordinates of the point on the projection are therefore:

$$b = 5.98 \sin 40° = 3.84 \text{ cm}$$

and

$$v = 5.98 \cos 40° = 4.58 \text{ cm}$$

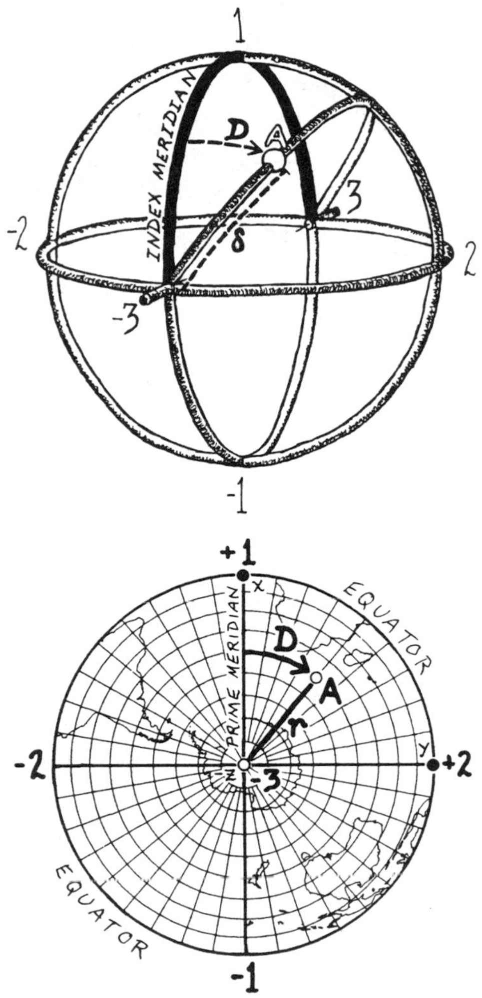

Figure 3-17.
Fixed reference frame in polar orientation.

Constructing Equatorial Projections

We now consider how to construct an equatorial equal area projection with the **y** axis $(0°, 90°)$ at the center of the projection (Figure 3-18). On the sphere we need to find the angles D and δ as before, which can be done using the graphical techniques described earlier in this chapter or algebraically as follows. For point **A** with coordinates (λ, ϕ)

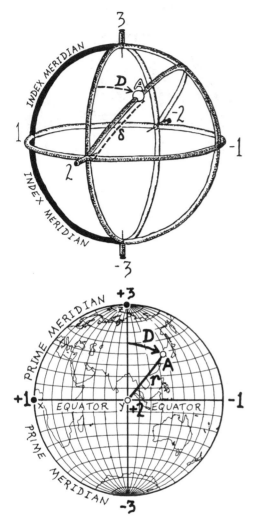

Figure 3-18.
Fixed reference frame in normal orientation. The **2** and **y** equatorial axes at the center of the projection are at the point of tangency of the projection with the sphere. The angular distance δ and the azimuthal angle D on the sphere again map to the polar coordinates r and D on the projection.

= (36°, 128°) for example, we first find the Cartesian coordinates of **A** from equations 3.3, 3.4, and 3.5:

$$A_x = \cos 36° \cos 128° = -.4981$$

$$A_y = \cos 36° \sin 128° = .6375$$

$$A_z = \sin 36° = .5878$$

Then δ and D are given by

$$\delta = \cos^{-1} A_y \qquad (3.23)$$
$$= 50.39°$$

$$D = \tan^{-1}(-A_x/A_z) \qquad (3.24)$$
$$= 40.28°$$

As before, let's set the scale factor $c = 14.14$ cm, so that (from 3.15):

$$r = 14.14 \sin 25.20° = 6.02 \text{ cm}$$

Then coordinates of the points on the projection corresponding to **A** are simply (from 3.19 and 3.20):

$$b = 6.02 \sin 40.28° = 3.89 \text{ cm}$$

and

$$v = 6.02 \cos 40.28° = 4.59 \text{ cm}$$

The Mercator Projection

Mercator projections are especially useful in plate tectonics because the small circles used to describe a set of transforms can be mapped as a set of parallel horizontal lines. Mercator projections, a type of cylindrical projection, are made by rolling the plane of the projection into a cylinder and placing it in contact with the sphere along a great circle. The grid lines on the globe are projected onto the cylinder and the latter then unrolled to form a plane. If the circle of contact was the equator, then meridians of longitude appear as horizontal lines and the circles of latitude as vertical lines. Therefore latitude and longitude coordinates can be transformed directly into Cartesian coordinates on all cylindrical projections. Since meridian circles that are equally spaced on the globe are mapped as equally spaced vertical lines, if

h is the horizontal axis on the projection, then for point **A** with coordinates (λ, ϕ), where $-180° \leq \phi \leq 180°$,

$$h = c\, \phi \, (\pi/180) \qquad (3.25)$$

where c is the scaling constant and the factor $\pi/180$ converts from degrees to radians. Taking the equator as the h axis of the projection along which $v = 0$, the value of the v coordinate for **A** is given by

$$v = c f (\delta) \qquad (3.26)$$

where c is the scaling constant and the function $f(\delta)$ determines the geometrical properties of the particular cylindrical projection that is used. For the Mercator projection,

$$f(\delta) = \ln \tan (45° + \lambda/2) \qquad (3.27)$$

As an example, we will find the coordinates of New York at $(40°, -74°)$ on a normal Mercator projection (having as the v and h axes the Greenwich meridian and the equator) having a width of 20 cm. The latter is achieved by setting $c = 10/\pi$ cm. Then (from 3.25, 3.26 and 3.27),

$$c = 10/\pi \text{ cm}$$
$$h = (10/\pi)\,(-74)\,(\pi/180) = -4.11 \text{ cm}$$

and

$$v = (10/\pi) \ln \tan (45 + 40/2) = 2.43 \text{ cm}$$

The 180° E and 180° W longitude circles plot as the vertical lines $h = 10$ cm and $h = -10$ cm. Since the North Pole and South Pole have coordinates $v = +\infty$ and $v = -\infty$, the regions near the poles are highly distorted and are not generally plotted.

Mercator projections are commonly used to determine how well an Euler pole fits the relevant plate tectonic data. For a perfect fit, if the cylinder of the projection contacts the globe along the great circle perpendicular to the Euler pole, the fracture zones will all be exactly parallel horizontal lines and the isochrons will approximate parallel vertical lines. The procedure used, therefore, is to rotate the globe until the Euler pole is aligned with the **3** axis of the projection and to then make a map of the fracture zones and isochrons using the procedure just described. An example for a hypothetical Euler pole on the equator is shown in Figure 3-19.

Figure 3-19.
Transforms on different projections. Concentric circles about a hypothetical Euler pole **E** show the locus of possible transforms (light lines), one possible set of which are shown with heavy lines. (*a*) Stereographic projection centered on the Euler pole. (*b*) Mercator projection with center at the Euler pole. (*c*) Mercator projection with equator perpendicular to the Euler pole.

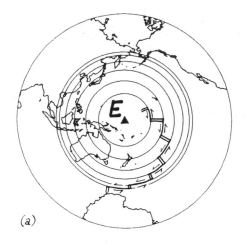

(*a*)

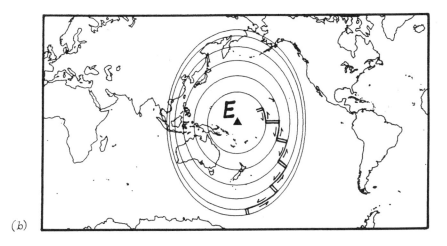

(*b*)

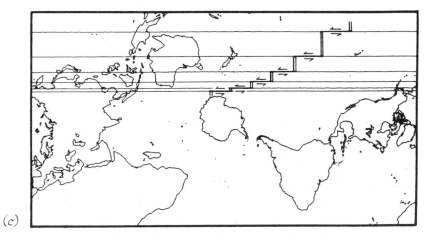

(*c*)

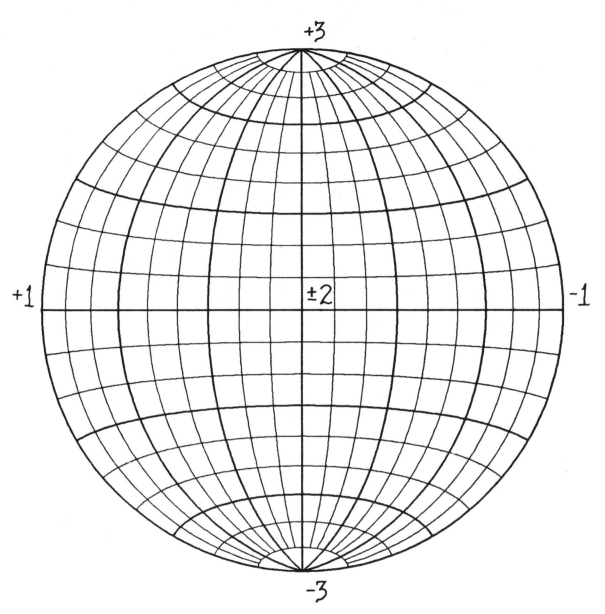

Figure 3-20.
Equatorial equal area projection.

Problems

3-1. With the North Pole **z** at $+3$ and $\phi = 0$ at the index meridian, draw on both hemispheres the great circle $<$**A**, **B**$>$ through **A** $= (0°, 30°)$ and **B** $= (-30°, 120°)$. Starting at **A**, plot points at intervals of $45°$ along this great circle. List the globe coordinates of these points.

3-2. Find the distance in kilometers from New York $(40°, -74°)$ to each of the following cities:
Cape Town $(-35°, 20°)$
Anchorage $(61°, -150°)$
Tokyo $(35°, 138°)$
Melbourne $(-37°, 145°)$
Buenos Aires $(-34°, -58°)$

3-3. What are the coordinates of New York after you have rotated it $30°$ about the **y** axis which has coordinates $(0°, 90°)$?

3-4. What are the coordinates of New York after you have rotated it $-30°$ about **x** $= (0°, 0°)$?

3-5. During the Permian the geographic South Pole had coordinates $(46°\ S, 279°\ E)$ as seen from North America. On a projection with **z** $= +3$ and $\phi = 310°$ aligned with the index meridian, plot the location of the Permian equator.

3-6. During the Cambrian the South Pole had coordinates $(7°\ S, 320°\ E)$ as seen from North America. Plot the Cambrian equator on the diagram of the previous problem and find the angle between the Permian and Cambrian equators.

3-7. (a) Find the point where the Permian equator intersects the meridian $\phi = 270°\ E$.

(b) At this intersection, find the local azimuth of the Permian equator.

3-8. Find the poles of the great circle connecting

(a) New York and Cape Town,

(b) Anchorage and Buenos Aires, and

(c) Cape Town and Tokyo.

3-9. An earthquake recorded at Stanford $(37°\ N, 238°\ E)$ arrives from a direction $15°$ east of south. The seismogram tells us that the distance to the epicenter is $80°$. What are the coordinates of the epicenter of the earthquake?

3-10. Later when we analyze seismic data we will need to know how to plot perpendicular great circles on a sphere. For practice, on the eastern hemisphere centered on $(0°, 90°)$, plot

(a) the great circle through Cape Town and Anchorage,

(b) the great circle through Cape Town perpendicular to the first great circle, and

(c) the great circle perpendicular to the two previous great circles. Next, find the poles of circles a, b, and c.

Can you generalize these results to a theorem about the poles of three mutually perpendicular great circles? Next, list the coordinates of the six points of intersection of these three great circles. Finally, plot and list the coordinates of the poles of the three great circles. Can you generalize these results to a theorem about the poles and points of intersection of three perpendicular great circles?

Suggested Readings

General

Map Projections Used by the U.S. Geological Survey, J. P. Snyder, U.S. Geological Survey Bulletin 1532, 313 pp., 1982. *Monograph explaining how projections are used to make maps.*

The Properties and Uses of Selected Map Projections, Tau Rho Alpha and J. P. Snyder, U.S. Geological Survey Map I-1402, 1982. *Single sheet describing main map projections used in geology.*

Also see Suggested Readings for Chapter 5

4

Wrapping Plate Tectonics Around a Globe

Now we're all set to go global. We'll see that plate tectonics on the globe is essentially the same as on a plane. In fact, if you've mastered the techniques described in Chapter 3, you will probably find plate tectonics on the globe simpler and more elegant.

First let's see what Euler poles and transform faults look like on the globe. Recall that an Euler pole is like the hinge of a pair of scissors and that a transform fault is like the arc swept out by the point of one of the blades. On the plane, transform faults are straight lines when the Euler pole is located an infinite distance away. On the globe, the analog of an Euler pole at infinity is one 90° away, and the corresponding transform lies along a great circle. Such transforms are rare on the globe because the distance δ from a point of observation to an Euler pole is rarely exactly 90°. Most transforms lie along small circles, which by definition have angular radii less than 90°.

Viewed as topographic features rather than as geometrical constructions, transforms and fracture zones are the longest linear features on the earth's surface, "linear" in this context meaning that they lie along small circles. What is this perfection of geometric form trying to tell us about the tectonic process that produces transforms? Some insight is provided by recalling that if constant force is applied to a plate or other object on a plane, all points on the plate will move along trajectories that are parallel straight lines. On the globe the analog of constant force is constant torque, which produces motion in which all points on a plate move along trajectories that are small circles concentric about a common point, the Euler pole. So motion about a fixed Euler pole on the globe is a direct analog of the simplest kind of motion on a plane, motion in a constant direction. Thus the tendency of fracture zones to lie along small circles tells us that on the globe the direction of plate motion doesn't shift around randomly. Once a plate starts moving, it continues for a long time to move in a constant direction, which is the direction of motion produced by a constant torque applied to the plate.

Although the concept of an Euler pole provides a very efficient and compact way of describing how two plates have been moving relative to each other over a period of millions of years, it is important to keep in mind that Euler poles are not real in the sense that transforms, ridges, and trenches are real. Euler poles are geometrical constructs and not geologic features on the earth's surface. When an oceanographic ship sails over an Euler pole, it can't detect it observationally because the pole is not a topographic or geologic feature. Euler poles exist only in the eye of the beholder of the plate tectonic process.

To calculate the location of an Euler pole, three types of data are used. The first is the observed trend of transforms, as determined from their topography and geology. The second is the direction in which a block on one side of a fault slips past the block on the other side during an earthquake along a fault boundary, as determined by analyzing earthquake waves. The azimuth of this "slip vector" gives the direction of relative plate motion and thus is analogous to the trend of a transform. The third type of data is the velocity of spreading across ridges, as determined from the spacings of magnetic isochrons. We will now consider ways of analyzing these three types of data, first using graphical methods and a qualitative approach, and then using a more precise (if not more accurate) mathematical analysis.

Transform Trends

In determining the location of an Euler pole using the trend of transforms, the basic data to be analyzed consist of the local azimuth angle T of the transform at each locality **B**.

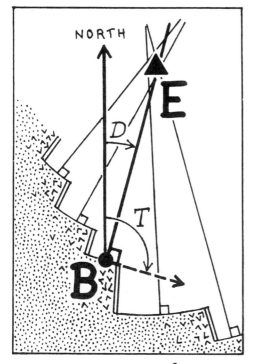

$$D = T \pm 90°$$

Figure 4-1.

Locating an Euler pole **E** from the trends T of transforms. Lines nearly intersecting at **E** are great circles perpendicular to the transforms.

More precisely, T is the azimuth or declination from north at **B** of the great circle locally parallel to the transform (Figure 4-1). Since the Euler pole lies on a great circle perpendicular to the transform, we know that ideally the Euler pole **E** should lie on a great circle through the point of observation **B**, with local azimuth $D = T \pm 90°$. The pole of this great circle may be found as the point 90° from **B** along the great circle defined by the transform (Figure 4-2). Ideally all of the great circles from the different sites **B** of observation should intersect at the exact position of the Euler pole **E**. However, if you have had any experience in conducting experiments of your own, especially experiments or observations of the earth, you'll not be surprised to learn that Nature is more complicated. For example, sometimes transforms are offset by short segments of ridge (Figure 4-3) with the result that the apparent azimuth of the transform-and-ridge system is different from the azimuths of the individual transform segments, which are the azimuths of interest. If data from only a few ship tracks across this area are available to you, which is usually the case, you will not be able to recognize the individual transform segments, and your final interpretive map will show a continuous transform with the general trend of the ridge-and-transform system. Because of this and other observational errors, great circles constructed perpendicular to transforms rarely intersect in one point, unless, of course, you have only two observations. When you start working with real data, you will always find at least a small amount of experimental scatter in your results, and so you will be faced with the problem of estimating the best location of your Euler pole from these somewhat scattered data.

Let's start by looking at the transforms that offset the Mid-Atlantic Ridge, and from them let's try to determine the Euler pole for the North America and Eurasia plates. These magnificent transforms are among the most striking physiographic features of the earth's surface. Trying to explain them was, in fact, one of the challenges that led J. Tuzo Wilson and W. Jason Morgan to invent plate tectonics. The narrow mountain ranges along the fracture zones that mark these transforms yield the four trends listed in Table 4-1. Included are the poles of the great circles perpendicular to the transform trends, which we will refer to as "Euler pole great circles." These great circles, which ideally should pass through the Euler pole, are shown in Figure 4-4. Note that the Euler pole great circles are nearly parallel and therefore intersect

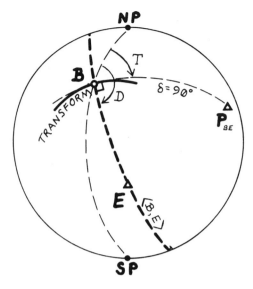

Figure 4-2.
Euler pole **E** is on the great circle perpendicular to the trend of the transform. $\mathbf{P}_{BE}$ is the pole of the great circle $<$**B**, **E**$>$.

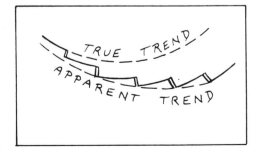

Figure 4-3.
Apparent and true trends of transform system offset by short ridge segments.

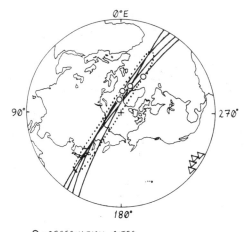

O – OBSERVATION SITES
△ – POLES OF EULER POLE GREAT CIRCLES
     PERPENDICULAR TO TRANSFORMS
⠂⠂ – REGION OF ACCEPTABLE EULER POLES

Figure 4-4.
Euler pole great circles perpendicular to transform trends along the Mid-Atlantic Ridge.

Table 4-1. Trends of transforms along Mid-Atlantic Ridge.

Site **T**		D_T	**P** $_{<T, E>}$	
λ	φ		λ	φ
79.0°	2.5°	308°	6.7°	235.0°
71.0°	352.0°	295°	7.9°	238.2°
66.5°	340.0°	278°	3.2°	242.7°
52.5°	326.5°	275°	3.0°	232.5°

D_T, trend of transform; $\mathbf{P}_{<T, E>}$, pole of great circle from transform to expected Euler pole.

at very small angles, so their points of intersection are widely scattered along the general trend of the package of great circles. It is obvious that any Euler pole within the dotted region of Figure 4-4 would provide an almost equally acceptable fit to the data. Because the observation sites along a spreading center tend to be located on a great circle perpendicular to the transforms, similar results are commonly obtained from other fracture zones, and the great circles defining the location of the Euler pole are closely grouped. Therefore, the location of the Euler pole is determined only within a narrow band, greatly elongated along the direction of the Euler pole great circles.

Slip Vectors

Can slip vectors help us determine the location of the Euler pole for the North America and Eurasia plates more accurately? Slip vectors are found from the radiation pattern of seismic waves from earthquakes along transforms (Chapter 6). They describe the relative motion of the plates on opposite sides of the transforms. Ideally, slip vectors are tangent to small circles centered on the Euler pole and therefore give us the same kind of information as transform trends. They can be analyzed using the same technique that we used for transforms. The azimuths of the slip vectors for the Mid-Atlantic Ridge are given in Table 4-2. The Euler pole great circles perpendicular to these slip vectors are plotted in Figure 4-5. Ideally the Euler pole should be at the intersection of these circles. As was true for the transforms (and for the same reasons), the Euler pole could lie anywhere within an elongate region along a package of nearly parallel, Euler

Table 4-2. Slip vectors along Mid-Atlantic Ridge.

Site S		A_S	$\mathbf{P}_{<S, E>}$	
λ	ϕ		λ	ϕ
79.8°	2.6°	317°	7.4°	226.1°
80.2°	353.0°	309°	6.0°	231.4°
70.9°	352.4°	295°	7.9°	238.6°
66.7°	341.8°	295°	9.6°	228.6°
66.3°	340.2°	287°	6.7°	234.6°
52.8°	325.2°	275°	3.0°	231.9°

A_S, azimuth of slip vector; $\mathbf{P}_{<S, E>}$, pole of great circle from observation site to expected Euler pole.

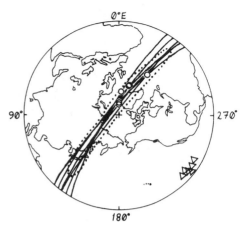

O—OBSERVATION SITES
△—POLES OF EULER POLE GREAT CIRCLES PERPENDICULAR TO SLIP VECTORS
⋯—REGION OF ACCEPTABLE EULER POLES

Figure 4-5.
Euler pole great circles perpendicular to slip vectors from earthquakes on transforms along the Mid-Atlantic Ridge.

pole great circles. Therefore, the slip vectors constrain the location of the Euler pole in about the same way the transforms did, and do not add much information.

To try to get a better fix on the North America-Eurasia Euler pole, we will next use our last source of information about plate motions, namely the observed variations in the spreading velocity along the Mid-Atlantic Ridge. Before turning to these data, we will discuss how such velocity variations are analyzed.

Velocities Due to Rotation about an Euler Pole

In Figure 4-6a we show plate B fixed and plate A rotating relative to plate B about an Euler pole **E** with antipole **E'** located 180° away on the opposite side of the globe. The rate of angular rotation of plate A relative to B is ω (omega), usually expressed in degrees per year (°/yr), in radians per year (rad/yr), in degrees per million years (°/my), or in radians per million years (rad/my). For simplicity we assume that the boundary between plates A and B is a ridge along a half great circle extending from **E** to **E'**. If we imagine an isochron pair forming along the ridge at time t, then today these two isochrons will be great circles meeting at **E** and **E'** and the maximum separation between the isochrons will be at the "equator" located 90° from **E** and **E'**. The maximum angular distance between the isochrons will be equal to ωt, where t is the age of the isochrons. If ω is in radians, the

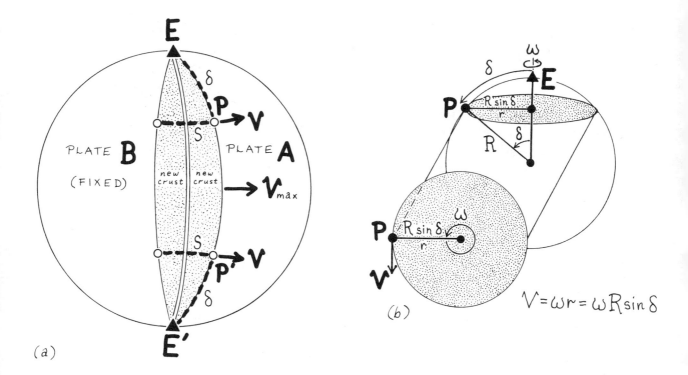

$$V = \omega r = \omega R \sin \delta$$

(b)

(a)

Figure 4-6.
E and **E′** are the Euler poles for the motion of plate A relative to plate B. (*a*): Stippled region is new crust added during time *t*, *S* is the distance between the isochrons that formed at time *t*, and $V = S/t$ is the velocity of plate A relative to plate B. (*b*) The velocity varies with distance δ from the point of observation **P** to the Euler pole **E**.

maximum spacing in kilometers is equal to $\omega t R$, where R is the earth's radius in kilometers (Figure 4-6). This equation simply says that as a wheel of radius R rotates an angle (in radians) of ωt, a point on the rim of the wheel moves a distance $\omega t R$. Closer to the Euler pole the spacing between isochrons will be $\omega t R \sin \delta$, where δ is the distance from the point of observation to the Euler pole. The "$R \sin \delta$" term reflects the fact that closer to the Euler pole, a point rotates along a circle with a smaller radius equal to $R \sin \delta$ (r in Figure 4-6b). The velocity of a point **P** at distance δ from **E** is $V = \omega R \sin \delta$ (Figure 4-6b).

In making marine geophysical surveys, we do not measure the angular velocity of the plates directly but infer this from the measured spacing between magnetic isochrons. If the isochron spacing is S km at a given point on the ridge, then the inferred linear velocity V is S/t. Noting that a velocity V of 10 mm/yr is the same as a velocity of 10 km/my, that 1 degree of angular distance equals 111 km of linear distance along a great circle, and that the earth's radius R is 6.38×10^3 km $= 6.38 \times 10^9$ mm, we have:

$$
\begin{aligned}
V \text{ (mm/yr)} = V \text{ (km/my)} &= \sin \delta\ R \text{ (mm)}\ \omega \text{ (rad/yr)} \\
&= \sin \delta\ 6.38 \times 10^{9}\ \omega \text{ (rad/yr)} \\
&= \sin \delta\ 6.38 \times 10^{3}\ \omega \text{ (rad/my)} \\
&= \sin \delta\ 1.11 \times 10^{8}\ \omega \text{ (°/yr)} \\
&= \sin \delta\ 111\ \omega \text{ (°/my)} \\
&= \sin \delta\ V_{max} \qquad\qquad (4.1)
\end{aligned}
$$

where V_{max} (mm/yr) $= 111\ \omega$ (°/my) is the maximum spreading rate, which is found at a point 90° from the Euler pole. V is sometimes called the local **full spreading velocity** of plate A relative to plate B. More commonly the **half spreading velocity** $V_{\frac{1}{2}} = \frac{1}{2}\ V$ is used. This is the velocity of either plate relative to the ridge, assuming (as is usually done) that spreading was symmetrical. If you stand astride the ridge with one foot on either diverging plate, $V_{\frac{1}{2}}$ is the velocity of one foot relative to your navel, which you as observer would regard as fixed. The half velocity is commonly cited in the literature because it is the half velocity that you determine by dividing the distance between two adjacent anomalies on one plate by the age difference between the anomalies. In reading the literature you will want to be alert because the expression "spreading velocity" sometimes refers to V and sometimes to $V_{\frac{1}{2}}$.

Suppose you have determined the spreading velocity V at a number of points along a ridge by measuring the distance between matching isochrons of known age on either side of the ridge (Figure 4-7a), but you don't have any information about fracture zones or slip vectors. Can you still find the location of the Euler pole and the rate of angular rotation? Recall that V (km/my) $= \sin \delta\ \omega$ (°/my)111(km/°). This equation tells you that if you plot velocity as a function of the distance δ from the Euler pole, your data points will fall on a sine curve passing through the origin. Since you don't know the location of the Euler pole, to determine the distance you will have to start with a trial pole **E** and then find the distance from **E** to each site. You will also need to pick a trial value of ω. If you've been lucky (Figure 4-7), then on your V versus δ plot your observed points will all fall on the theoretical sine curve (Figure 4-7a). If your trial pole $\mathbf{E}_1$ is wrong (Figure 4-7a), then your data points won't fit (Figure 4-7c). You can then try other Euler poles and other values of ω, until you're satisfied that you've found as good a fit to the data as possible.

A slight improvement in the accuracy with which Euler poles can be located using velocities is possible if account

(a)

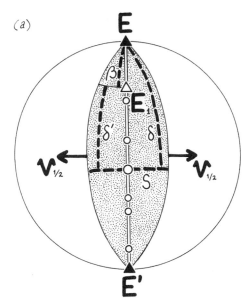

(b)

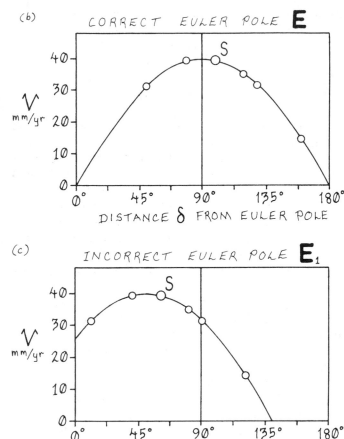

Figure 4-7.
(*a*) The phase shift β of a wrong Euler pole **E**₁. (*b*) If δ is measured from the true Euler pole **E**, then on a plot of velocity *V* versus δ the observed values of *V* will fall on a sine curve through the origin. (*c*) If an incorrect Euler pole like **E**₁ is used, the data points will not fall on a sine curve through the origin.

is taken of the observation that, whereas transforms are by definition exactly parallel to the direction of relative motion between two plates, ridges and the magnetic isochrons parallel to them are not always exactly perpendicular to the fracture zones. Therefore the apparent velocity V_{iso} determined by isochron spacing measured in a direction perpendicular to the isochrons is not always equal to the true velocity V_{true} measured parallel to the fracture zone (Figure 4-8). In the ideal case, if a ridge is adjacent to an adjacent transform at some point of observation **B**, then the trend of the ridge will be along the great circle <**B**, **E**> from **B** to the Euler pole **E**. On the other hand, if the trend of the ridge is not perpendicular to the transform, then the true velocity V_{true} is related to the apparent velocity V_{iso} by

$$V_{true} = V_{iso}/\cos (D_R - D_{BE}) \qquad (4.2)$$

where D_R is the declination of the trend of the ridge, and D_{BE} is the declination of <**B**, **E**> (Figure 4-8). If the trends of the ridges are substantially different from a direction perpendicular to the transforms, the use of true velocities permits the Euler pole and rotation rate to be determined somewhat more accurately. Along the Mid-Atlantic Ridge, D_{BE} is observed to differ from D_R by as much as 20°. The corresponding corrections in V_{iso} are 6 percent or less, which is comparable with the errors in measuring V_{iso}. We will not use this correction in our examples.

Spreading Velocities on the Mid-Atlantic Ridge

We are now intellectually equipped to determine whether variations in spreading velocity along the Mid-Atlantic Ridge can help us determine a better location for the Euler pole describing motion between the North America and Eurasia plates. In Figure 4-9 the dotted outline shows the region of poles that we earlier found to be in reasonable accord with the trends of transforms and slip vectors. Along the median line of this region are shown a set of trial Euler poles **A**, **B**, **C**, **D**, and **E**. We want to determine which of these poles provides the best fit to the observed velocities of seafloor spreading as determined by marine geophysicists during the past decade.

A summary of these velocities is given in Table 4-3, together with a list of the calculated distances between the trial Euler poles and the sites of observation. The velocities are plotted as a function of the calculated distance δ in Figure 4-10 for each of the trial poles, along with theoretical sin δ curves adjusted to pass through the observed velocity (28 mm/yr) for Site 5, the most distant site from all the trial poles, by solving for ω:

$$\omega \ (°/my) = V \ (mm/yr)/111 \sin \delta$$
$$= 28 \ (mm/yr)/111 \sin \delta$$

For trial pole **A**, most of the calculated velocities are greater than the observed velocities, indicating that the trial pole is too far from the observation sites. For trial pole **D** the converse is true, indicating that the trial pole is too close. Trial pole **E** is obviously impossible. Since it is located between sites, it would predict convergence at some of the sites, whereas spreading (extension) was observed at all of the sites. The best fit is found at or near trial pole **C**.

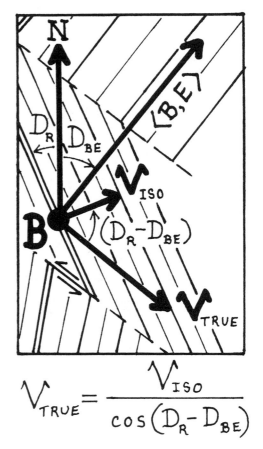

$$V_{TRUE} = \frac{V_{ISO}}{\cos\left(D_R - D_{BE}\right)}$$

Figure 4-8.
The spreading velocity V_{iso} measured perpendicular to isochrons is less than the true velocity, V_{true}, if the isochrons are not perpendicular to the transforms.

Table 4-3a. Observed velocities along Mid-Atlantic Ridge.

	Observation Site		
Site	λ	φ	V_{obs}
1	72.1°	1.0°	16 mm/yr
2	70.0°	345.0°	18 mm/yr
3	60.0°	330.6°	20 mm/yr
4	46.0°	332.5°	27 mm/yr
5	40.0°	330.0°	28 mm/yr

Table 4-3b. Trial Euler Poles.

	Trial Euler Pole		
Pole	λ	φ	V_{max}
A	30°	142°	30 mm/yr
B	50°	142°	28 mm/yr
C	70°	135°	30 mm/yr
D	85°	52°	37 mm/yr
E	69°	339°	57 mm/yr
F	70°	191°	31 mm/yr

Table 4-3c. Distance from trial Euler poles.

Site	Trial Euler Pole					
	A	B	C	D	E	F
1	74°	55°	35°	15°	8°	38°
2	79°	59°	39°	19°	2°	39°
3	90°	70°	50°	29°	10°	47°
4	103°	84°	63°	43°	23°	61°
5	110°	90°	70°	50°	29°	66°

Table 4-3d. Expected velocities; sums of squared errors.

Site	V_{obs}	Trial Euler Pole					
		A	B	C	D	E	F
1	16	29	23	17	10	8	19
2	18	29	24	19	12	2	20
3	20	30	26	23	18	10	23
4	27	29	28	27	25	22	27
5	28	28	28	28	28	28	28
$\Sigma (V_{ex} - V_{obs})^2$		394	142	11	80	445	22

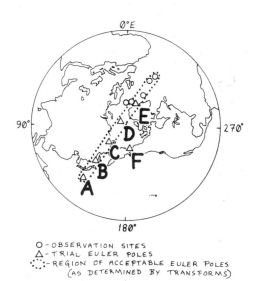

O – OBSERVATION SITES
△ – TRIAL EULER POLES
∴ – REGION OF ACCEPTABLE EULER POLES
(AS DETERMINED BY TRANSFORMS)

Figure 4-9.
Trial Euler poles used to make the V versus δ plots shown in Figure 4-10.

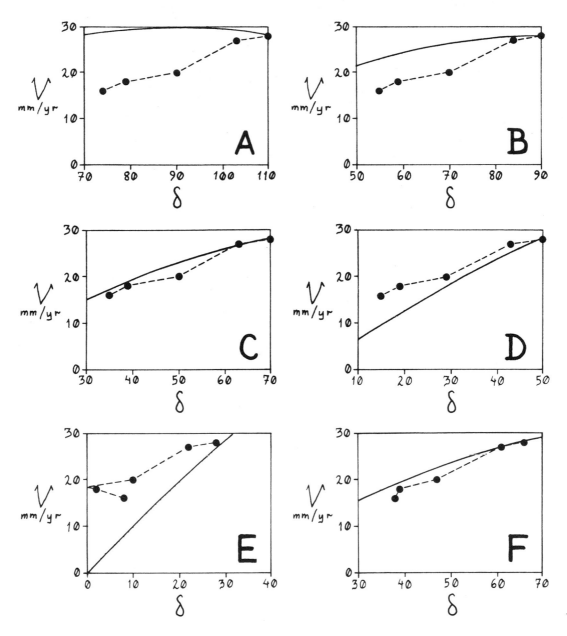

Note that trial pole **F**, near **C** and to one side of the arc of **A**, **B**, **C**, **D** and **E**, provides almost as good a fit to the data as does trial pole **C**. This illustrates an important difference between the ways in which fracture zones and spreading velocities constrain the location of Euler poles. Fracture zones and slip vectors constrain the pole to a region elongate in the direction of the Euler pole great circles, between the

Figure 4-10.
Plots of observed velocity *V* versus δ as determined for each of the trial poles in Figure 4-9. Solid curve is the calculated velocity V_{ex} versus δ. Trial pole **C** gives the best fit.

Figure 4-11.
Determining Euler poles from transforms and velocities gives regions of uncertainty that are respectively elongate along or perpendicular to the trend of Euler pole great circles.

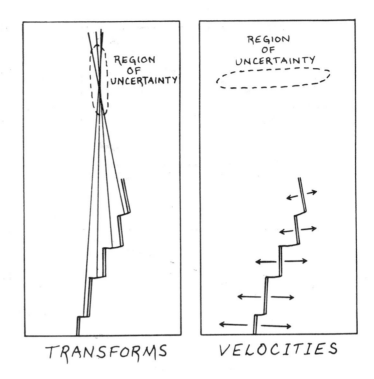

TRANSFORMS   VELOCITIES

pole and the sites where the observations were made. Velocities constrain the Euler pole to a region elongate in a direction perpendicular to the Euler pole great circles (Figure 4-11). The two types of data are complementary and together they constrain the location of the Euler pole much more accurately than either type of data does separately.

Best Fit Determined by Least Squares

Much of the power of plate tectonics stems from its ability to describe and predict earth movements quantitatively. Since Euler poles are the key quantitative element in plate tectonics, you will want to develop a good feel for the reliability of Euler poles based on knowledge about how they are calculated. Consider for a moment exactly how you would go about calculating the best Euler pole from a set of measurements of the local azimuth T_i of transforms along the Mid-Atlantic Ridge. One obvious approach would be to find the points of intersection of all the great circles drawn perpendicular to the transforms and then find the average location of all these points. This approach often gives disappointing results because the great circles commonly converge at very small angles, as we saw for the Mid-Atlantic Ridge data, so the location of the point of intersection is very sensitive to small errors in the observations. A second approach would be to pick a trial pole location for **E** and then find for each great circle the distance ϵ_i from the great circle to **E** (Figure 4-12). A measure of the overall misfit is then provided by the sum $\Sigma \, \epsilon_i^2$ of the values of ϵ_i^2 at all sites. You can then try different trial positions of **E** until you're satisfied that you have found a position that minimizes $\Sigma \, \epsilon_i^2$, the sum of the squares of the errors.

A third approach, and one that is used by most researchers, is to calculate the expected theoretical value of each observation and then compare this with the actual observation, using the following procedure. (1) Select a trial Euler pole. (2) At a given point of observation, calculate the expected value T_{ex} that the transform would have in the absence of observational error (or reluctance on the part of Nature to obey the rules of our plate tectonic model). (3) Subtract the expected transform trend T_{ex} from the observed trend T_{obs} (Figure 4-12) and square the difference. This "squared error"

$$\epsilon_i^2 = (T_{obs} - T_{ex})^2 \qquad (4.3)$$

is commonly used in plate tectonics, as in other forms of experimental analysis, as a measure of the misfit between an observation and the predicted value of the observation, based on some theoretical model such as the Euler pole model. (4) As the combined measure of misfit at all points of observation, use the sum $\Sigma \, \epsilon_i^2$ of the squared errors at all sites, where ϵ_i^2 is as defined in equation 4.3. (5) Select other trial poles and repeat the process until you're satisfied that you have found the Euler pole for which $\Sigma \, \epsilon_i^2$ is a minimum. This is termed the best-fit pole in the sense of least squares. This approach is the one used in most contemporary analyses.

Let's first apply this quantitative analysis to our observations of transform trends along the Mid-Atlantic Ridge (Table 4-1) which we earlier analyzed qualitatively. In Figure 4-13 two sets of trial poles are shown, one along the trend of Euler pole great circles determined from transforms and slip vectors and the other along a great circle transverse to the Euler pole circles. The sum of the squared errors for each of these trial poles is displayed in Figure 4-14, along with smooth curves sketched through these points. Two important observations are to be noted. For the trial poles along

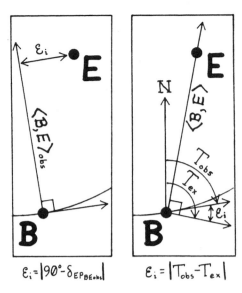

$\epsilon_i = \left| 90° - \delta_{EP_{BE_{obs}}} \right|$ $\qquad$ $\epsilon_i = \left| T_{obs} - T_{ex} \right|$

Figure 4-12.
Alternative measures of error in determining Euler poles from transform data. On the left, a great circle $<B, E>_{obs}$ is constructed perpendicular to the transform; the measure of error is the distance ϵ_i from the great circle to the trial Euler pole **E**, which is equal to the absolute value of 90° minus the distance $\delta_{EP_{BE \, obs}}$ from the Euler pole **E** to the pole $P_{BE \, obs}$ of the great circle. On the right, the expected trend of the transform T_{ex} is found as the direction perpendicular to the great circle $<B, E>$ from the point of observation **B** to the trial Euler pole **E**; the measure of error is then the absolute value of the difference between the expected trend T_{ex} and the observed trend T_{obs}.

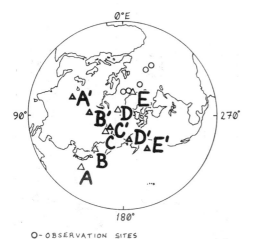

O – OBSERVATION SITES
△ – TRIAL EULER POLES ALONG TREND
▲ – TRIAL EULER POLES ALONG TRANSVERSE

Figure 4-13.
Trial Euler poles used to generate the curves showing $\Sigma\, \epsilon_i^2$ versus pole locations in Figures 4-14 and 4-15. Primed and unprimed poles are along and transverse to the trend of the Euler pole great circles, respectively.

the Euler circles, the summed squared errors $\Sigma\, \epsilon_i^2$ remain small except for trial poles located near points of observation. As trial poles approach transforms, the calculated expected trends change rapidly and the squared error soars. Secondly, the error curve for the set of poles transverse to the Euler pole circles displays quantitatively what we saw intuitively from our earlier plot. Trial poles that fit the data reasonably well are confined to a narrow zone along the great circle through the sites of observation.

Results of a similar analysis of the slip vectors are also displayed in Figure 4-14. The measure of misfit used for the slip vector data is

$$\epsilon_i^2 = (S_{obs} - S_{ex})^2 \qquad (4.4)$$

These results are remarkably similar to those for the transforms. The best-fit minimum along the transverse profile is somewhat better defined for the slip vectors, mainly because more data were used in this analysis.

Now let's turn to the velocity data and apply the same least-squares analysis, using the same two sets of trial poles. As seen in Figure 4-15, this time the results are exactly the opposite: For trial poles along the transverse great circle, the location of the best-fit pole is poorly defined, whereas for trial poles along the trend defined by the transforms and slip vectors, the location of the best-fit pole is very well defined by a minimum in the summed squared errors of the velocities. While neither the transform-slip vector nor velocity data determine the location of the best-fit pole separately, together they determine it very well.

Figure 4-14.
The sum of the squares of the errors for the azimuths of transform trends T and slip vectors S for trial poles along and transverse to the trend of Euler pole great circles. Best-fit pole is near $\mathbf{C'}$.

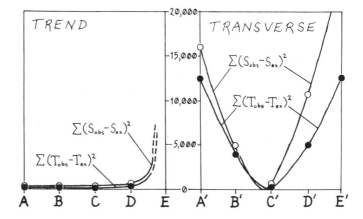

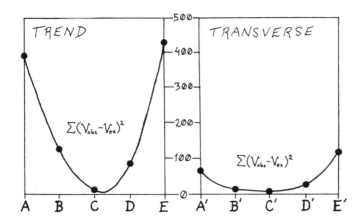

Figure 4.15.
The sum of the squares of errors of spreading velocities for the trial poles along and transverse to the trend of the Euler pole great circles. Best-fit pole is near **C**.

A refinement of the least-squares technique is used when the accuracy of the data varies from site to site. If σ is the standard deviation describing the experimental uncertainty at a site, then the measure of misfit at that site is taken to be

$$\epsilon_i^2 = (\frac{V_{obs} - V_{ex}}{\sigma})^2 \qquad (4.5)$$

in the case of velocity measurements and similarly for transforms and slip vectors. This has the desirable effect of giving less weight to data with large errors. As before, the best-fit Euler pole and rotation rate are the values that minimize ϵ_i^2 summed over sites for all types of data.

Using an eyeball estimate of the minima of the curves shown in Figures 4-14 and 4-15, we obtain the following parameters for the angular rotation of North America and Eurasia:

Euler Pole		Rotation Rate	Description
λ	ϕ	ω	
73°	121°	0.278° /my	Eyeball fit
65.85°	132.44°	0.258° /my	Computer least-squares
±6.17°	±5.06°	±0.019° /my	Standard error of computer least-squares

In the second line is listed the result of a computer analysis and in the third line are listed the standard errors of the parameters determined by the computer fit. The two analyses are in fairly good but not perfect agreement. One reason is that the computer analysis weighted each datum according to the inverse of its standard deviation, whereas we didn't.

The second reason is that we used only data from sites along the North America-Eurasia plate boundary, whereas the computer used not only these data but also data from other plates to determine the North America-Eurasia pole. At first glance this doesn't make much sense; however, we will soon see how the motion between any pair of plates is partially constrained by the motion between other plates to which the pair is linked.

If the amount of marine geophysical data available for determining Euler poles and rotation rates is large, finding the best-fit pole becomes very tedious and it is easy to make computational errors, as you will appreciate if you solve some of the problems at the end of this chapter. Therefore computers are generally used to analyze the large data sets currently available. The vocabulary used in this modeling research describes the plate tectonic interpretation of the observational data as a "model." The parameters of this model are the latitude λ and longitude ϕ of the Euler pole and the rate of rotation ω of the two plates about this pole. As the parameters λ, ϕ, and ω vary over their possible range, they define a "parameter space" within which the best-fit model is located. The computer algorithm carries out some strategy for searching this parameter space for the model that best fits the data in the sense of least squares, weighting each observed datum inversely with its standard deviation. A great concern of modelers is that in finding a minimum value of $\Sigma \, \epsilon_i^2$, they will inadvertently zero in on a local "best-fit pole" in some region of parameter space and completely miss a better fit located some distance away from the region searched. Modelers take pains to convince their readers that they have not fallen into this classic pitfall of data analysis.

Angular Velocity Vectors

Recall that on a plane, if you want to find the velocity $_A\mathbf{V}_C$ of plate C relative to plate A and you know the velocities $_A\mathbf{V}_B$ and $_B\mathbf{V}_C$, you can use vector addition to find the desired velocity $_A\mathbf{V}_C = {}_A\mathbf{V}_B + {}_B\mathbf{V}_C$. We now need to find an analogous technique to use on the globe. We do this by using a three-dimensional rotation vector $\boldsymbol{\omega}$ to describe the relative motion between any two plates, where the boldface $\boldsymbol{\omega}$ indicates that this is a vector quantity. In contrast, the symbol ω, which is a simple number or scalar quantity, gives the speed of angular rotation but not the direction. The angular rotation vector **ω** is a vector originating at the earth's center and pointing toward the Euler pole **E**. Its length is equal to the scalar angular velocity ω, which is found, as we've just seen, from spreading velocities. So an angular velocity vector **ω** can be thought of as consisting of two parts: a unit vector, corresponding to the Euler pole plotted on a sphere of unit radius; and a scalar velocity ω which, when multiplied into the Euler pole unit vector, gives the rotation vector. If λ and φ are the coordinates of **E**, we can describe the parameters of **ω** as the triad (λ, φ, ω). We will now review briefly some of the basic physics that describes how rotation vectors are added.

The rotational motion of any object can be described most simply by a rotational velocity vector **ω**. In the case of a rotating wheel, **ω** is directed along the axis of rotation. If you wrap the fingers of your right hand around the wheel in the direction of motion, your thumb will point in the direction of the vector—the right-hand rule. The length of the vector is determined by the rate of rotation of the wheel, which may be expressed in revolutions, degrees, or radians per unit time. In the case of the earth's rotation, the direction of **ω** is from the center of the earth toward the North Pole and the length of the vector is one revolution per day or 15°/hr.

To visualize how rotations add, consider a fan sitting at the center of a phonograph turntable rotating in a clockwise direction at an angular velocity of 33 rpm (revolutions per minute). We will describe this rotation by the vector $_C\boldsymbol{\omega}_T$, where T refers to the turntable, C refers to the console in which the phonograph is mounted, and the velocity vector specifies the rotation of T with respect to C. The orientation of $_C\boldsymbol{\omega}_T$ is downward and the length of $_C\boldsymbol{\omega}_T$ is 33 rpm. Now let's imagine that we're moving with the turntable and observing the rotation of the fan blades. The vector $_T\boldsymbol{\omega}_F$ describes the angular motion of the fan blades F relative to the turntable T. This vector has a length of 120 rpm and is directed horizontally, perpendicular to the face of the fan (Figure 4-16). To find the angular velocity of the fan blades relative to the console C we can simply find the vector sum (Figure 4-17):

$$_C\boldsymbol{\omega}_F = {_C\boldsymbol{\omega}_T} + {_T\boldsymbol{\omega}_F} \qquad (4.6)$$

In order to do the vector addition, we generally resolve **ω** into a north component ω_n, an east component ω_e, and a

Figure 4-16.
Fan on turntable. Angular rotation vector of fan blade is horizontal. Angular rotation of turntable is vertical.

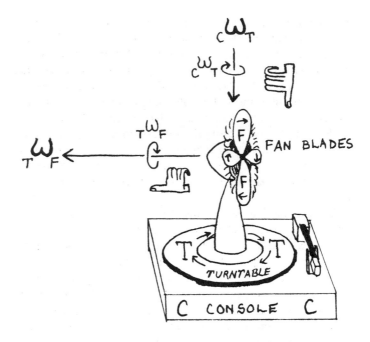

downward component ω_d of angular velocity. At the moment with the fan facing to the east, $_T\boldsymbol{\omega}_F$ is directed toward the west, so we have the following components:

	ω_n	ω_e	ω_d
$_C\boldsymbol{\omega}_T$	0	0	33 rpm
$_T\boldsymbol{\omega}_F$	0	-120 rpm	0
$_C\boldsymbol{\omega}_F$	0	-120 rpm	33 rpm

The length of $_C\boldsymbol{\omega}_F$ is $[120^2 + 33^2]^{1/2} = 124.5$ rpm. The direction is toward the west and inclined downward at an angle $\tan^{-1}(33/-120) = 15.4°$ below the horizontal. The vector $_C\boldsymbol{\omega}_F$ describes the instantaneous rotation of the fan blades F relative to the console C. This vector itself rotates about the vertical vector $_C\boldsymbol{\omega}_T$. If it bothers you that the net rotation vector is itself rotating, recall that any fixed rotation axis in

Figure 4-17.
Vector addition gives the instantaneous angular rotation vector of the fan blade relative to the console on which the turntable sits.

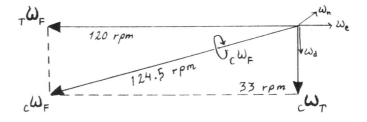

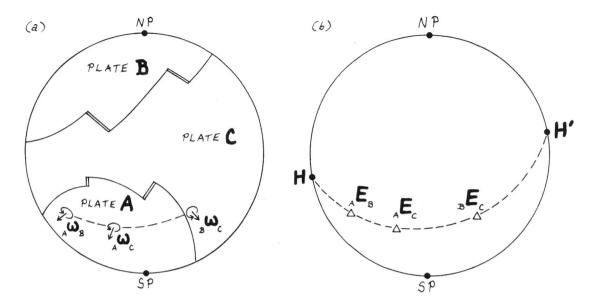

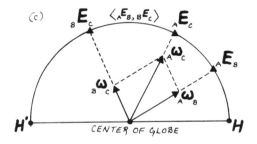

Figure 4-18.
The relative angular velocity vectors for three plates (*a*) are coplanar and therefore lie along a great circle (*b*). The vectors can be added in two dimensions on the plane through the vectors (*c*).

our everyday reference frame on the earth's surface is actually rotating once a day about the earth's rotation axis.

Velocity Space on the Globe

Recall that the Euler pole $_A\mathbf{E}_B = (\lambda, \phi)$ that describes the motion of plate B relative to plate A, together with ω, the rate of angular rotation, comprise the vector $_A\boldsymbol{\omega}_B$. The length of the vector describes the rotation rate and the direction is from the center of the earth toward the Euler pole. Similarly the relative motion of a third plate, C, relative to plate B can be described by the angular motion vector $_B\boldsymbol{\omega}_C$. We can find the velocity of C relative to A by vector addition, just as we did on the plane, using the vector equation $_A\boldsymbol{\omega}_C = {_A\boldsymbol{\omega}_B} + {_B\boldsymbol{\omega}_C}$. In this case, both $_A\boldsymbol{\omega}_B$ and $_B\boldsymbol{\omega}_C$ can be visualized as originating at the earth's center and lying in a plane which cuts the earth's surface along the great circle $<_A\mathbf{E}_B, {_B\mathbf{E}_C}>$ through the Euler poles $_A\mathbf{E}_B$ and $_B\mathbf{E}_C$. Figure 4-18 shows how to find the resultant vector by vector addition, plotting the vectors in the plane $<_A\mathbf{E}_B, {_B\mathbf{E}_C}>$.

To do the same operation analytically, it is useful to decompose the rotational vectors into three components along the **x**, **y**, and **z** axes, which originate at the center of the earth and end at the following points on the Earth's surface: $\mathbf{x} = (0°, 0°)$, $\mathbf{y} = (0°, 90°)$, and $\mathbf{z} = (90°, *)$. **x** and **y** lie in the plane of the equator and **z** is directed toward the North Pole. An angular rotation vector with length ω and an Euler pole at (λ, ϕ) has the following three Cartesian components:

$$\omega_x = \omega \cos \lambda \cos \phi \qquad (4.7)$$

$$\omega_y = \omega \cos \lambda \sin \phi \qquad (4.8)$$

$$\omega_z = \omega \sin \lambda \qquad (4.9)$$

To convert back to **E** and ω, we use:

$$\phi = \tan^{-1}(\omega_y/\omega_x) \qquad (4.10)$$

$$\lambda = \tan^{-1}(\omega_z/[\omega_x^2 + \omega_y^2]^{1/2}) \qquad (4.11)$$

$$\omega = [\omega_x^2 + \omega_y^2 + \omega_z^2]^{1/2} \qquad (4.12)$$

In working on a projection it's easier to use the parameters (λ, ϕ, ω) in spherical coordinates whereas using a computer it's easier to use the components $(\omega_x, \omega_y, \omega_z)$ in Cartesian coordinates. The sum of any two vectors **A** and **B** with components (A_x, A_y, A_z) and (B_x, B_y, B_z) is simply

$$\mathbf{A} + \mathbf{B} = (A_x + B_x, A_y + B_y, A_z + B_z)$$

The following example shows how to find the velocity of the India plate relative to the Antarctica plate, given the angular velocities of Africa relative to Antarctica and of India relative to Africa.

$$_{AN}\boldsymbol{\omega}_{IN} = {}_{AN}\boldsymbol{\omega}_{AF} + {}_{AF}\boldsymbol{\omega}_{IN}$$

First we find the Cartesian components of

$$_{AN}\boldsymbol{\omega}_{AF} = (\lambda, \phi, \omega) = (9.4°, -41.7°, 0.15°/my)$$

$$\omega_x = 0.15 \cos(9.4°)\cos(-41.7°) = .1105$$

$$\omega_y = 0.15 \cos(9.4°)\sin(-41.7°) = -.0984$$

$$\omega_z = 0.15 \sin(9.4°) = .0245$$

For $_{AF}\boldsymbol{\omega}_{IN} = (17.3°, 46.0°, 0.64°/my)$ we have

$$\omega_x = 0.64 \cos(17.3)\cos(46.0) = .4245$$

$$\omega_y = 0.64 \cos(17.3)\sin(46.0) = .4396$$

$$\omega_z = 0.64 \sin(17.3) = .1903$$

Add vectorially,

	ω_x	ω_y	ω_z
$_{AN}\boldsymbol{\omega}_{AF}$	.1105	−.0984	.0245
$_{AF}\boldsymbol{\omega}_{IN}$	.4245	.4396	.1903
$_{AN}\boldsymbol{\omega}_{IN}$	.5350	.3412	.2148

And convert back to spherical coordinates

$$\phi = \tan^{-1}(.3412/.5350) = 32.5°$$

$$\lambda = \tan^{-1}(.2148/[.5350^2 + .3412^2]^{1/2}) = 18.7°$$

$$\omega = [.5350^2 + .3412^2 + .2148^2]^{1/2} = .67°/my$$

Keeping the plus and minus signs straight can be tricky until you've become accustomed to the conventions. Here are some helpful tips, examples, and rules of the game.

Rules of Angular Velocity Vectors

1. $_A\boldsymbol{\omega}_B$ means "the rotation of plate B relative to plate A."

2. An angular rotation vector can be expressed as the Euler pole (λ, ϕ), which can be regarded as a unit vector, and the scalar velocity ω, or alternatively as the vector $\boldsymbol{\omega}$ with parameters λ, ϕ, and ω, which is not a unit vector. As shorthand, we sometimes write $\boldsymbol{\omega} = (\lambda, \phi, \omega)$. The value of the scalar quantity ω is usually positive, corresponding to a vector directed from the earth's center toward the Euler pole **E**. If the sign of the scalar ω is negative, $\boldsymbol{\omega}$ is directed from the earth's center away from the Euler pole **E**, toward the antipole **E′**.

3. $_B\boldsymbol{\omega}_A = -_A\boldsymbol{\omega}_B$. The vector $-\boldsymbol{\omega}$ is directed 180° from $+\boldsymbol{\omega}$. If $\mathbf{E} = (\lambda, \phi)$ and ω corresponds to $\boldsymbol{\omega}$, then $\mathbf{E}'$ and ω correspond to $-\boldsymbol{\omega}$, where the antipole $\mathbf{E}' = (-\lambda, \phi + 180°)$. In shorthand, $-\boldsymbol{\omega} = (-\lambda, \phi + 180°, \omega)$. An alternative way to express $-\boldsymbol{\omega}$ is by retaining the original Euler pole and changing the sign of the scalar ω, so that $-\boldsymbol{\omega} = (\lambda, \phi, -\omega)$. In Cartesian coordinates if $\boldsymbol{\omega} = (\omega_x, \omega_y, \omega_z)$ then $-\boldsymbol{\omega} = (-\omega_x, -\omega_y, -\omega_z)$.

4. Subtracting the vector $\boldsymbol{\omega}$ is the same as adding the vector $-\boldsymbol{\omega}$. Thus $_A\boldsymbol{\omega}_B - _C\boldsymbol{\omega}_B = _A\boldsymbol{\omega}_B + _B\boldsymbol{\omega}_C = _A\boldsymbol{\omega}_C$. Note how adjacent subscripts cancel when vectors are added.

As an example, consider $_{NA}\omega_{EU}$ describing the rotation of plate EU relative to plate NA, with Euler pole $\mathbf{E} = (66°, 132°)$ and $\omega = 0.23°$ /my. Regarding NA as fixed and looking down on the Euler pole from above, the positive sign of ω means that points on EU are moving **counterclockwise** around $\mathbf{E}$. According to rule 3, the corresponding rotation vector for the rotation of points on plate NA relative to EU is:

$$_{EU}\omega_{NA} = -_{NA}\omega_{EU} = (-66°, 312°, 0.23° \text{/my})$$
$$= (66°, 132°, -0.23° \text{/my})$$

The first equation says that points on NA are rotating counterclockwise around the antipole $\mathbf{E'} = (-66°, 312°)$. The second equation says that the points on NA are rotating clockwise around the original Euler pole $(66°, 132°)$. Of course the two statements are saying the same thing. According to rule 4, if we add these two vectors they should cancel:

$$_{NA}\omega_{EU} + {}_{EU}\omega_{NA} = {}_{NA}\omega_{EU} - {}_{NA}\omega_{EU}$$

which, by the procedure we followed before:

	ω_x	ω_y	ω_z
$_{NA}\omega_{EU}$	$-.0626$	$.0695$	$.2101$
$_{EU}\omega_{NA}$	$.0626$	$-.0695$	$-.2101$
$_{NA}\omega_{NA}$	$.0000$	$.0000$	$.0000$

we see is true.

Checking Internal Consistency

A demonstration of the power of plate tectonics is provided by its ability to calculate the relative motion between plates with no common boundary between them. For example, in order to understand the tectonics of western North America during the early Tertiary, geologists need to know the relative velocities of the North American and Pacific plates. During the early Tertiary these plates were not yet in contact, so there are no data from which an early Tertiary Euler pole can be directly determined. However, ridges existed during the early Tertiary on the boundaries between the North America and Africa plates, the Africa and India plates, the India and Antarctica plates, and the Antarctica and Pacific plates, so angular velocity vectors can be determined for spreading along all of these boundaries, which can be thought of as links in a chain connecting the North America plate with the Pacific plate. Therefore, the angular velocity vector for the Pacific plate relative to North America may be found by adding all of the relevant vectors:

$$_{NA}\omega_{PA} = {}_{NA}\omega_{AF} + {}_{AF}\omega_{IN} + {}_{IN}\omega_{AN} + {}_{AN}\omega_{PA} \quad (4.13)$$

Of course each link added to such a chain increases the accumulated error. Not only do the errors of each vector contribute to the total error, but it is always possible that one or more of the plates has undergone internal deformation, which would invalidate one of the most important assumptions of the plate tectonic model. For example, if hidden beneath the ice of Antarctica there is a geologically undetected boundary between two Antarctica subplates that were undergoing relative movement during the early Tertiary but subsequently were welded together, as is quite possible, then the above equation is incorrect because one of the links in the chain is broken.

The laws of vector addition can be used not only to find an unknown plate velocity but also to check the internal consistency of known plate velocities. Suppose, for example, that you want to know the angular velocity vector for the motion of Africa relative to Antarctica. As Chief Scientist on the *R. V. Vema,* you cruise a zigzag pattern across the ridges and transforms between the Africa and Antarctica plates, gathering data. However, because of stormy seas and short fuel supplies you have to turn back with an incomplete data set. Quite commonly in marine geophysics you don't have all the data you'd like. Nevertheless you decide to analyze the available data using the techniques described above and you obtain the following results for the angular rotation parameters (λ, ϕ, ω):

$$_{AN}\omega_{AF} = (15°, -48°, 0.19°/my)$$

In view of the sparseness of the data, you would like to obtain an independent check on this result. Is it possible to do this without gathering more data from the Africa-Antarctica boundary? Yes, if you have independent information about the motions of any chain of plates linking Africa and Antarctica, say Africa-India-Antarctica. Let's assume, for example, that these angular velocities are known to have the following parameters:

$$_{AN}\omega_{IN} = (18.7°, 32.5°, 0.67°/my)$$
$$_{AF}\omega_{IN} = (17.3°, 46.0°, 0.64°/my)$$

Using vector addition you find:

$$_{AN}\boldsymbol{\omega}_{AF} = {}_{AN}\boldsymbol{\omega}_{IN} + {}_{IN}\boldsymbol{\omega}_{AF}$$
$$= {}_{AN}\boldsymbol{\omega}_{IN} - {}_{AF}\boldsymbol{\omega}_{IN}$$

Converting to Cartesian coordinates and subtracting,

	ω_x	ω_y	ω_z
$_{AN}\boldsymbol{\omega}_{IN}$	.5352	.3410	.2148
$- {}_{AF}\boldsymbol{\omega}_{IN}$	.4245	.4396	.1903
$_{AN}\boldsymbol{\omega}_{AF}$	.1107	$-.0986$	.0245

and converting to spherical coordinates, you arrive at

$$_{AN}\boldsymbol{\omega}_{AF} = (9.4° - 41.7°, 0.15°/\text{my})$$

Comparing the two angular velocity vectors, the Euler poles are displaced by 8° and the velocities are different by 25 percent. Why the discrepancy? The answer is that the two results were obtained from two completely independent sets of data and therefore have independent errors. Moreover, the errors that produced this discrepancy reside in all three angular velocity vectors $_{AN}\boldsymbol{\omega}_{AF}$, $_{AN}\boldsymbol{\omega}_{IN}$, and $_{AF}\boldsymbol{\omega}_{IN}$. The procedure used to make the vectors internally consistent mathematically is to adjust all of the three angular velocity vectors, weighting the individual vectors inversely with their individual errors, until the vectors have perfect internal consistency in the sense that

$$_{AN}\boldsymbol{\omega}_{AF} = {}_{AN}\boldsymbol{\omega}_{IN} + {}_{IN}\boldsymbol{\omega}_{AF} \qquad (4.14)$$

When the three sets of observational data along the three boundaries are analyzed by least-squares analysis, an additional constraint requires that the final angular velocity vectors can be added in any combination with perfect consistency. This consistency can be described as the requirement that around any closed chain of three plates taken in the proper sequence, the vector sum must equal zero:

$$_{AN}\boldsymbol{\omega}_{IN} + {}_{IN}\boldsymbol{\omega}_{AF} + {}_{AF}\boldsymbol{\omega}_{AN} = {}_{AN}\boldsymbol{\omega}_{AN} = \mathbf{0} \qquad (4.15)$$

When the analysis is extended to include all of the plates on the globe, all of the angular velocity vectors must be inter-

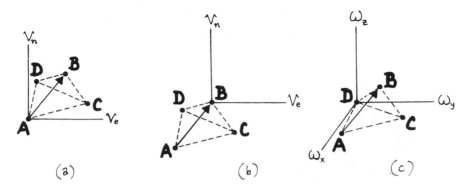

nally consistent in the above sense. All researchers who publish tables of global plate velocities check their results to be sure that they satisfy this criterion.

Angular Velocity Space

In analyzing plate velocities on a plane, we found it useful to plot the velocities in a two-dimensional coordinate system in which the axes were north velocity V_n and east velocity V_e. Similarly we can plot the angular velocity vectors in a three-dimensional coordinate system in which the axes are components of angular velocity in the usual sense, with ω_x and ω_y in the plane of the equator directed toward longitudes 0° and 90°, respectively, and ω_z directed toward the North Pole. As was true on the plane, the velocity of each plate is described by a point in this space, and the relative velocity between any two plates, say of plate B relative to plate A, is described by a vector arrow beginning at point **A** and going to point **B** (Figure 4-19). In both the two- and three-dimensional diagrams the relative velocity is determined by the orientation and length of this vector arrow and not by its position relative to the origin. Because of this convention, the same relative velocities will be found for all plates, regardless of which point is placed at the origin—the relative plate velocities are exactly the same in Figures 4-19a, b, and c. Angular velocity space is thus a simple and direct extension of plane velocity space to three dimensions.

If you have determined the velocities for several pairs of plates and decide to plot these velocities in velocity space, you will quickly discover any inconsistencies in your relative velocities. In the earlier example, if you decide for conve-

Figure 4-19.
Two-dimensional (a) and (b) and three-dimensional (c) plate velocities. In all cases the vector arrow from **A** to **B** represents the velocity of plate B relative to plate A. The length and orientation of a vector do not depend on the choice of origin.

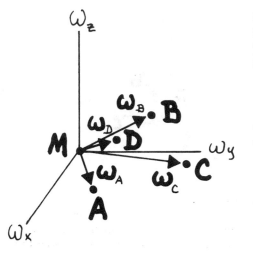

Figure 4-20.
Arrows represent the angular velocities of plates A, B, C, and D relative to M, which may represent either a real or a fictitious plate M, or perhaps an absolute reference frame fixed relative to the lower mantle or mesosphere.

nience to plot the angular velocity of Antarctica at the origin, then the coordinates in angular velocity space for Africa which you found from direct measurements over the Africa-Antarctica Ridge are

$$\omega_x = .1228, \omega_y = -.1364, \omega_z = .0492$$

On the other hand, the coordinates for the angular velocity of Africa using the Antarctica-India-Africa circuit are

$$\omega_x = .1107, \omega_y = -.0986, \omega_z = .0245$$

You would need to reconcile these different values by introducing a suitable least-squares adjustment into your original calculations for each angular velocity vector, or if you were in a hurry, you might pick a value intermediate between the two inconsistent points, using a weighting factor determined by the quality of each datum.

Your freedom to choose any origin you like in representing a set of relative plate motions has some interesting implications. Let's assume that for reasons of your own you have plotted points representing the relative angular velocities of plates A, B, C, and D in such a way that none of the angular velocities is at the origin in angular velocity space (Figure 4-20). You then plot a point **M** at the origin corresponding to an imaginary plate M, so all of the angular velocity vectors describing the motions of plates A, B, C, and D relative to plate M would radiate out from point **M** to the other points. Then all of the information about the relative velocities between all of the plates can be specified simply by listing in a table the relative velocities $_M\omega_A$, $_M\omega_B$, $_M\omega_C$, and $_M\omega_D$. As shorthand we might shorten this to ω_A, ω_B, ω_C, and ω_D where by ω_B we mean the angular velocity of plate B relative to plate M. An advantage of using this representation is that from a list of the values of ω_A, ω_B, ω_C, and ω_D the relative velocities of any two plates, say plate D relative to plate B, can be found by simple vector subtraction:

$$_B\omega_D = -\omega_B + \omega_D = \omega_D - \omega_B \qquad (4.16)$$

Plate M might be regarded as a fictitious plate chosen for mathematical convenience. For example, the velocity $V = 0$ of plate M might be regarded as describing a reference frame that remains fixed in the sense that we choose to

Table 4-4. Rotation vectors relative to plate "M."

Plate	Code	Spherical			Cartesian		
		λ	ϕ	ω	ω_x	ω_y	ω_z
Africa	AF	18.8°	338.2°	0.139	.1222	−.0489	−.0448
Antarctica	AN	21.8°	75.6°	0.054	.0125	.0486	−.0201
Arabia	AR	27.3°	356.1°	0.388	.3440	−.0235	.1780
Caribbean	CB	−42.8°	66.8°	0.129	.0373	.0870	−.0876
Cocos	CC	21.9°	244.3°	1.422	−.5722	−1.1889	.5304
Europa	EU	0.7°	336.8°	0.038	.0349	−.0150	.0005
India	IN	19.2°	35.6°	0.716	.5498	.3936	.2355
North America	NA	−58.3°	319.3°	0.247	.0984	−.0846	−.2102
Nazca	NZ	48.0°	266.2°	0.585	−.0259	−.3906	.4347
Pacific	PA	−61.7°	97.2°	0.967	−.0575	.4548	−.8514
South America	SA	−82.3°	75.7°	0.285	.0094	.0370	−.2824

Note: ω, ω_x, ω_y, and ω_z are listed in °/my.

assign it zero velocity. Some researchers refer to an absolute reference frame that remains fixed with respect to the lower mantle, or mesosphere, and they suggest different ways for finding the velocity of plates relative to this reference frame (see Chapter 10). "Plate M" would then be replaced by "lower mantle" or "mesosphere" in the above discussion. Regardless of whether you find this approach useful, listing the velocity of each plate relative to a common plate or reference frame with an assumed velocity of zero is a very compact and self-consistent means of presenting all of the relative velocity information. The vectors in such a listing can be regarded as points in a velocity space which has the fictitious plate M at its origin. Table 4-4 presents a current evaluation of the angular velocities of the major plates in such a velocity space.

To dramatize how this calculation of relative plate velocities does not depend on the choice of origin, suppose that a colleague were to tell you that she has caught on to your use of "plate M" to thinly disguise your presentation of angular velocities in some secret "absolute" reference frame, the definition of which is known only to you. Further, your colleague tells you that she knows you are wrong because she has just determined a better absolute reference frame, and in her reference frame yours is rotating about the North Pole at a rate of 1°/my. In other words, the velocity of your metaphorical plate M should be plotted with velocity coordinates (0, 0, 1.0000) instead of (0, 0, 0). Your colleague's table of plate velocities is exactly like yours, except that the values of ω_z are all increased by adding exactly 1.000. When you subtract any two lines from her table to obtain a relative plate velocity, the resultant vector is exactly the same as when you used your own table. In other words, the disagreement about the meaning of "absolute" plate velocity does not affect the values you determine for the relative velocities of plates. Conversely, determinations of the relative velocities between plates does not determine their absolute velocities until some additional information or assumptions are brought to bear on the problem.

Finding the Local Velocity V from the Angular Velocity ω

Finding local velocities on the globe requires techniques that you already know how to use. Here's how to proceed. Let's say that you are standing at point **P** on plate B and that you know the angular velocity $_A\boldsymbol{\omega}_B$. To find your local velocity at **P**, recall that your motion relative to plate A will be along a small circle centered on the Euler pole **E** corresponding to $_A\boldsymbol{\omega}_B$. The great circle tangent to this small circle at **P** describes your direction T of motion. If δ is the distance from **P** to **E** and D is the local azimuth of the great circle $<$**P**, **E**$>$ from **P** to **E**, then your direction of motion is $T = D + 90°$ and your speed is V (mm/yr) $= \omega$(°/my) 111 sin δ.

If you prefer algebra to spherical geometry, you can find the local velocity from

$$\mathbf{V} = \boldsymbol{\omega} \times (R\ \mathbf{P}) = \omega R \mathbf{E} \times \mathbf{P} \qquad (4.17)$$

where R is the radius of the earth. The coordinates of **P** (the position unit vector for the point of observation) and **E** (the Euler pole for the rotation) are usually given in global (x, y, z) coordinates, where **x** and **y** are in the plane of the equator and **z** is along the rotation axis. You will probably want to know the direction of **V** in a local (n, e, d) coordinate system with axes of northward, eastward, and downward. Box 4-1 shows how to carry out the cross product multiplication of **ω** and **P** and Box 4-2 shows how to convert the

---

Box 4-1. Velocity Due to Angular Rotation: An Algebraic Approach.

If you prefer algebra to geometry, you may want to use the cross product of the angular velocity vector and the position vector to find the linear velocity of a point on plate B relative to plate A. Let **P** = (P_x, P_y, P_z) be the components of the unit vector describing the position of point **P** located on plate B, and let **E** = (E_x, E_y, E_z) be the coordinates of the Euler pole for the motion of plate B relative to plate A. Then if ω is the scalar angular rotation rate, the velocity of point **P** relative to plate A is given by

$$\begin{aligned}\mathbf{V} &= (\omega\,\mathbf{E})\times(R\,\mathbf{P}) = \omega\,R\,(\mathbf{E}\times\mathbf{P})\\ &= \omega\,R\,(E_x, E_y, E_z)\times(P_x, P_y, P_z)\\ &= \omega\,R\,(E_y\,P_z - P_y\,E_z,\; E_z\,P_x - P_z\,E_x,\; E_x\,P_y - P_x\,E_y)\end{aligned}$$

where R is the earth's radius. Since $V = \omega\,R\sin\delta$, the length of the vector defined in parentheses is obviously equal to $\sin\delta$, as it should be for the cross product of the two unit vectors **E** and **P**. In the above expression we have found the components of the velocity in the equatorial plane (x and y components) and along the rotation axis (z component). We're usually interested in the northward, eastward, and downward (n, e, and d) components. Box 4-2 shows how to convert between the two coordinate systems.

---

results from global components to local components. When the dust settles, for mathematical and physical reasons the vector **V** should be tangent to the sphere, and V_d should equal zero.

Box 4-2. How to Change from Global (x, y, z) Coordinates to Local (n, e, d) Coordinates.

If we know the global Cartesian components of a vector **V** = (V_x, V_y, V_z), the local Cartesian components, northward (n), eastward (e), and downward (d), at some point **A** will depend upon the coordinates (λ, ϕ) of **A**.

To find the local Cartesian coordinates from our global coordinates, we define the 3×3 matrix **T** as follows.

$$\mathbf{T} = \begin{bmatrix} T_{nx} & T_{ny} & T_{nz}\\ T_{ex} & T_{ey} & T_{ez}\\ T_{dx} & T_{dy} & T_{dz} \end{bmatrix}$$

where $T_{nx} = \mathbf{n}\cdot\mathbf{x}$, $T_{ey} = \mathbf{e}\cdot\mathbf{y}$, $T_{dz} = \mathbf{d}\cdot\mathbf{z}$, etc. Note that the vectors **x**, **y**, and **z** define the columns, and the vectors **n**, **e**, and **d** define the rows.

Because **e** is tangent to the small circle of latitude λ, and circles of latitude are defined by planes perpendicular to **z**, we know that $\mathbf{e}\cdot\mathbf{z} = 0$. By similar reasoning we arrive at the following table:

$T_{nx} = -\sin\lambda\cos\phi$	$T_{ny} = -\sin\lambda\sin\phi$	$T_{nz} = \cos\lambda$
$T_{ex} = -\sin\phi$	$T_{ey} = \cos\phi$	$T_{ez} = 0$
$T_{dx} = -\cos\lambda\cos\phi$	$T_{dy} = -\cos\lambda\sin\phi$	$T_{dz} = -\sin\lambda$

(continued)

Box 4-2. *(continued)*

Operation of this matrix is written in the form

$$\mathbf{V}_L = \mathbf{T}\,\mathbf{V}$$

and is defined as follows.

$$\begin{bmatrix} V_n \\ V_e \\ V_d \end{bmatrix} = \begin{bmatrix} T_{nx} & T_{ny} & T_{nz} \\ T_{ex} & T_{ey} & T_{ez} \\ T_{dx} & T_{dy} & T_{dz} \end{bmatrix} \begin{bmatrix} V_x \\ V_y \\ V_z \end{bmatrix} = \begin{bmatrix} T_{nx}V_x + T_{ny}V_y + T_{nz}\,V_z \\ T_{ex}V_x + T_{ey}V_y + T_{ez}\,V_z \\ T_{dx}V_x + T_{dy}V_y + T_{dz}\,V_z \end{bmatrix}$$

This means that

$$V_n = \mathbf{n}\cdot\mathbf{V} = T_{nx}\,V_x + T_{ny}\,V_y + T_{nz}\,V_z$$
$$V_e = \mathbf{e}\cdot\mathbf{V} = T_{ex}\,V_x + T_{ey}\,V_y + T_{ez}\,V_z$$
$$V_d = \mathbf{d}\cdot\mathbf{V} = T_{dx}\,V_x + T_{dy}\,V_y + T_{dz}\,V_z$$

Note that the first component equals the first row dotted with the vector.

To convert a vector expressed in local (n,e,d) components to one in global (x,y,z) components, we simply use the inverse matrix

$$\mathbf{T}^{-1} = \begin{bmatrix} T_{nx} & T_{ex} & T_{dx} \\ T_{ny} & T_{ey} & T_{dy} \\ T_{nz} & T_{ez} & T_{dz} \end{bmatrix}$$

so that

$$\mathbf{V} = \mathbf{T}^{-1}\,\mathbf{V}_L$$

where

$$V_x = \mathbf{x}\cdot\mathbf{V}_L = T_{nx}\,V_n + T_{ex}\,V_e + T_{dx}\,V_d$$

etc. Note that in the inverse matrix $\mathbf{T}^{-1}$, $\mathbf{x}$, $\mathbf{y}$, and $\mathbf{z}$ define the rows.

Problems

4-1. Find the rotation vector $\boldsymbol{\omega} = (\lambda, \phi, \omega)$ for the following pairs of plates.

(a) Cocos-Pacific.

(b) North America-South America.

4-2. A point located in the Mid-Atlantic Ridge in the South Atlantic has coordinates $(-30°, -15°)$.

(a) What azimuth would you expect for a nearby fracture zone?

(b) What is the azimuth and length (in mm/yr) of the velocity vector of a point on the east scarp of the rift valley relative to a point on the west scarp?

4-3. The data in the following table are observations of the azimuths of transforms and velocities of spreading along the Mid-Atlantic Ridge. Find the angular rotation vector most nearly consistent with both types of data and make a rough estimate of your error. Include a plot of variations of velocity as a function of distance from the Euler pole.

Site	Locality λ	Locality φ	Azimuth of Transform	Spreading Rate
1	80°	−1°	308°	
2	71°	−8°	295°	
3	67°	−18°	295°	
4	66°	−20°	278°	
5	52°	−35°	276°	
6	85°	90°		10 mm/yr
7	70°	−18°		16 mm/yr
8	60°	−29°		24 mm/yr
9	45°	−28°		28 mm/yr

4-4. The Galapagos Islands with coordinates $(0°, -90°)$ are located adjacent to the Nazca-Cocos spreading center. How far apart (in km) would you expect the 0.7 Ma isochrons to be at that locality?

Suggested Readings

General

The North Pacific: An Example of Tectonics on a Sphere, Dan P. McKenzie and Robert L. Parker, Nature, v. 216, p. 1276–1280, 1967. *Classic article introducing concept of plotting plate velocities in linear and angular velocity space.*

Rises, Trenches, Great Faults, and Crustal Blocks, W. Jason Morgan, Journal of Geophysical Research, v. 73, p. 1959–1982, 1968. *Classic article describing how plate tectonics works on a sphere; first calculation of global set of Euler poles.*

Sea Floor Spreading and Continental Drift, Xavier Le Pichon, Journal of Geophysical Research, v. 73, 3661–3705, 1968. *Gives global set of Euler poles and tests them by plotting plate boundaries on Mercator projections projected from the Euler poles.*

Sources of Data

The N-Plate Problem of Plate Tectonics, Clement Chase, Geophysical Journal Royal Astronomical Society, v. 29, p. 117–122, 1972. *Classic paper on obtaining globally consistent set of Euler poles.*

Present-Day Plate Motions, J. B. Minster and Thomas Jordan, Journal of Geophysical Research, v. 83, p. 5331–5351, 1978. *Source of data summarized in Table 4-4; gives lists of data used to calculate Euler poles and confidence limits for poles.*

Plate Kinematics: The Americas, East Africa and the Rest of the World, Clement Chase, Earth and Planetary Science Letters, v. 37, p. 355–368, 1978. *Globally consistent set of Euler poles.*

Statistical Test of Intraplate Deformation from Plate Motion Inversions, Seth Stein and Richard Gordon, Earth and Planetary Science Letters, v. 69, p. 401–412, 1984. *Contains discussion of least-squares techniques for obtaining Euler poles from observations.*

5

Plotting Planes and Vectors in Local Coordinates

In analyzing vectors and planes we have casually shifted back and forth between two coordinate systems, one global and one local. In the global system we have described vectors either in terms of the spherical coordinates of latitude λ and longitude ϕ or, alternatively, in terms of Cartesian components x, y, and z, where $\mathbf{x}$ and $\mathbf{y}$ are in the plane of the equator and $\mathbf{z}$ is directed toward the North Pole. This system has been especially useful in analyzing vectors of global interest such as the angular velocities of plates. In discussing vectors and planes of local interest we have sometimes found it more useful to employ a local coordinate system with three perpendicular axes in the north, east, and downward (n, e, d) directions. For example, in Chapter 4 we used (n, e, d) coordinates to describe the angular rotation of a fan on a turntable. In the next few chapters we will want to use local coordinates to analyze several different types of local vectors that are important in plate tectonics, including earthquake slip vectors, magnetic field vectors, and the velocity of a point on a plate. Before tackling these subjects, we will explore some of the techniques that are useful for analyzing vectors and planes in a local reference system using both Cartesian and spherical coordinates. Since the underlying ideas are the same as the ones you have already mastered for global coordinates, you will probably find this chapter to be as easy as it is short.

159

Referring to the two coordinate systems as global and local reflects our habit of thinking of vectors as arrows located in some particular place. We think of global vectors as being located at the center of the earth. We think of local vectors as being located at some point on the earth's surface. Mathematically, however, a vector is completely described by specifying its orientation and length. These are independent of its location and do not change if the vector is translated from the earth's center to the surface. We took advantage of this translatability of vectors in constructing velocity diagrams. Admittedly it is sometimes useful in developing one's intuition to think of a vector as being attached to some particular point in the earth. For example, a point **R** on the earth's surface can be represented by the vector **R**, which we can visualize as an arrow with its tail at the earth's center and its head at the point **R** on the surface. However if it suits our purposes, we can regard the vector **R** as being located in outer space (or on the earth's surface). We can always translate a vector conceptually, provided it keeps the same orientation and length.

Another preliminary point to keep in mind is that there are as many different local (n, e, d) coordinate systems as there are points on the earth's surface. For example, the "down" direction at the equator is perpendicular to the "down" direction at either of the geographic poles. In fact, the spatial orientations of **n**, **e**, and **d** unit vectors are generally different at any two points on the earth's surface. The n axis is only constant along the equator, the e axis is only constant along the same meridian of longitude, and no two d axes are the same. This is in contrast to the global (x, y, z) coordinate system, the axes of which are always the same. In solving problems, it's important to remember not to mix coordinate systems: (n, e, d) coordinates for different points on the earth's surface cannot be analyzed together, nor can (n, e, d) and (x, y, z) coordinates. Always transform the vectors to one coordinate system by the equations listed in Box 4-2, remembering that the equations for the conversion yield different values for each point on the earth's surface where the n, e, and d components are determined.

Inclination and Declination

To describe the orientation of a local vector **A**, first pass a vertical plane through the vector (Figure 5-1). The **declination** D is the azimuth or trend of this vertical plane. D is

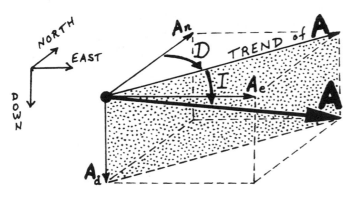

D: AZIMUTH OR DECLINATION CLOCKWISE FROM NORTH
I: INCLINATION OR PLUNGE BELOW HORIZONTAL

Figure 5-1.
Inclination *I* and declination *D*. *I* is measured downward from the horizontal in the vertical (shaded) plane passing through the vector **A**.

measured in degrees clockwise from north, with a range of values from 0° to 360°. The **inclination** *I* is the dip or plunge of the vector below the horizontal, measured in the vertical plane through **A** (Figure 5-1). The range of values of *I* is from −90° to +90°, with **positive** values indicating plunge **below** the horizontal. Vectors with **negative** inclinations are pointed **above** the horizontal.

We can regard a local vector as consisting of two elements: a scalar length that is simply a number, and a unit vector that specifies the orientation of the vector. Imagine a local unit vector in a sphere of unit radius (1 meter, for example) with its tail at the center of the sphere and its head touching the surface of the sphere. The point of contact can be used to describe the orientation of the unit vector. Therefore, if we can find a suitable grid system to describe the location of points on the unit sphere in local coordinates, this same grid system can be used to describe the orientation of unit vectors. A very useful grid consists of curves of constant inclination and curves of constant declination, which are analogous to the latitude and longitude grid lines on a globe. The curve corresponding to *I* = 0 is a great circle formed by the intersection of the sphere with a horizontal plane passing through the center of the sphere, analogous to the equator. Curves of constant inclination *I* correspond to small circles formed by the intersection of other parallel horizontal planes with the sphere. These small circles are concentric about two points on the sphere corresponding to the downward +**d** and upward −**d** directions. They are thus analogous to circles of latitude, concentric about the geographic poles.

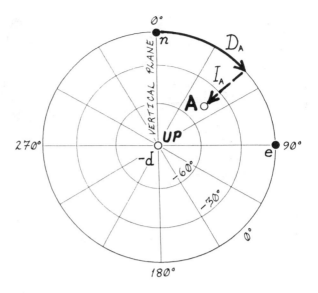

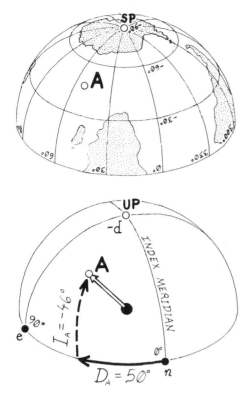

Figure 5-2.
On an inverted globe surrounding the local unit vector **A**, the latitude and longitude circles can be used to describe the inclination and declination of the unit vector. The equator of the globe corresponds to the set of unit vectors lying in a horizontal plane, and the index meridian of the globe corresponds to unit vectors with varying inclinations that trend northward. The inclination and declination of vector **A** can be read from the old latitude and longitude lines.

Curves of constant declination are great circles that correspond to the intersection with the sphere of vertical planes passing through the center of the sphere. All of these great circles intersect at two points on the sphere, one defining the downward or $+\mathbf{d}$ direction and one defining the upward or $-\mathbf{d}$ direction. Circles of constant declination are thus analogous to meridians of longitude, all of which intersect at the geographic poles.

It may help you to visualize this grid system by taking a transparent globe of the earth, flipping it upside down so that the South Pole is pointing straight up, and erasing the continents but not the grid lines (Figure 5-2). You now have a ready-made grid system with which to describe the points corresponding to unit vectors. Simply orient the globe with its old **x** axis (0°, 0°) pointed north. With the globe oriented in this manner, its old Greenwich meridian, $\phi = 0°$, corresponds to declination $D = 0°$. Positive (down) inclinations are plotted on the lower (down) hemisphere. In Figure 5-2, the vector **A** is inclined upward to the northeast. In the old coordinates of the globe, the geographic coordinates of **A** are $\lambda = -46°$ and $\phi = 50°$. We can translate this vector directly to the orientation of $I = -46°$ and $D = 50°$.

An advantage of plotting unit vectors on a sphere of unit radius, as you have just done, is that you can then use all of the tricks you learned earlier in working with a globe representing the earth. It's important to remember that in doing this, you are simply taking advantage of the fact that in both

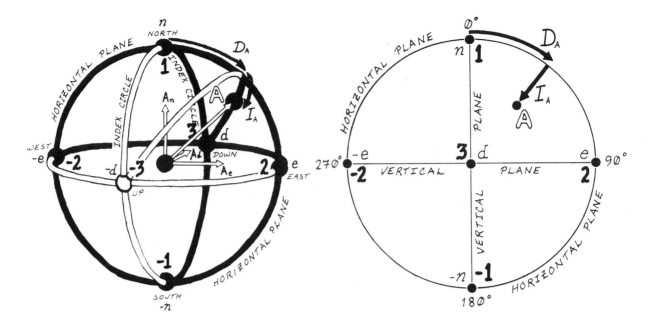

Figure 5-3.
Fixed reference frame for local (*n, e, d*) coordinates showing the inclination *I* and the declination *D;* **A** plotted on the lower hemisphere of a polar projection.

types of analysis you are using a spherical grid system consisting of (1) a set of great circles passing through a common point at the center of the projection and (2) a set of small circles concentric about the point of intersection of the great circles. To emphasize the essential identity of the globe and vector grid systems, Figure 5-3 shows a "fixed reference frame" that describes the local polar projection grid system. We imagine vectors as points plotted on a globe nestled within this frame. Points may be rotated about the ± **1** and ± **3** axis, as with a global polar projection. The grid system of Figure 5-3 is, in fact, exactly the same as that of the polar projection we used in plotting points on the globe. The center of the projection now corresponds to the up or −**d** direction rather than the −**z** axis or South Pole. As before, solid symbols and continuous lines are used to plot on the lower hemisphere, and open symbols and dashed lines are used to plot on the upper hemisphere.

Local Cartesian Components

Sometimes it is convenient to analyze a unit vector using its northward, eastward, and downward components. Noting that (Figure 5-1) the horizontal component of unit vector **A** is equal to cos *I,* it is easy to see that the Cartesian components of **A** are

$$A_n = \cos I \cos D \qquad (5.1)$$

$$A_e = \cos I \sin D \qquad (5.2)$$

$$A_d = \sin I \qquad (5.3)$$

If you know the Cartesian components, the angles I and D can be found from

$$I = \sin^{-1} A_d \; ; \; -90° \leq I \leq 90° \qquad (5.4)$$

$$D = \tan^{-1} (A_e/A_n) \; ; \; 0° \leq D < 360° \qquad (5.5)$$

Boxes 5-1 through 5-4 show how to plot vectors and planes in local coordinates using a stereographic or equal area projection.

Faults and Slip Vectors

Faults along plate boundaries are especially important in plate tectonics. Two fault parameters are of particular interest: The orientation of **fault planes** and the orientation of **slip vectors** for individual earthquakes. The orientation of fault planes, like that of other planar surfaces, can be specified by giving the strike and the dip of the plane. Box 5-4 defines these terms and shows how they're used. An alternative description is the orientation of a line perpendicular to the plane, which is usually called the **pole** of the plane. (Keeping track of all the different kinds of poles used in plate tectonics will be a recurring theme as we progress.) Boxes 5-3 and 5-4 review the ways of plotting planar surfaces and their poles.

Normal faults, which occur along diverging boundaries, are characterized by (1) steeply dipping fault planes and (2) downward motion of the block on the upper side of the fault (Figure 5-4). This is the type of faulting you would expect in a region of tension where plates are being pulled apart. The slip vector can be visualized as a scratch made on one block during an earthquake by a sharp rock embedded in the block on the other side of the fault plane. More formally, if block B is regarded as fixed during an earthquake (Figure 5-4), then the slip vector $_B\mathbf{S}_A$ describes the motion of block A during the earthquake. Similarly $_A\mathbf{S}_B$ describes the motion of block B relative to block A. Slip vectors are similar to relative velocity vectors in the sense that the vectors $_A\mathbf{S}_B$ and $_B\mathbf{S}_A$ are antiparallel, having the same length but exactly opposite directions (Figure 5-4), so that $_B\mathbf{S}_A = -_A\mathbf{S}_B$. In normal faulting, the orientation of the slip vector in the plane of the fault is perpendicular to the horizontal strike of the fault.

Thrust faults, which occur along converging boundaries, are characterized by fault planes with upward motion

---

Box 5-1. Plotting Vectors as Points on Polar Projections.

Although polar projections are almost always used as the "base map" upon which vectors are plotted in publications, equatorial projections, because of their greater flexibility, are generally used for the actual plotting. We will use an equatorial projection to plot the following two unit vectors on a polar projection:

$$\text{Vector } \mathbf{A}: \quad I_A = 35°, D_A = 290°$$
$$\text{Vector } \mathbf{B}: \quad I_B = -55°, D_A = 150°$$

• Label axes on globe

Place over the grid representing the fixed reference frame a piece of tracing paper representing the globe upon which the vectors will be plotted as points. Draw on the tracing paper the points **n**, **e**, and **d** representing north, east, and down axes attached to the globe. Pin the paper at the $\pm\mathbf{d}$ axis at the center of the fixed reference frame.

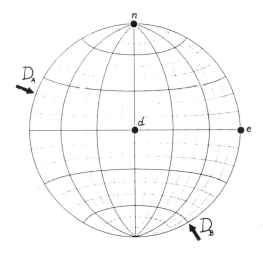

• Draw arrows showing declinations of **A** and **B**

Draw arrows on the outer circle showing the declination of the two vectors, with declination angles measured clockwise from **n** at the top of the projection.

(continued)

Box 5-1.   *(continued)*

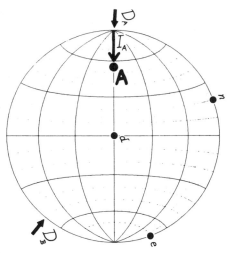

- Rotate D_A arrow to top of projection

Rotate the tracing paper clockwise until the D_A arrow is aligned with the vertical axis along which inclination angles can be measured.

- Plot vector **A**

Starting at the outer circle of the projection, which represents a horizontal plane, count 35° downward from the location of the D_A arrow at the top of the grid system. Because inclination is positive, plot vector as solid circle on lower hemisphere.

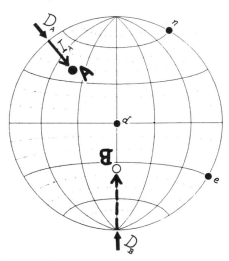

- Rotate D_B arrow to top of projection

See the third step, above.

(continued)

Box 5-2. *(continued)*

- Plot vector **B**

Count 55° from the location of the D_B arrow at the top of the grid system. Because inclination is negative, plot vector as open circle on upper hemisphere.

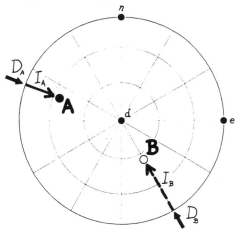

- Rotate **n** to top of projection

Vectors **A** and **B** are now in correct positions on a standard polar projection.

Box 5-2. Measuring the Angle Between Two Vectors.

To find the angle between two vectors, we use the same procedure we used to find the angular distance between two points on the globe (Box 3-2). This is one of the many operations that are impossible on polar projections but are easy on equatorial projections. Here we find the angle between vectors **A** and **B** of Box 5-1.

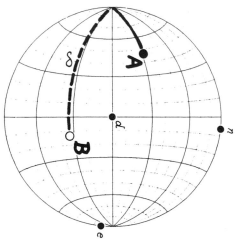

(continued)

Box 5-2.   *(continued)*

• Rotate about center of projection

Rotate the tracing paper until **A** and **B** lie on the same great circle. Note that the traces of the great circle on the upper (dashed line) and lower (solid line) hemisphere are symmetrical about the vertical axis of the projection.

• Read angle δ = 146°

Count the number of degrees along the great circle between **A** and **B**.

Box 5-3.   Plotting a Plane If You Know the Pole.

This is analogous to plotting a great circle on a globe if you know its pole, as described in Box 3-3. We will plot the plane with the pole of inclination $I = 40°$ and declination $D = 190°$.

• Plot pole

Mark arrow showing the declination D_P of the pole on the outer circle. Proceed as in Box 5-1 to plot pole **P**.

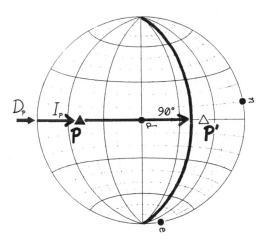

• Rotate about center of projection

Rotate the tracing paper until the arrow marking D_P and the pole **P** are aligned with the horizontal axis through the center of the projection.

(continued)

Box 5-3. *(continued)*

● Draw plane

Plot the desired plane along the great circle that passes through the point 90° from the pole along the horizontal axis through the center of the projection. Note that **P** and **P′** are 180° apart and that both are poles of the plane.

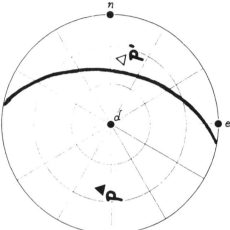

● Rotate about center of projection

Rotate tracing paper until **n** is at the top of the projection. The circle is now in its correct position on the lower hemisphere of a polar projection. The plot of the circle on the upper hemisphere (not shown) is also 90° from **P** and **P′**.

Box 5-4. Plotting Planes If You Know the Dip and Strike.

A geologist in the field describes the orientation of a bedding plane or fault plane by recording the strike and dip of the plane. The **strike** is the azimuth or trend of a horizontal line lying in the plane. If the plane dips into a lake, the strike line is the water line. The expression "strike N 20° E" and the expression "strike S 20° W" both describe a strike line with an azimuth of 20° or 200°, which are equivalent. The **dip** describes the inclination of the plane below (+) or above (−) the horizontal. Dip is measured in a direction perpendicular to the strike. Usually a positive dip is recorded, along with a description of the quadrant toward which the plane dips downward. For example, "strike N 40° W, dip 30° SW" describes a plane dipping 30° to the southwest. The equivalent expression "dip −30° NE" is rarely used. We will now plot a plane corresponding to the orientation of a fault with strike N 80° W, dip 40° SW.

(continued)

Box 5-4. *(continued)*

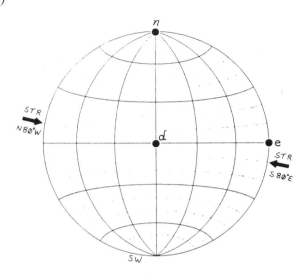

● Plot strikes

Plot two arrows along the outer circle of the projection at the two equivalent strikes for the plane, which are 180° apart.

● Plot dip quadrant

Plot the quadrant symbol *"SW"* along the outer circle of the projection 90° from the strike arrows.

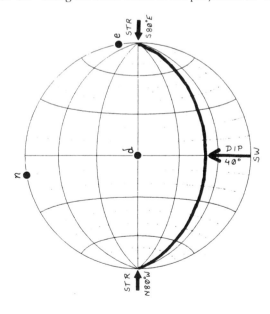

(continued)

Box 5-4. *(continued)*

- Rotate

Rotate tracing paper until the strike arrows are at the top and bottom of the projection.

- Draw plane

Starting at the symbol "*SW*," count 40° downward along the horizontal line through the center of the projection and draw the plane through this point. This plane dips 40° below the horizontal.

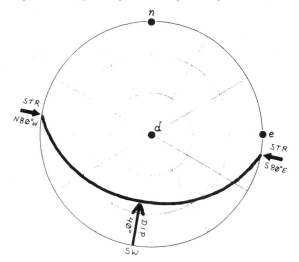

- Rotate

Rotate the tracing paper until **n** is back at the top of the projection. The plane describing the fault plane is now plotted properly on a polar projection.

of the block on the upper side of the fault (Figure 5-5). This is the type of faulting you would expect in a region undergoing compression, where plates are converging. If block B is regarded as fixed during an earthquake (Figure 5-5), then the slip vector $_BS_A$, which describes the motion of block A relative to block B, is directed up the plane of the fault. In thrust faulting as in normal faulting, the orientation of the slip vector in the plane of the fault is perpendicular to the strike of the fault, but the sense of the motion is opposite between normal and thrust faults. In thrust faulting, the slip vector for the upper block is directed up the plane of the fault.

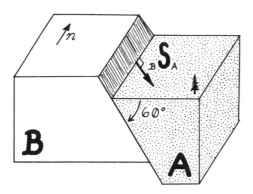

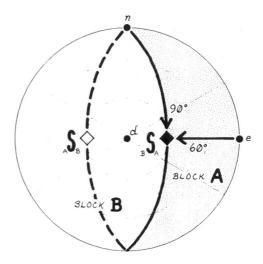

Figure 5-4.
Normal fault striking due north. The projection can be visualized as a unit sphere with its center located in the plane of the fault. The shaded region shows the part of the lower hemisphere of the unit sphere that is occupied by block A. On the projection, the fault plane is the great circle through **n**, shown by a solid line on the lower hemisphere and a dashed line on the upper hemisphere. Solid square (lower hemisphere) is the slip vector showing the motion of block A relative to block B. Open square (upper hemisphere) is the slip vector for the motion of block B relative to block A.

Transform faults are an example of the more general class of **strike-slip faults,** along which blocks simply slide past each other without converging or diverging. Strike-slip faults are characterized by (1) steeply dipping fault planes and (2) horizontal motion in a direction parallel to the strike of the fault (Figure 5-6). Examples of strike-slip faults include transforms like those that offset the Mid-Atlantic Ridge and the San Andreas fault between the North America and Pacific plates. The slip vectors $_A\mathbf{S}_B$ and $_B\mathbf{S}_A$ are antiparallel.

A fault with a given dip, say 70°, can be either normal, thrust, or strike-slip, depending on the orientation of the slip vector in the plane of the fault. In nature, the orientation of the slip vector is rarely exactly perpendicular or parallel

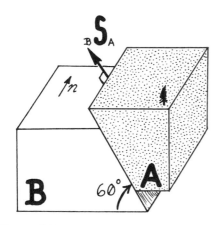

Figure 5-5.
Thrust fault striking due north. Conventions are the same as in Figure 5-4. Note that the subscripts of the slip vectors are reversed from those for the normal fault in Figure 5-4.

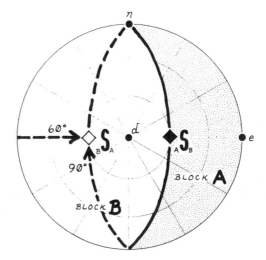

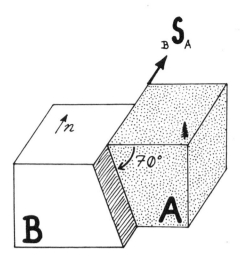

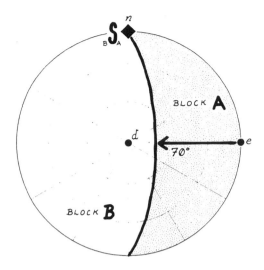

to the strike of the fault, as specified in the above definitions for ideal cases; however, the terms normal, thrust, and strike-slip are still used if the slip vectors are oriented within 20° or so of the ideal directions. Otherwise the earthquake is said to have two components of motion, for example, strike-slip and normal or strike-slip and thrust. (Faults never have components of both normal and thrust motion for obvious reasons.) Figure 5-7 shows a fault with strike-slip and normal components of motion.

In Chapter 4 we used the azimuth or trend of slip vectors in exactly the same way that we used the trend of transforms. In both cases we are interested in only the horizontal component of the motion between two plates. The trend of the horizontal component of the slip vector is simply equal to the declination D of the slip vector.

Figure 5-6.
Strike-slip fault striking due north. Conventions are the same as in Figure 5-4.

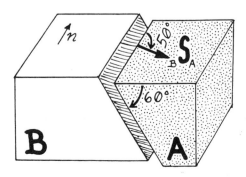

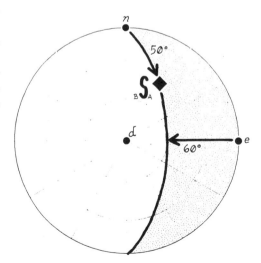

Figure 5-7.
Fault with components of normal and strike-slip motion. Conventions are the same as in Figure 5-4. The slip vector $_B\mathbf{S}_A$ lies in the plane of the fault.

Problems

5-1. The vectors described in the following table include five that are in the same plane and two that are not.

(a) Plot all of the vectors as points on an equal area projection.

(b) Draw a great circle through the five points corresponding to the coplanar vectors.

(c) Plot the poles for the plane on both hemispheres.

(d) List the coordinates of the two poles.

(e) Find the angle between vector **A** and each of the other vectors.

Vector	I	D	Angle to A
A	25°	230°	0°
B	43°	259°	30°
C	43°	274°	
D	50°	288°	
E	58°	303°	
F	49°	310°	
G	43°	335°	

5-2. The following list gives the inclinations and declinations of poles of planes.

(a) Plot all of the poles.

(b) Draw the planes corresponding to the poles.

(c) If you're good at recognizing patterns, you will be intrigued by your plot. Try to make as general a statement as you can describing the pattern you see.

Pole	I	D
A	18°	55°
B	57°	68°
C	75°	77°
D	68°	103°
E	40°	122°

5-3. (a) Plot the following poles and the corresponding planes.

(b) Give in words any generalizations you can make about the relationships of the poles and planes.

Pole	I	D
A	28°	43°
B	36°	157°
C	40°	285°

5-4. The orientation of a fault plane is strike N 40° E, dip 60° SE.

(a) Plot the fault plane on an equal area projection.

(b) Plot the pole of the fault plane.

(c) Draw four planes perpendicular to the fault plane.

(d) On the equal area plot, write the dip and strike of each of these four planes.

5-5. In the following list of fault planes, find the angle between fault plane A and each of the other fault planes. (Recall that the angle between two planes is equal to the angle between their poles.)

Fault Plane	Strike	Dip	Angle with A
A	N 60° W	30° SW	0°
B	N 60° E	70° SE	58°
C	S 15° W	20° SE	
D	S 70° E	60° NE	
E	N 45° E	80° SE	
F	-	0°	

5-6. For each of the following fault planes and slip vectors, assume that along the fault surface block A is above block B and assume that the slip vector listed gives the motion of block B relative to block A. For each of the fault planes and slip vectors:

(a) Determine the angle in the fault plane between the slip vector and the horizontal.

(b) State whether the fault is normal, thrust, strike-slip, or composite.

Fault Plane		Slip Vector		Answer
Strike	Dip	I	D	
S30° W	30° SE	−30°	−60°	90°, normal
S70° E	20° NE	10°	85°	
N30° W	40° NE	60°	170°	
N45° W	40° NE	30°	0°	
S20° W	70° SE	0°	20°	
N45° E	45° SE	−45°	−45°	
340°	50° SW	−23°	0°	

5-7. In analyzing the radiation pattern of earthquake waves, we will need to fit data on a sphere with pairs of perpendicular planes that divide the unit sphere into four equal quadrants.

To get a feeling for this kind of analysis, plot the points listed in the following table on a projection and try to find two perpendicular planes oriented so that all points fall in opposing quadrants that are separated by quadrants without points. List the strikes and dips of the two planes and the inclination and declination of their poles. State briefly how you made sure that the two planes are perpendicular to each other.

I	D	I	D
72°	73°	45°	160°
43°	95°	70°	160°
4°	138°	7°	170°
32°	142°	45°	287°
17°	154°	7°	358°

Suggested Readings

Structural Geology—An Introduction to Geometrical Techniques, 3d ed., Donal Ragan, John Wiley, New York, 393 pp., 1984. *Introduction to use of stereographic and equal area projections in structural geology.*

The Techniques of Modern Structural Geology, v. 1, Strain Analysis, Johan Ramsay and Martin Huber, Academic Press, London, 307 pp., 1983. *Use of stereographic and equal area projections to plot the orientation of faults, lineations, and other elements of structural geology.*

The Use of the Stereographic Projection in Structural Geology, Frank Phillips, Edward Arnold, London, 1971. *Classic text.*

6

Earthquakes and Plates

Most of the world's earthquakes originate along the boundaries between plates (Figure 6-1). In fact, if you were given a map of the world showing only the locations of large earthquakes, you would probably be able to guess where most of the plate boundaries are located. Earthquakes occur along trenches, ridges, and transforms. Those generated along each of the three types of boundary are distinctly different. Small to moderate earthquakes are generated along ridges at depths of 10 kilometers or less. Larger earthquakes are generated along transforms at depths up to 20 km. The 1906 San Francisco earthquake is an example of a large earthquake along a transform. The very largest earthquakes occur along subduction zones. These are also the deepest earthquakes, occurring at depths up to 700 km. The 1964 Alaskan earthquake was a subduction zone earthquake.

Birth of an Earthquake

Why do earthquakes occur along plate boundaries? Wouldn't it be reasonable to suppose that a subducting slab would slide quietly into its trench, like a toboggan sliding down a hill? The answer would be yes if it weren't for friction. This makes living near plate boundaries slightly more dangerous and much more interesting than living in the interior of a plate. The friction along the fault between two plates appears to be highly variable. In places the friction is low and plates

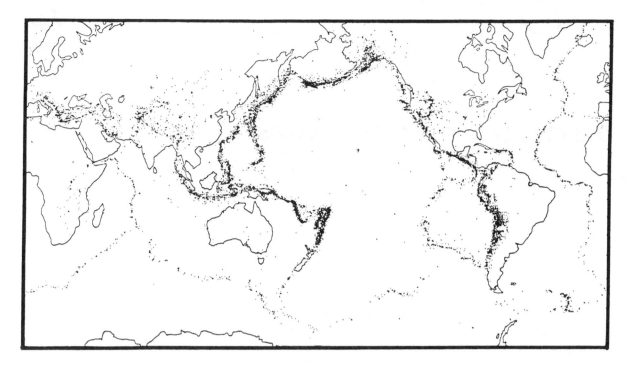

Figure 6-1.
Earthquake epicenters for the period 1961–1968 are shown by small dots. The boundaries of the major plates are clearly outlined with narrow belts of smaller earthquakes along spreading centers. Concentrations of large earthquakes occur along transforms, and broad belts of large earthquakes occur along zones of convergence.

move past each other quietly, a process sometimes described as aseismic creep. Along other parts of a fault where friction is high, plates tend to get stuck temporarily. Continued plate motion produces distortion in the form of elastic strain in the vicinity of the rough spots (Figure 6-2). Rocks, since they're elastic, resemble springs in being capable of storing up elastic energy when stretched or compressed. Prior to an earthquake, the rocks along a fault can be regarded as a spring-loaded system waiting to go off. Eventually one more millimeter of plate motion provides the straw that breaks the camel's back, the frictional forces are exceeded, and the rocks on either side of the fault suddenly leap past each other in opposite directions.

A rupture typically starts as a small crack which grows rapidly as it propagates along the fault. As the rupture propagates, elastic energy stored in the rocks on either side of the fault is radiated in the form of seismic waves. The size of the accompanying earthquake depends on two factors. The first is the amount of elastic energy per unit area of the fault released at the time of the rupture. The second is the area of the rupture. The latter is the most variable of the two factors and largely determines the size of the earthquake. In the largest earthquakes the rupture propagates for several

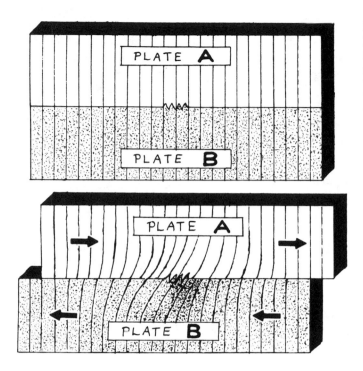

Figure 6-2.
Friction between plates A and B is concentrated at the rough spot near the center of the fault. Straight lines painted on the two blocks prior to plate movement (above) are bent as plates move past each other (below). Elastic energy is stored in the rocks adjacent to the rough spot. Crowding together of lines indicates regions of compression.

hundred kilometers whereas in the smallest detectable earthquakes the rupture propagates for only a few kilometers or less.

What is the true place of origin of an earthquake? As a rupture propagates, it releases many local patches of stored elastic energy, so in a sense a large earthquake originates from the entire zone of rupture. However, the action really starts at the first point of rupture, which is termed the **focus** or **hypocenter** of the earthquake (as opposed to the **epicenter**, the point on the surface of the ground directly above the focus). As an earthquake builds up, seismic waves radiated from different parts of the fault are superimposed on each other and on echoes of the earthquake reflected from different layers in the earth. After large earthquakes like the one in Alaska in 1964, the earth rings like a bell for about a day. The result is a complex signal containing a rich store of information about the earth's interior. Seismologists spend their professional lives trying to decipher such signals.

Seismologists are remarkably patient and persistent scientists. They keep hundreds of very sensitive seismometers in operation all over the globe, waiting for earthquakes to occur. By analyzing the records from these instruments, seismologists are able to determine three main facts about a large earthquake: the **epicentral location**, which they determine by measuring the exact time of arrival of the earthquake signal at different stations; the **focal depth**, which they determine from the timing of arrivals and also from the shape of the recorded waveform; and the **magnitude**, a measure of the size of the earthquake as determined from the amplitude of the ground motion recorded by the seismometers.

The locations of the epicenters can be combined with focal depths to provide a striking three-dimensional picture of earthquake foci along a given boundary. The results from subduction zones are particularly interesting. Invariably the foci are found to lie along one plane or two parallel planes inclined beneath the upper of the two converging plates (Figure 6-3). These planes are called Benioff-Wadati zones, after the two seismologists who discovered that the foci of deep earthquakes often lie along such inclined planes.

One of seismology's greatest challenges has always been to understand why earthquakes occur where they do. Before plate tectonics, seismologists knew that the distribution of earthquakes throughout the earth is far from random. They knew that most earthquakes occur in narrow belts separating regions where earthquakes are rare. They knew that deep earthquakes occur in only a few narrow belts located beneath zones of intense volcanic activity. They knew that outside these narrow zones of deep seismicity, almost all earthquakes occur at shallow depths of 50 kilometers or less.

When scientists encounter a pattern as regular as the distribution of earthquakes, their hearts beat faster, their intellectual juices start to flow, and they become very restless until they find an explanation for the pattern. Many a seismologist felt this way about the global distribution of earthquakes, yet none was able to explain the pattern until the theory of plate tectonics was advanced to explain the relative youth of the ocean floor, the topography over the mid-oceanic rises, and the existence of magnetic stripes. Seismologists then realized with amazement that this theory, which had been advanced to explain non-seismic data, provided a framework within which they could, for the first time, understand many of the first-order features of global seismicity.

The explanation is remarkably simple and involves two key ideas. The first is that because of natural radioactivity, the earth's temperature increases with depth at a rate of about 30° per kilometer. At a depth of 20 km the temperature

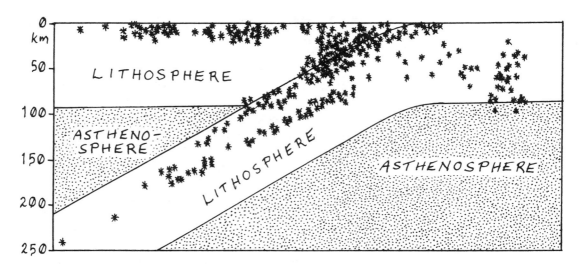

is about 600°C and at a depth of 30 km it is about 900°C, which is red hot. The second key idea is that in the temperature range 600–900°C, the lithosphere becomes soft or ductile like tar and deforms smoothly by plastic flow or creep. It doesn't store up a lot of elastic energy and then snap, as it does at lower temperatures. This loss of elasticity and increase of plasticity with increasing temperature and depth explains why earthquakes along transform faults do not occur at depths greater than 20 or 30 km. Although the lithosphere may be as much as 100 km thick, the lower part of the lithosphere deforms by ductile flow and only the upper part is cool and brittle enough to generate earthquakes. Along spreading centers, where the heat flow is higher and where high temperatures occur at shallower depths than in older parts of the lithosphere, earthquake foci are confined to very shallow depths.

The explanation of why deep earthquakes occur along descending slabs at subduction zones is equally simple. If a slab of lithosphere is being subducted fast enough, say at a rate of 100 mm/yr or so, the interior of the slab remains cool and brittle enough to generate earthquakes. During fast subduction the lithosphere carries its surface isotherms down with it. As was discussed in Chapter 2, where subduction is slower than about 30 mm/year, as it is along the coast of Oregon and Washington, for example, the lithosphere heats up as it descends and earthquakes are not generated. It is easy to appreciate why earthquakes are so important in plate tectonics: they tell us where rocks are undergoing lots of deformation while remaining cold enough to snap.

Figure 6-3.
Foci of earthquakes beneath Japan define a double Benioff-Wadati zone. Some earthquakes have loci along the top of the subducting slab and some have loci near its center. Foci also occur near the hinge point of the slab and in the upper part of the overlying plate to the west of the trench.

First Motion

One might imagine that seismic energy would move out from the focus of an earthquake in a spherically symmetrical pattern, like light radiated from a candle. However, the pattern of radiation from an earthquake is more like the beam of light from a lighthouse in the sense that energy is beamed in certain directions. Moreover, the directions of these "beams" of seismic energy are directly related to the stresses released at the time of an earthquake and indirectly to the direction of plate motion that gave rise to the stresses. From the viewpoint of plate tectonics, the radiation patterns from earthquakes are very important. We will now discuss the techniques used to analyze these radiation patterns.

During an earthquake a seismometer that records the vertical component of ground motion will produce a record showing a series of upward and downward pulses. Seismologists routinely determine whether the first recorded pulse was upward or downward. By analyzing the so-called **first motions** of the ground recorded at many different stations distributed widely over the globe, a seismologist can determine the orientation of the fault along which the earthquake was produced. (We are fortunate that seismologists in almost all countries have the tradition of exchanging records freely, even during times of political tension.) From each seismographic record, we will use only one bit of information: a determination of whether, when the earthquake started, the ground at the station first moved away from the earthquake source (a push) or toward the source (a pull). It is surprising that from just this one binary bit of information determined at a dozen or so seismograph stations surrounding an earthquake we are able to determine the direction of the elastic forces that were released during the earthquake.

We might begin by asking: In what direction would you expect the ground to move as the first shock arrives from an earthquake? Let's start with a common-sense example. Acting presumably for the common good, various governments have detonated numerous nuclear fission and fusion explosive devices in underground cavities. These always produce artificial earthquakes. What is the initial direction of ground motion? The pressure from the explosion pushes outward on the walls of the cavity, and this first push continues along the advancing spherical front of the seismic wave. So it's not surprising that the first motion of the ground

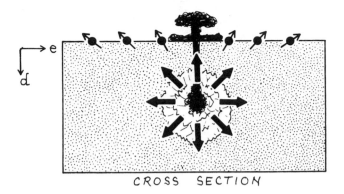

CROSS SECTION

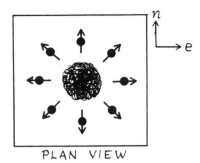

PLAN VIEW

is a push directly away from the focus in the cavity (Figure 6-4). If the earth responds like an ideal uniform elastic solid, the first motion recorded by seismographs will be perfectly symmetrical around the epicenter.

Consider now what happens when a cavity collapses. If the collapse is a radially symmetrical implosion, the effect will be the exact opposite of an explosion. The first motion of the ground surface will be a pull downward and toward the cavity (Figure 6-5).

In the early days of seismology it was thought that earthquakes were generated by some type of implosion or explosion deep in the earth. If this were correct, we would expect the first motions to be either all pushes or all pulls. However in the radiation pattern of a natural earthquake we always observe both pushes and pulls in different quadrants. Let's see if we can figure out why.

To start with, we know that earthquakes don't occur continuously but as irregular series of jerks. The two sides of a

Figure 6-4.
Hydrogen bomb detonated in an underground cavity. The shock wave from the bomb exerts a push on the walls of the chamber. The first motion of the ground surface is also a push directed radially away from the cavity.

Figure 6-5.
A collapsing cavern. The imploding walls produce a pull toward the cavity.

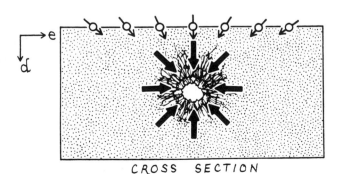

CROSS SECTION

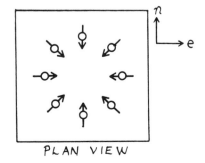

PLAN VIEW

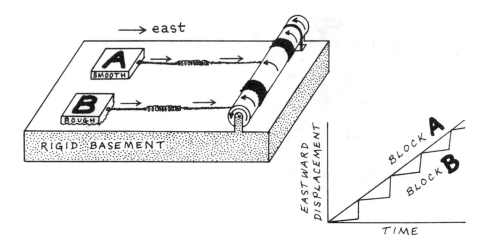

Figure 6-6.
Block B is rough and moves in jerks. Block A is smooth and moves continuously. The total displacement is the same for both blocks.

fault seem to stick together for a while as elastic energy is stored due to plate motion, then suddenly break loose. A simple analogy that comes to mind is a block of wood pulled with a rubber band over a rough surface (Figure 6-6). Static friction keeps the block from moving until the tension in the rubber spring reaches a critical value. The block then leaps ahead until friction again takes over. On the other hand a smooth, well-lubricated block pulled at the same rate will slide along continuously without jerks. The two blocks are good analogs of different segments of a fault. Along some segments the rocks glide (or creep) silently past each other, deforming plastically without generating earthquakes. Along other segments the rocks stick for a while due to frictional forces and then suddenly release their elastic energy in an earthquake. A pretty good analog of a fault is thus provided by a series of rough and smooth blocks connected by springs and pulled by a common string. Box 6-1 shows how even a simple one-dimensional model like this is able to generate both pushes and pulls as first motions.

Since a fault surface is two-dimensional, we can gain even more insight by considering the following two-dimensional experiment. Allow a bowl of hot jelly to cool and congeal with a coin embedded in the middle of its upper surface (Figure 6-7). Now gently pull the coin toward the east parallel to the surface of the jelly. The work you do in pulling the string is stored in the jelly in the form of elastic energy. Note in Figure 6-7 the two ants who are standing on the surface of the jelly, watching this strange experiment. Suddenly the coin tears loose. Elastic waves are generated as

---

Box 6-1. First Motions of One-Dimensional Earthquake.

1. All of the blocks are frictionless except block 4, on which a static frictional force **F** is acting.

2. The earthquake originates in the focal region when block 4 breaks loose.

3. At all locations the first motion of the block when the seismic wave arrives is toward the **east**.

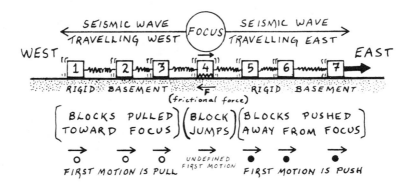

West of the focus at blocks 1, 2, and 3 the direction of propagation of the seismic wave is toward the **west**. The effect is to pull the block toward the focus and to expand or dilate the spring. First motion that is opposite to the direction of seismic wave propagation is called a **pull** or **dilatation** (open circles).

East of the focus at blocks 5 and 6 the seismic wave is propagating eastward and the first motion is eastward. The effect is to push the blocks away from the focus and to compress the spring. This type of first motion is called a **push** or **compression** (solid circles).

The direction of motion of block 4 during the earthquake is parallel to the regional eastward motion and could be found by drawing an arrow from the open circles to the solid circles.

---

the hole where the coin was attached races back toward the center of the bowl. The ant standing near the east rim of the bowl feels the first motion as a violent pull toward the center. For the ant standing near the west rim, the first motion is a push away from the center. A line along which neither push nor pull occurs extends from the north to the south rim and is called a **nodal line**. An ant standing exactly on that line would feel no motion during the first part of the jellyquake.

Our second jelly experiment is exactly like the first one, though it may not seem the same at first. Cut out a small circle of sandpaper and nail it, rough side up, to a smooth, well-greased table top. Center a cylindrical tube that will serve as a mold over the patch of sandpaper (Figure 6-8).

Figure 6-7.
First jelly experiment. A coin congealed into the surface of a bowl of jelly is pulled eastward by a string. When the coin tears loose, the ensuing elastic wave has first motions that are pushes along the west rim and pulls along the east rim of the bowl.

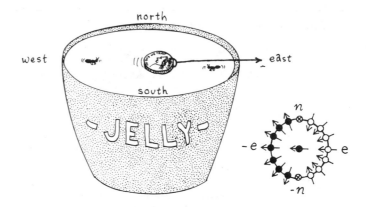

Now pour the cylinder full of hot jelly and let it set. Then remove the tube and push the cylinder of jelly smoothly and gently toward the west. The lower surface of the jelly will slip smoothly along the table top except where it's attached to the sandpaper. Sooner or later another jellyquake will occur as the jelly tears loose from the sandpaper. The result will be exactly the same as before: a pull along the east side of the container and a push along the west. If we define a first motion **direction vector** as one pointing from the pulls toward the pushes, then this west-pointing vector is parallel to three other important vector quantities: (1) the forces exerted on a parcel of jelly around the patch of sandpaper by the rest of the jelly at the moment of release, (2) the initial direction of motion of this parcel of jelly, and (3) the westward direction in which you were pushing the entire cylinder of jelly. Box 6-2 shows how this would work for rocks and introduces some useful symbols.

Figure 6-8.
Second jelly experiment. A small disk of sandpaper is nailed to a smooth table top. A cylinder is centered on the sandpaper disk and filled with hot jelly. After the jelly sets up, the cylinder is moved slowly westward until the jelly breaks loose from the sandpaper. The first motions are the same as in the first jelly experiment.

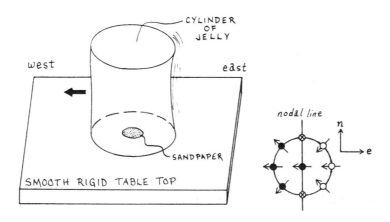

Box 6-2.   First Motions of Two-Dimensional Earthquake with One Elastic Plate and One Rigid Plate.

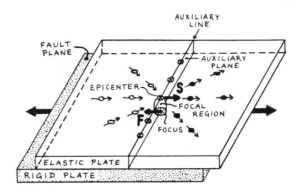

→ DIRECTION OF FIRST MOTION
● FIRST MOTION IS PUSH (COMPRESSION)
○ FIRST MOTION IS PULL (DILATATION)
⊗ NO FIRST MOTION (UNDEFINED)

1. The contact between the elastic plate and a hypothetical undeformable rigid plate is frictionless except at a circular rough patch (shaded), where a local frictional force **F** balances the elastic force due to the deformation of the elastic plate in the focal region.

2. When an earthquake occurs, **F** vanishes (or, more realistically, drops to a much lower value of sliding friction). The elastic force **S** is now free to accelerate the rock in the focal region, sending out seismic waves.

The directions of first motions are **not parallel** in an east-west direction. They are directed **radially** away from (push) or toward (pull) the focus. The **amount** of radial motion is greatest along an east-west line through the focus and drops to zero along a north-south line through the focus. This line is perpendicular to **S** and is termed the **auxiliary line**. The auxiliary line divides the plane of observation into two regions with different types of first motion: pushes or compressions (solid circles) to the east; and pulls or dilatations (open circles) to the west.

The direction of **S** can be found from observations of first motions as follows.

1. Plot on a map the locations of seismographic stations and the observed sense of first motion.

2. Draw the auxiliary line separating the open and closed circles.

3. Draw **S** perpendicular to the auxiliary line from the region of pull (open circles) toward the region of push (solid circles).

S is sometimes called the **slip vector** of the earthquake. It is the motion of the rocks in the focal region relative to the rigid plate. The slip vector **S always lies in the fault plane** between the two plates. If a diamond were embedded in the elastic plate, it would scratch a line in the rigid plate parallel to the slip vector.

What is the analog along an actual fault of our rigid table top? There isn't any. The rocks on both sides of the fault undergo deformation and store elastic energy. We could make a model of this by taking a large slab of jelly, cutting it, and lubricating the cut with oil except at one place, where the jelly on both sides is fastened to a double-faced disk of sandpaper. If the two slabs are displaced in a direction parallel to the lubricated cut, the sandpaper will eventually break loose but this time the pattern of pushes and pulls will be markedly different because equal and opposite forces are now acting on the parcels of jelly on opposite sides of the sandpaper. The result is that there are now four areas of alternating pushes and pulls separated by two nodal lines. Box 6-3 shows how this works for two rectangular slabs of jelly (or rock), and Box 6-4 shows how to find the direction of a fault from first-motion data.

Box 6-3. First Motions of Two-Dimensional Earthquake with Two Elastic Plates.

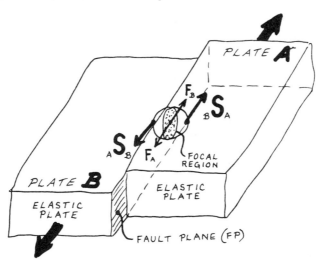

1. The contact between plates A and B is frictionless except at the circular patch (shaded), where plate A exerts a frictional force $\mathbf{F}_B$ on plate B and plate B exerts frictional force $\mathbf{F}_A$ on plate A. The elastic force $_B\mathbf{S}_A$ acting on the hemispherical focal region in plate A is balanced by $\mathbf{F}_A$, and similarly $\mathbf{F}_B$ balances $_A\mathbf{S}_B$.

2. When an earthquake occurs $\mathbf{F}_A$ and $\mathbf{F}_B$ drop to much lower values of sliding friction. The elastic forces $_B\mathbf{S}_A$ and $_A\mathbf{S}_B$ are now free to accelerate the rocks in the focal region, sending out seismic waves.

(continued)

Box 6-3. *(continued)*

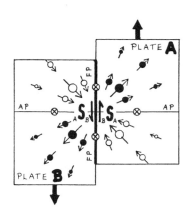

→ DIRECTION OF FIRST MOTION
● FIRST MOTION IS PUSH (COMPRESSION)
O FIRST MOTION IS PULL (DILATATION)
⊗ NO FIRST MOTION (UNDEFINED)
$_A\mathbf{S}_B$ SLIP VECTOR SHOWING MOTION OF PLATE B RELATIVE TO PLATE A

3. The waves from the two focal hemispheres cancel along the fault plane (F.P.) because the two hemispheres are equidistant from all points on F.P. As before, along an auxiliary line the first motion has zero amplitude. The auxiliary line is perpendicular to the slip vectors $_A\mathbf{S}_B$ and $_B\mathbf{S}_A$ and to the fault plane. When we generalize to three dimensions, the auxiliary line will become an **auxiliary plane** perpendicular to the slip vectors and fault plane. The maximum motion occurs midway between the fault and auxiliary planes.

The forces $_B\mathbf{S}_A$ and $_A\mathbf{S}_B$ form a **force pair** or **couple** and have a right-lateral (dextral) or left-lateral (sinistral) sense. Left-lateral plate motion as shown above produces a left-lateral force couple at the time of an earthquake, and conversely, for right-lateral plate motion a right-lateral force couple is produced.

Box 6-4. How to Interpret an Array of First Motions.

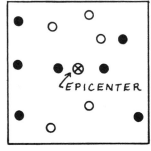 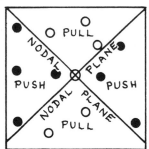

Seismographic stations distributed around the epicenter of an earthquake observe the first motions shown above. What pattern of faulting would be consistent with these observations? Here's how to proceed.

1. By trial and error, divide the map into four quadrants by two lines representing the "strikes" of the two nodal planes. If the slip vector **S** is horizontal, the two lines will be perpendicular. If one nodal plane is horizontal, or if both nodal planes have the same "strike" (i.e., they intersect in a horizontal line), only one line will be able to be drawn.

(continued)

Box 6-4. *(continued)*

2. Arbitrarily select one of the lines and label it "fault plane"; label the other line "auxiliary plane."

3. Draw arrows representing the force couple, with each arrow pointing from a quadrant of pull (open circles) toward a quadrant of push (closed circles).

4. Now switch the labels on the nodal planes. Note that this changes the sense of the faulting from left-lateral to right-lateral. The two interpretations are equally consistent with the observed array of first motions.

5. Compare your solutions with a geologic map of the area. Geologic constraints usually eliminate the intrinsic ambiguity of a first motion solution.

ZONE OF COMPRESSION
ZONE OF DILATATION

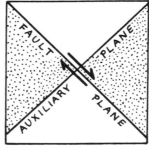

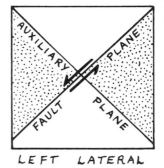

RIGHT LATERAL   LEFT LATERAL

Going Three-Dimensional

So far we've thought about earthquakes in terms of one-dimensional or two-dimensional analogs. We've now reached a stage where we need to start thinking in three dimensions. In doing this, we'll find it especially useful to use as our frame of reference a sphere centered on the focus of an earthquake. This is called the **focal sphere**. There's no need to specify its radius because we use the focal sphere mainly to keep track of directions. A sphere is a natural frame of reference for an earthquake because the advancing seismic front is a spherical surface of constantly increasing radius, similar to the advancing front of photons from a flash bulb. We carry the analogy with light farther by using **seismic ray paths** perpendicular to the wave fronts to describe the direction of wave propagation (Figure 6-9). Seismic rays radiate from the earthquake focus like the spines of a sea urchin. Later we will see that the seismic rays are refracted and curve when they travel over long distances because the

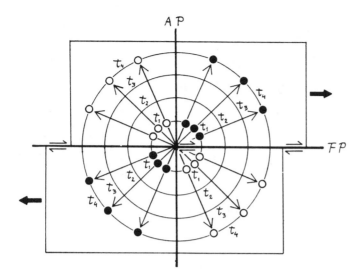

Figure 6-9.
Wave fronts at times t_1, t_2, t_3, etc., are spheres shown here in cross section as circles. Ray paths shown by arrows are straight lines perpendicular to the wave fronts which radiate outward from the focus. The arrows depict the direction of ray propagation. Ray paths along which the first motion is a push are shown with solid circles and those along which the first motion is a pull, with open circles. The focal sphere can be considered to be the wave front at time t_1. The focal plane (FP) and the auxiliary plane (AP) are perpendicular.

earth's seismic velocity increases with depth. For the present we'll stick to constant velocities and straight rays. We will always consider the focal sphere to be small enough so that the ray paths from the focus are essentially straight within it. Later we'll consider curved rays.

Our standard procedure is to represent the focal sphere by an equal area projection. To get started, label the three perpendicular axes as $\pm\mathbf{n}$ (north), $\pm\mathbf{e}$ (east), and $\pm\mathbf{d}$ (down). Let's track an advancing ray as it pierces the focal sphere at a point with spherical coordinates (I, D), where I is the inclination below (+) or above (−) the horizontal and D is the azimuth measured clockwise from north. A small volume of rock located on the focal sphere at this point will, as the wave arrives, move either radially outward parallel to the ray, in which case the first motion is a push or compression, or it will move radially inward, in which case it is a pull or dilatation. We place a solid or open circle on the projection at this point to indicate compression or dilatation and then proceed to analyze another first motion. Generally the directions of compression and dilatation are grouped in four regions which can be separated by two planes perpendicular to each other. These are called **nodal planes** because rays directed along them have zero first-motion amplitudes. One of these nodal planes is the **fault plane** of the earthquake and the other is called the **auxiliary plane**. The dip and strike of the two planes can be easily read from the equal area projection. The intersection of the two planes is called the **B** or **null axis** (Figure 6-10).

Figure 6-10.
First motion of strike-slip fault. **T** is axis of tension; **P** is axis of compression; **B** (null) axis is intersection of fault plane and auxiliary plane; $_BS_A$ is slip vector showing the velocity of plate A relative to plate B.

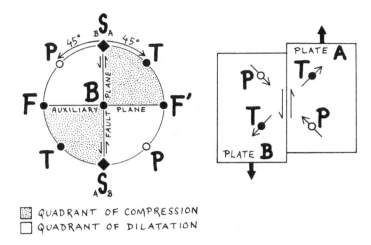

☐ QUADRANT OF COMPRESSION
☐ QUADRANT OF DILATATION

The two slip vectors **S** describe the direction of relative motion of the two plates on opposite sides of the fault. To visualize the slip vector, imagine a sharp rock in the fault plane embedded in and moving with plate B. Then the scratch made by the rock in plate A is parallel to the slip vector **S**. The two slip vectors are the poles of the auxiliary plane. The motion of plate A relative to plate B is given by the vector $_BS_A$ and that of B relative to A by $_AS_B$. Note carefully the word "relative": exactly the same pattern of first motions is produced by movement of A relative to a fixed B as is produced by movement of B relative to a fixed A. First motion analysis, like plate tectonics, does not provide an absolute frame of reference.

The amplitude of the first few pulses on a seismogram dies off toward the nodal planes and reaches its maximum along two axes inclined 45° from the nodal planes. To find these directions, first construct the great circle <**F**, **S**> where **F** is the pole of the fault plane (Figure 6-10). The null axis **B** is the pole of <**F**, **S**>. The two **T** axes are the directions where the compressional first motion outward from the focal sphere is greatest. The **P** axes are the directions where dilatation is greatest. The **P** and **T** axes are in the <**F**, **S**> plane, where they are 45° from both slip vectors **S** and 90° from each other (Figure 6-10).

The planes, axes, and vectors of first-motion analyses are highly symmetrical. You'll quickly notice that every element of a focal plane analysis which appears on the lower hemisphere projection has an antipodal counterpart on the upper hemisphere. For example, if a slip vector **S** is inclined 10° along the horizontal to the east on the upper hemisphere

($I = -10°, D = 90°$), a second slip vector is inclined 10° downward to the west on the lower hemisphere ($I = 10°$, $D = 270°$). Similarly, a fault plane through the focus intersects both hemispheres in two half-circles. Moreover, each compressive ray with coordinates (I, D) has an antipodal compressive ray with the same amplitude at ($-I, D + 180°$) on the opposite hemisphere, and so do dilatational rays. Although using plots of both hemispheres sometimes helps one understand what's going on, most experienced seismologists prefer to condense all of the information onto a lower hemisphere projection, plotting data from rays that pierced the upper half of the focal sphere as the corresponding antipodal directions. It is customary in seismology to specify the inclination of a ray vector by giving the **angle of incidence** i of the ray relative to the vertical. If i is measured as the angle from the downward ($+\mathbf{d}$) axis, then $i = 90° - I$, where I is the familiar inclination of a vector below the horizontal plane.

By now we've named just about every possible axis and pole formed by the intersection of the two perpendicular nodal planes. In Boxes 6-5 to 6-9 you can see how these definitions and labels can be used to describe real faults.

Box 6-5. Focal Sphere for a Vertical Normal Fault.

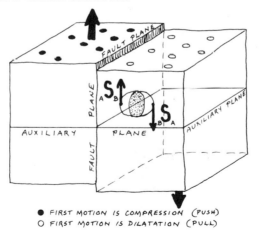

● FIRST MOTION IS COMPRESSION (PUSH)
○ FIRST MOTION IS DILATATION (PULL)

Vertical movement of plate relative to plate B produces slip vector $_B\mathbf{S}_A$. The auxiliary plane is perpendicular to the slip vectors and thus is horizontal. The northeast-southwest-trending vertical fault plane and the horizontal auxiliary plane are mutually perpendicular and together define four quadrants in which the first motion is alternately compression (shading and solid symbols in diagram) and dilatation (unshaded regions and open symbols in diagram).

(continued)

Box 6-5. *(continued)*

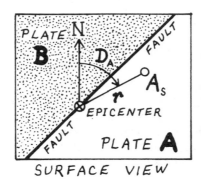

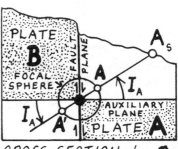

☑ *ZONE OF COMPRESSION (PUSH)*

☐ *ZONE OF DILATATION (PULL)*

SURFACE VIEW   CROSS-SECTION along **r**

Near the epicenter **E**, the zones of compression and dilatation as observed at sites on the surface of the ground can be separated by a single line (see map to the left). The cross section to the right shows the quadrant distribution of regions of compression and dilatation separated by the fault and auxiliary planes.

The azimuth or declination angle D_A can be seen on the map to the left to be the angle relative to north at which the ray leaves the focus and epicenter. The inclination angle I_A can be seen in the cross section to the right to define the angle above the horizontal at which the ray leaves the focus.

The **focal sphere** may be imagined as a spherical shell centered on the focus. The seismic ray that reaches the surface at point A_S may be represented by point **A** where the ray passes through the sphere. The first motion is a pull at point **A** on the focal sphere, just as it is at point A_S on the earth's surface. For reasons of symmetry, a ray leaving the focus in exactly the opposite direction will also have a dilatational first motion. The point **A'** where this "image" ray passes through the focal sphere can also be used to represent on the lower hemisphere the ray from the focus to A_S.

Box 6-6. Plotting First Motions for a Normal Fault on a Projection.

☑ COMPRESSION
☐ DILATATION

For each seismographic station like the one at A_S (Box 6-5), find the azimuth D_A of a line *r* from the epicenter **E** to the station.

(continued)

Box 6-6.   *(continued)*

Find the inclination angle I_A relative to the horizontal of the ray as it leaves the focus. Since the direct ray to station $\mathbf{A}_S$ is inclined above the horizontal, the inclination angle for point $\mathbf{A}$ on the focal sphere is negative. In order to be able to plot all points on the lower hemisphere of the focal sphere, it is traditional in seismology not to plot points like $\mathbf{A}$ with negative inclinations ($I < 0$), but instead to plot their "image" points $\mathbf{A}'$, for which the corresponding inclination is $I_{A'} = -I_A > 0$ and the corresponding declination or azimuth is $D_{A'} = D_A + 180°$ (see Box 6-5).

On a projection with axes $\mathbf{n}$ (north), $\mathbf{e}$ (east), and $\mathbf{d}$ (down), representing the lower hemisphere of the focal sphere, plot $\mathbf{A}'$ at coordinates $(-I_A, D_A + 180°) = (I_{A'}, D_{A'})$.

The auxiliary plane plots as the outer circle ($I = 0$) of the projection. The quarter-sphere or quadrant to the southeast, with the $\mathbf{T}$ axis at its center, undergoes a compression downward during the first motion of the earthquake. The quadrant to the northwest with the $\mathbf{P}$ axis at its center undergoes a dilatation or upward pull. The slip vector $_B\mathbf{S}_A$ is directed straight down ($+\mathbf{d}$) and is the pole of the auxiliary plane. Note that this vector describes the motion of plate A relative to plate B.

Box 6-7.   Nonvertical Normal and Thrust Faults.

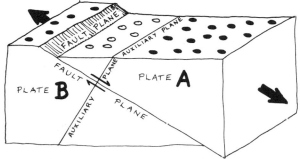   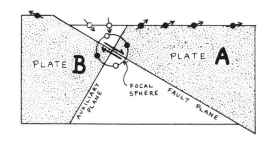

The stations recording pushes (solid circles) can be separated from those recording pulls (open circles) by the lines of intersection at the earth's surface of the fault plane and auxiliary plane.

Solid and open symbols show pushes and pulls and arrows show directions of first motion toward or away from the focus.

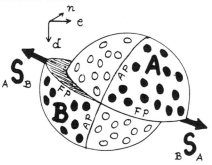   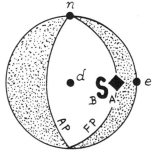

NORMAL FAULT   *(continued)*

Box 6-7. *(continued)*

To the left is an oblique view of the focal sphere sliced like a grapefruit along the fault plane. To the right is a view of the same focal sphere looking directly down on the lower hemisphere. Note that for a normal fault like the one shown here, at the center of the focal sphere the first motion is dilatational (no shading). This is true of all normal faults. Note that the slip vector $_BS_A$ is the pole of the auxiliary plane.

If the relative motion of plates A and B is reversed by 180° to generate a thrust fault as shown above, the locations of the fault plane (F.P.) and auxiliary plane (A.P.) remain the same. However, the directions of the slip vectors change by 180°, $_BS_A$ and $_AS_B$ interchanging directions. The quadrants of compression and dilatation are also interchanged. Note that at the center of the projection the first motion is compressional. This is true of all thrust faults.

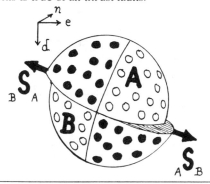

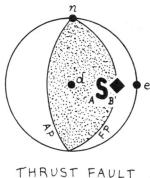

THRUST FAULT

Box 6-8. How to Plot Fault Motion with Both Normal and Strike-Slip Components.

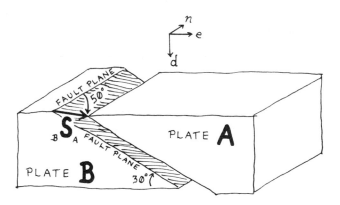

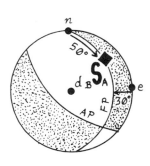

We want to plot the fault and slip vector for a north-south-trending fault that dips 30° to the east. The trend of the slip vector is not horizontal, as it would be for strike-slip motion, nor does it trend down dip at an azimuth 90° from the trend of the fault, as it would for pure normal or thrust motion. So the motion has both strike-slip and normal components which can be plotted as follows.

1. Plot the fault plane on the lower hemisphere of a projection.

(continued)

Box 6-8. *(continued)*

2. The angle between the slip vector $_BS_A$ and a horizontal line trending north is given as 50° measured in the plane of the fault. The strike-slip component of the slip vector is given by $_BS_A \cos 50°$ and the normal or down-dip component is given by $_BS_A \sin 50°$.

3. Plot the slip vector $_BS_A$ 50° from the $+\mathbf{n}$ axis measuring the angle along the great circle that describes the fault plane.

4. Plot the auxiliary plane perpendicular to the slip vector $_BS_A$.

Box 6-9. How to Plot a Strike-Slip North-South Fault Dipping 30° E.

1. Plot the fault plane as before.

2. Plot $_BS_A = \mathbf{n}$ and $_AS_B = -\mathbf{n}$.

3. Plot the A.P. perpendicular to $_BS_A$ and $_AS_B$. It is a vertical east-west plane.

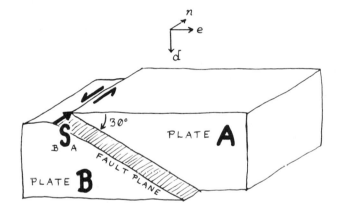

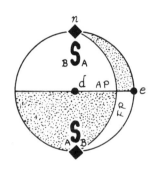

Directions of Compression and Tension

Let's put a rock in a vice and squeeze it (Figure 6-11). When it breaks, will the cracks have a preferred orientation or will they be random? First let's watch what happens before the rock breaks. Let's imagine that there are two small spheres with the same radius at the exact center of our rock sample. One sphere (dotted line in Figure 6-11) is attached to the atoms of the rock and moves with them as the rock is strained; the other sphere (not shown) remains fixed in space and will represent the focal sphere when the rock breaks. As we squeeze the sample, material moves in toward the center

Figure 6-11.
Rock squeezer. As the rock is squeezed in the vertical (+ **d**) direction, material moves toward the center of the sample from the top and bottom and away from the center toward the sides. Dotted line shows deformation of an initially spherical surface at the center of the sample.

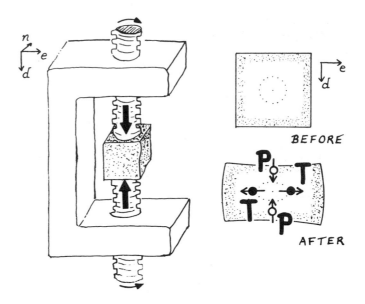

through the top and bottom of the focal sphere and out through its equator. The dotted sphere gets shorter in the direction of compression and gets wider in the perpendicular direction. The direction of maximum shortening is also the direction of compression and is termed the **P** axis (for pressure). We will see that it is the same **P** axis described earlier. All directions in the plane perpendicular to **P** are directions of tension, as shown in cross section by the **T** axis in Figure 6-11.

Let's continue to squeeze the rock until it breaks. Geophysicists have done this to thousands of rocks and have found both systematic and random elements to the pattern of breaking. The breaks are in the form of minifaults inclined at an angle of 45° or slightly less from the vertical direction **P**. This is the systematic element. The random element is that the strikes or azimuths of minifaults in successive experiments occur in different directions. Two typical breaks with different minifault azimuths are shown in Figure 6-12.

When the rock breaks, the elastic strain relaxes and the rock specimen becomes shorter and wider because of displacements along the minifault. Because shortening is in the vertical direction **P** in which the rock was squeezed, rocks on the two sides of the minifault move in toward the focal sphere in the two antipodal **P** directions. These are thus axes of pull or dilatation (open circles). If this sounds contradic-

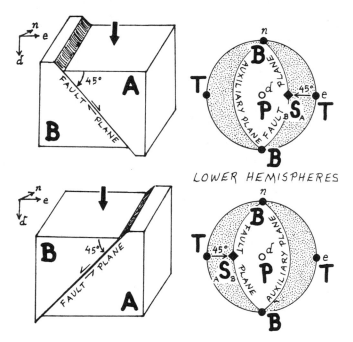

LOWER HEMISPHERES

Figure 6-12.
Rocks squeezed in the experiment of Figure 6-11 have broken along two different mini-faults. The **P** (compression) axis for both breaks is the same. In most experiments an angle less than 45° is observed between the axis of compression and the minifault. So the true compression axis may be 10° or so closer to the slip vector **S** than is the **P** axis formed by analyzing first motions.

tory (which it does at first), just remember that if we subject a subterranean spherical cavity to enough pressure it will eventually implode and the resulting first-motions will all be pulls or **dilatations**. So compression can and usually does cause dilatational first motions because it causes movement toward the center of the focal sphere in the direction of compression.

On breaking the rock becomes wider in the antipodal horizontal directions **T** perpendicular to the azimuth of the minifault. As rocks on either side of the fault move away from the focal sphere in these two directions, they produce pushes or **compressions** (solid circles). The **P** and **T** axes lie in a plane perpendicular to the two nodal planes at an angle of 45° from either plane (Figure 6-12). The **P** axes are at the midpoint of the dilatation quadrants and the **T** axes are at the midpoint of the compression quadrants.

In effect, when the rock breaks, faulting transforms the elastic shortening stored in the rock before the break into a permanent displacement. (Recall that the same kind of transformation occurred in the experiments with the wooden blocks and the jelly.) In the equal area plots for the two minifaults, the **P** axes calculated from the orientation of the nodal planes are vertical (±**d**) for both breaks, which makes sense because this is the direction along which compression

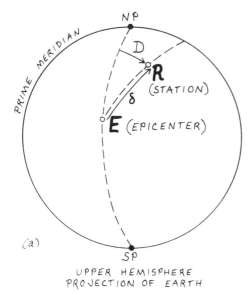

(a)

UPPER HEMISPHERE
PROJECTION OF EARTH

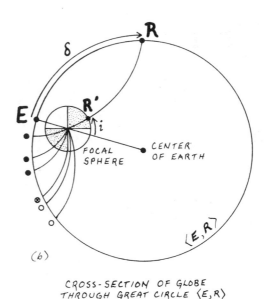

(b)

CROSS-SECTION OF GLOBE
THROUGH GREAT CIRCLE ⟨E,R⟩

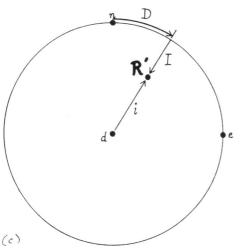

(c)

LOWER HEMISPHERE
PROJECTION OF FOCAL SPHERE

Figure 6-13.
Ray paths are curved because of the increase with depth of seismic velocity. After finding the azimuth *D* and epicentral distance δ using a projection of the globe (*a*), seismologists are able to calculate the paths of individual rays and find the angle of incidence *i* of a ray as it leaves the focus (*b*). Knowing *i*, the ray may be plotted as a compression or dilatation point at (*I, D*) on the focal sphere (*c*), where $I = 90° - i$.

was applied by the vice. The two nodal planes in this particular example are also the same except that the fault plane and the auxiliary plane are interchanged in the two experiments. When this experiment is repeated many times using homogeneous rocks, the breaks are observed to have random strikes or azimuths, as expected for reasons of symmetry. However, all of the **P** axes are observed to be vertical or nearly so.

Curved Ray Paths Through a Spherical Earth

As a seismic wave travels downward in the earth, it encounters faster seismic velocities and its ray path is bent upward. A ray that leaves the focus at a steep angle *I* below the horizontal (that is, at a small angle of incidence *i* from + **d**) will travel a long distance δ before it reaches the surface (Figure 6-13). A ray that leaves the focus at a large angle of incidence *i* from + **d** will reach the surface at a shorter distance. Knowing the distance δ of a seismographic station from the epicenter and knowing the depth of the focus *h* of an earthquake, the angle of incidence *i* can be read from standard seismological tables. The ray path from the epicenter **E** to a seismographic station **R** lies in a plane through **E**, **R**, and the earth's center **O**. This plane intersects the earth's surface along the great circle <**E**, **R**> (Figure 6-13). The direction

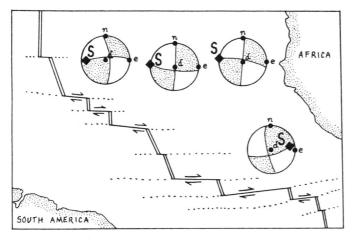

Figure 6-14.
First motions of earthquakes located on transforms that offset the Mid-Atlantic Ridge. Shown below are the first-motion data used to determine the two most easterly earthquakes. Dotted lines, inactive fracture zones. Solid circles, compressions. Open circles, dilatations. Crosses, station near nodal plane.

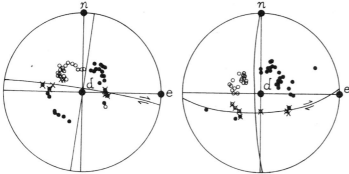

or azimuth D at which the ray leaves the epicenter may be found on a projection of the globe as shown in Figure 6-13. Converting angle of incidence i to inclination below the horizontal by $I = 90° - i$, the direction of the ray may be plotted on a projection of the focal sphere as a solid circle with direction coordinates (I, D). Data from widely separated seismographic stations may be reduced in this way to a set of points on the focal sphere, and from this plot the nodal planes and slip vector can be determined.

Earthquakes at Transforms

During the late 1960s, first-motion analysis provided a great breakthrough in establishing the validity of the theory of plate tectonics. The crucial test of the theory was made along the Mid-Atlantic Ridge where the ridge is offset by fracture zones (Figure 6-14). The usual way to interpret a feature which is offset by faults is to assume that it was once straight. In that case, since the ridges are clearly offset to the left on opposite sides of the fault, the faulting should be left-lateral strike-slip faulting. Plate tectonic analysis, for reasons discussed in Chapter 1, came to the startling conclusion that the rocks on opposite sides of the fracture zones are moving right-laterally between the ridges, exactly the opposite direction. Since many geologists found this interpretation hard to swallow, an independent method was needed to determine whether the motion is right-lateral or left-lateral. Many large earthquakes occur along the segments of fracture zones between the offset ridges, and analysis of the first motions of these earthquakes has answered unequivocally the question of how the fracture zone faults are moving.

Four examples of first-motion solutions based on worldwide data are shown for mid-Atlantic fracture zones in Figure 6-14. The null axes or **B** axes where the nodal planes intersect are steep, so the faulting is clearly of the strike-slip type. Theoretically either the north-south-trending nodal planes or the east-west-trending nodal planes could be the fault planes, but north-south faults wouldn't make much sense because the fracture zones obviously run east-west. So let's select the east-west nodal planes as our fault planes. We can then draw pairs of arrows as shown to describe the relative motion on either side of the fault. The arrows are directed from quadrants of dilatation to quadrants of compression (shaded). The motions are all right-lateral, as predicted. It's hard to imagine more convincing support from the field of seismology for the plate tectonic theory.

Two examples of the data used for the first-motion analyses are shown in Figure 6-14. It might seem that the number of data is rather small to serve as the basis for so far-reaching a conclusion and that other nodal planes might fit the data equally well. If you're a doubter (as most of us were at first), try to find other nodal planes equally consistent with the observations. Remember, the two nodal planes must be perpendicular. You'll find that it's hard to move the present nodal planes more than 5° without creating more inconsistencies than are shown in Figure 6-14.

Note how closely the strikes of the nodal planes in Figure 6-14 parallel the local strikes of the fracture zones. There is some question, in fact, as to whether ocean floor topography or first-motion solutions provide a better measure of the direction of relative plate motions. Commonly both are used. Ideally the motion along a transform fault should be pure strike-slip, which means that the slip vector **S** should be horizontal and the null axis vertical, although variations of 10° or so can be expected because of local geologic irregularities. If the slip vector is not perfectly horizontal, its horizontal component is used to define the direction of relative plate motion. If the slip vector has the orientation **S** = (*I, D*), the orientation of the horizontal component of **S** is simply *D*.

Earthquakes at Ridges

The pattern of seismic energy radiated from earthquakes along spreading centers is completely different from that of transforms—so different, in fact, that one set of earthquakes would never be mistaken for the other. This difference is not surprising because the relative plate motions are completely different at the two types of boundaries: the plates on opposite sides of a ridge are moving away from each other and not tangentially, as was true for transforms. A typical cross section through the Mid-Atlantic Ridge is shown in Figure 6-15. At first glance it might appear that the entire rift valley is a collapse feature such as one might expect if magma had been withdrawn at depth. However, the true story of what's happening along rifts, which was worked out in the late 1970s by geologists visiting the rift valleys in submarines, is much more complicated and interesting than a simple collapse. Formed by volcanic processes at the center of the rift, new seafloor first moves up a distance of 1500 meters over the walls of the rift and then begins a slow descent to the deep ocean floor as it migrates outward on the flank of a cooling plate. The nature of the earthquakes generated by new oceanic crust near ridges has provided much insight into the way this submarine conveyer belt works.

To visualize this process, let's imagine that we are standing in diving suits or, more realistically, sitting in a submarine on the ocean floor at the center of a rift. We're going to be making observations for a million years, so let's get comfortable and start taking notes. We quickly notice that the inner floor of the rift valley is a flat plain about 10 kilometers wide, with scattered small volcanoes several hundred meters high. A new volcano erupts near the center of the valley floor every 10,000 years or so, a fairly rapid rate of eruption by terrestrial standards. Cruising over the inner valley floor, we notice lots of small faults oriented parallel to the axis of

Figure 6-15.
Idealized cross-section of the Mid-Atlantic spreading center. Typical focal mechanism solutions are shown for foci on normal faults at points 1, 2, and 3.

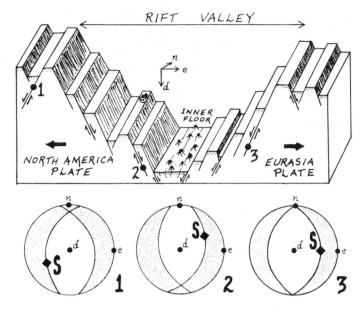

the rift valley. The rocks on either side of these faults have pulled apart to form cracks with openings a few centimeters to a few meters wide. Water circulating through these cracks driven by thermal convection is geologically important for two reasons: the seawater helps cool the initially molten volcanic rocks that make up the oceanic crust, and it interacts chemically with the hot rocks. Some of the minerals initially present in the seawater are left behind in the volcanic rocks whereas some of the minerals initially present in the volcanic rocks are dissolved. We become aware of the dynamic nature of this process when our submarine encounters a column of extremely hot water (375°C) shooting up like a jet from a nozzle made of minerals precipitated when the hot water rising from the interior of the oceanic crust cools.

During a period of several tens of thousands of years we encounter several of these jets and learn to recognize their life cycles. Initially they are extremely hot and carry such a heavy load of dissolved minerals that the hot jet appears black with precipitating minerals. We give these jets the working nickname "black smokers." As the jets cool, the amount of precipitation decreases although the rising hot seawater is still rich in dissolved sulfide minerals. During this stage the ocean floor around the spring becomes an oasis of animal life of a type usually not found on the deep ocean floor. We're excited to discover animals never seen before, including giant clams and tube worms more than a meter long. At the bottom of the food chain are bacteria that metabolize sulfides contained in the hot water—an unusual example of a biological economy based on rock energy rather than solar energy. Cruising along the inner floor of the rift valley, we learn to identify submarine hot springs at all stages of their development from black smokers where the water is hottest, to animal colonies where the water is warm, to abandoned sites where flow has stopped, marked by the presence of sulfide ores and the shells of dead clams. Any doubts that we may have had about the reality of seafloor spreading vanish in the face of such clear evidence for the formation of new seafloor by volcanic processes.

Let's anchor our submarine to the seafloor just west of a black jet near the center of the rift valley and turn off our engines. After attaching some seismometers to the sea bottom, we begin a long series of observations. As the hot spring goes through its life cycle and dies, we record many small, shallow earthquakes around us on the floor of the rift valley and at the same time we observe long cracks opening up all around us, oriented parallel to the trend of the valley. Soon the valley floor around us is riddled with faults. When we initially anchored, we were located midway between the east and west walls of the rift valley, which rise up steeply from the flat inner valley floor. Soon we notice that the eastern wall is getting farther away, as are hot springs and small volcanic cones that continue to erupt along the center of the rift valley. Quite clearly the volcanic eruptions are forming a growing belt of new ocean floor between us and the east wall of the rift.

We also notice that the west wall of the rift is getting closer to us and that this is happening by a curious process. The floor of the rift valley is cut into a number of slivers by the north-south-trending faults. As we approach the west wall, we notice that from time to time the sliver of ocean floor immediately adjacent to the wall is suddenly uplifted to become part of the wall. Eventually we find that the sliver of ocean crust to which we are anchored has migrated to a position next to the west wall of the rift valley. A short time later we are jolted by an earthquake and when the dust settles, we find that the sliver to which we're anchored has risen above the flat inner floor, from which it is separated by a steeply dipping fault. We are now perched on a step on the west wall. Since the sliver of crust to the west appears to have moved upward with us, the only displacement during the earthquake appears to have occurred along the fault between us and the flat inner valley floor. The next earthquake in our vicinity occurs along the fault to our west and is accompanied by upward displacement relative to us of the sliver of crust to the west, increasing the height of the fault scarp between the two slivers. By a series of small steps like this we are slowly jacked up the face of the west wall of the rift valley.

Let's consider for a moment whether we should expect the motion along the faults to be normal (extensional) or reversed (compressional). We note that the faults dip eastward toward the rift valley and that during earthquakes the block toward the east always moves downward relative to the block to the west. Therefore by definition the faulting is normal. In agreement with this model, the observed pattern of first motions from earthquakes along rifts invariably shows normal faulting with nodal planes that trend parallel to the rift. This is consistent with a tensional regional stress regime, which in turn is consistent with the concept of plate divergence.

Whether the lithosphere near a ridge is under tension or compression tells us something important about the nature of the driving forces that move plates, a subject that will be discussed more fully in the final chapter. One of the main candidates for a driving force is "ridge push," which is a hypothetical stress (force per unit area) acting over the interface between the lithosphere and the asthenosphere in the vicinity of a ridge. Ridge push can also be regarded as an equivalent force per unit length along the ridge acting in a direction perpendicular to the ridge. If one were to visualize ridge push as a giant hand pushing against a ridge wall, then one would expect a compressional stress regime to be reflected in the earthquake focal mechanisms. Since this is not observed, one might conclude that ridge push is not an important driving mechanism.

However, deeper consideration of the nature of ridge push leads to a different conclusion. Ridge push arises because the lithosphere is heavier than the asthenosphere. Analysis of the stress within the lithosphere and asthenosphere reveals that the asthenosphere exerts a "ridge push" stress all along the inclined lower surface of the lithosphere, which slopes downward away from the ridge as the lithosphere thickens (Figure 6-16). Ridge push in a direction away from the ridge

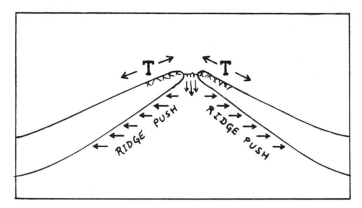

Figure 6-16.
The ridge push force is produced by a stress distributed over the sloping lower surface of the lithosphere. The net result is tension in the zone around the rift valley.

is thus distributed over a broad band hundreds of kilometers wide running parallel to the ridge. Although the net effect at the center of the plate is probably an increase in compressional stress, within the narrow zone of active seismicity adjacent to the ridge the stress regime is tensional.

What happens when slivers of oceanic crust that originated in the floor of the rift valley eventually reach the "corner" at the top of the rift valley? One might imagine that the direction of motion along the normal faults would be reversed to bring the tops of the slivers into alignment. If this explanation is correct, some of the first-motion solutions should indicate thrust faulting; however, essentially all of the solutions indicate normal faulting. Geological observations on the ocean floor have revealed a surprisingly simple explanation. When the slivers of ocean floor rise to the top of the wall, a new set of normal faults develops, dipping away from the center of the rift valley. In first-motion diagrams, the fault planes for these earthquakes have the same orientation as the auxiliary planes for earthquakes originating on the wall of the rift (Figure 6-15), with the result that the zones of compression and dilatation for the two types of earthquakes are the same. Earthquakes along both sets of faults are consistent with regional extension in a direction perpendicular to the trend of the rift zones.

Earthquakes at Trenches

What state of stress would you expect in a slab being subducted beneath a trench, compression or tension? The answer isn't obvious. First let's think about the descending slab itself.

To keep things simple, let's imagine that the slab is detached from the rest of the plate and is sinking under its own weight. Eventually the slab will reach an equilibrium free-fall velocity in which the viscous drag exerted by the asthenosphere on the sides of the slab exactly balances the gravitational force due to the difference in density between the slab and the asthenosphere. When the slab is falling freely, the stresses within the slab are negligible except for a small amount of compression along the leading edge of the slab and a small amount of tension along the trailing edge.

Now let's change our conceptual experiment and hang the slab suspended from its upper edge so that it hangs down into the asthenosphere under its own weight. The state of stress throughout the slab is obviously one of tension. (If this isn't obvious, make a horizontal slice through the slab and you'll find that the bottom piece will drop down into the asthenosphere, which means that the upper piece was exerting upward stress or tension on the lower piece.)

Now let's free the upper edge of the slab and allow it to support its own weight, resting on its lower edge, which butts up against the less-deformable lower mantle (mesosphere) or a layer of high viscosity deep in the asthenosphere. In this case the state of stress throughout the slab will be one of compression.

Turning our attention to the plate, if the plate were not attached to the sinking slab let's hypothesize that it would have a velocity V_p due to various non-slab forces acting on it. Now let's reattach the slab to the plate. If the free-fall velocity V_f of the slab is greater than V_p, then the slab will be pulling the rest of the plate into the trench and the state of stress in the slab will be tension. On the other hand if the plate velocity V_p is greater than V_f, the plate will be pushing the slab into the asthenosphere at a velocity greater than V_f and the state of stress in the slab will be one of compression.

First-motion studies of earthquakes along a typical Benioff-Wadati zone show that faulting in the sinking slab is along fault planes inclined about 45° from the inclined surface of the slab. At depths greater than 300 kilometers or so the direction of compression as measured by the **P** axes is in the plane of the slab and oriented in the direction of slab motion. The **T** axes are perpendicular to the plane of the slab. The slab appears to be undergoing compression in the direction of its downward motion, as if the slab were butting up against a layer of great strength or high viscosity deep in the earth (Figure 6-17). However, in the upper part of the

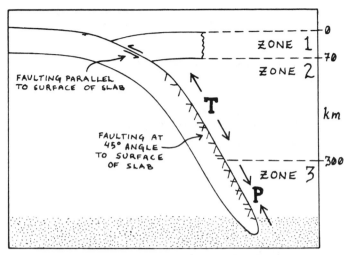

Figure 6-17.
Earthquakes in subducting slab. Zone 1, complex faulting including faulting parallel to surface of slab. Zone 2, tension parallel to surface of slab. Zone 3, compression parallel to surface of slab.

slab the orientations of the **T** and **P** axes are reversed, indicating tensional stress in the direction of slab motion.

In some subduction zones the orientations of **T** and **P** axes are more complicated than that just described, especially near the hinge line where the plate is bending downward. However, in general the focal mechanism analyses support the idea that slab-pull is one of the main plate driving forces. The slab-pull model predicts the presence at moderate depth of tension axes aligned with the direction of slab motion, and these are commonly observed in first-motion studies. The observed tension axes support theoretical models in which the lithosphere is denser than the underlying asthenosphere. Such models are consistent with much of what we know about plate tectonics. In passing it is interesting to ask why, if old lithosphere is thick and heavy, it doesn't sink everywhere? The answer may be that like a ship made of steel, it isn't able to sink because it doesn't leak. Occasional leaks are plugged by volcanoes, so in a sense the oceanic lithosphere is self-sealing. However, along trenches the lithosphere breaks up and sinks on a grand scale. This sinking constitutes a kind of engine which converts the gravitational potential energy of heavy lithosphere into the kinetic energy of plate motion. Whether or not this interpretation is correct, the first-motion studies provide a clear indication that in some sense the slab is pulling the adjacent plate into the trench.

Box 6-10. Examples of Normal (Extensional) and Reversed (Compressional) Focal Mechanism Solutions.

Circles are focal mechanism solutions. Shaded areas indicate compression, open areas indicate dilatation. Squares flanking the circles are cross sections showing the two possible geologic interpretations of the focal mechanism solution.

If the down (+**d**) axis is in a zone of compression, both geologic interpretations are reversed (or thrust) faults, indicating compression. If the (+**d**) axis is in a zone of dilatation, both geologic interpretations show normal faults (extension). If the fault plane is exactly vertical or horizontal, it is indeterminate whether the faulting is normal or reversed.

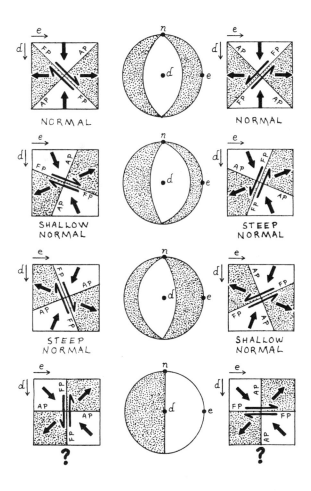

(continued)

Box 6-10. *(continued)*

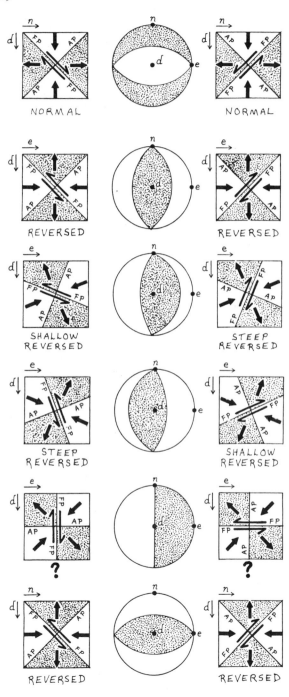

Problems

6-1. You are given the nodal planes for earthquakes a through e and the additional information that for each earthquake, the ray leaving the focus at an angle $I = 45°$, $D = 45°$ is a dilatation. For each earthquake plot on the lower hemisphere the nodal planes and show the slip vector(s) (**S**), the compression axis (**P**), the tension axis (**T**), and the poles of the nodal planes; then shade the quadrants of compression. Finally, describe the nature of the faulting (for example, right-lateral with a small component of thrust faulting toward the north).

Earthquake	Fault Plane		Auxiliary Plane	
	strike	dip	strike	dip
a	20°	30° SE	200°	60° NW
b	100°	80° SW	10°	90°
c	10°	90°	100°	80° SW
d	26°	70° SE	80°	32° NW
e	33°	70° SE	71°	25° NW

6-2. Two magnitude 6 earthquakes in the Gulf of California are located near the junction of a ridge and a transform. The uncertainties in the locations of the epicenters (both shown as one point **E** in the figure) would permit either of the two

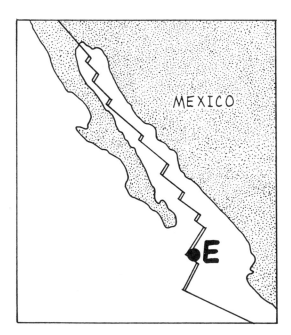

epicenters to be located on the ridge or the transform. As a seismologist you have obtained seismograms from 10 world-wide stations and have determined the first motions listed in the following table, where D is the azimuth from the epicenter to each seismographic station, I is the angle below the horizontal at which each ray left the focus, ● indicates first motions that were compressions, and ○ indicates those that were dilatations. Now you want to interpret these data.

(a) Plot the compressions and dilatations and try to find the nodal planes and the **P** and **T** axes.

(b) Give the strike and dip of the planes and the azimuth and inclination of the axes, along with a subjective estimate of the uncertainty in each.

(c) Based on the tectonic setting, try to decide which of the nodal planes is the fault plane and describe the nature of faulting (e.g., reversed with a small component of right-lateral strike-slip faulting).

(d) Based on your pick of the fault plane, plot the slip vector for each solution and list its azimuth and inclination.

(e) Would either of these slip vectors be useful in making global reconstructions of plate motions? How?

I	D	**EQ1**	**EQ2**
9°,	20°	○	●
15°	46°	●	○
62°	50°	●	○
34°	119°	●	●
56°	133°	●	○
46°	148°	○	●
22°	215°	○	○
9°	236°	●	●
22°	314°	○	●
75°	343°	○	○

6-3. Having learned what you can from seismology about spreading and transform tectonics in the Gulf of California, you now turn your attention to earthquakes occurring on the opposite side of the Pacific along the Kurile-Kamchatka Trench. You decide to study shallow earthquakes (focal depths less than 60 km) with epicenters immediately west of the trench. The results of your focal mechanism analyses are presented in the following table. These real data describe the large shallow earthquakes that occurred west of the trench between 1962 and 1979. Having done all this work, you have now reached the crucial stage of interpreting your data and trying to relate it to the big picture.

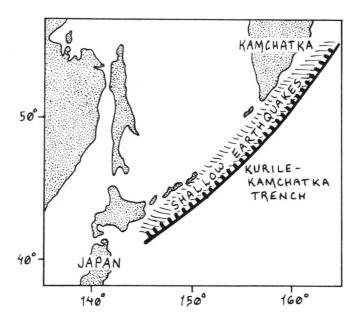

Focus			Nodal Planes				P Axis	
λ	ϕ	**Depth**	I_1	D_1	I_2	D_2	I_p	D_p
(44.7°,150.7°)		25 km	(3°,311°)		(84°,190°)		(42°,136°)	
(50.5°,156.6°)		27 km	(20°,310°)		(69°,144°)		(25°,134°)	
(54.8°,161.6°)		32 km	(25°,320°)		(64°,128°)		(20°,136°)	
(57.8°,163.5°)		33 km	(15°,302°)		(71°, 82°)		(29°,112°)	
(56.0°,163.3°)		33 km	(10°,300°)		(78°,150°)		(34°,125°)	
(46.5°,153.2°)		33 km	(12°,298°)		(74°,162°)		(32°,127°)	
(45.6°,150.5°)		33 km	(26°,300°)		(62°,146°)		(19°,129°)	
(44.3°,149.5°)		33 km	(16°,300°)		(70°,160°)		(28°,130°)	
(44.1°,148.0°)		33 km	(29°,312°)		(57°,102°)		(15°,121°)	
(45.8°,150.8°)		40 km	(29°,294°)		(56°,150°)		(14°,128°)	
(44.8°,149.0°)		40 km	(16°,302°)		(74°,132°)		(29°,124°)	
(43.6°,146.7°)		40 km	(23°,303°)		(62°,160°)		(20°,135°)	
(43.5°,146.8°)		48 km	(23°,319°)		(64°,114°)		(21°,131°)	
(43.2°,145.8°)		48 km	(18°,320°)		(72°,135°)		(27°,139°)	
(43.3°,146.4°)		50 km	(23°,305°)		(65°,150°)		(21°,133°)	
(50.9°,157.4°)		52 km	(25°,310°)		(64°,143°)		(19°,134°)	
(50.8°,157.1°)		54 km	(35°,310°)		(55°,140°)		(10°,134°)	
(44.8°,149.6°)		54 km	(16°,302°)		(74°,135°)		(29°,125°)	
(44.8°,149.5°)		60 km	(22°,313°)		(60°,179°)		(20°,148°)	

(a) To get an overview, plot the poles of the nodal planes and the **P** axes on an equal area projection. What generalization can you make about the grouping of these axes?

(b) For a typical earthquake, plot the nodal planes and the **P** and **T** axes. Considering the tectonic setting, which plane is most likely the fault plane?

(c) Is the tectonic regime one of tension or compression?

(d) On the basis of your pick of the fault plane, identify the slip vector in the diagram you made for (b) and the mean slip vector in the diagram you made for (a). You should end up with the azimuth of the mean slip vector for this set of earthquakes.

(e) The upper block of the subduction zone on which the Kurile Islands and Kamchatka Peninsula rest appears to be part of the North America plate. Using the data in Chapter 4, see how well your mean slip vector agrees with the relative motion of the North America and Pacific plates.

6-4. The earthquakes in the following table have intermediate and deep focal depths and epicenters to the west of the shaded zone of shallow earthquakes in the preceding figure. They are obviously different from the preceding set of shallow earthquakes. To analyze them, you might first plot the nodal planes and the **P** and **T** axes , and then consider the following. For each of these earthquakes, is there tension or compression in the down-dip direction in the plane of the slab? In analyzing these real data, you may find that not all earthquakes give the same answer. This is typical of real data from subduction zones. Some of the earthquakes at roughly the same focal depth in a slab may indicate compression and others tension. How would you interpret this? Try to wring as much tectonic information as you can from this data set.

Focus			Nodal Planes				P Axis	
λ	ϕ	Depth	I_1	D_1	I_2	D_2	I_p	D_p
(46.5°,151.1°)		101 km	(5°,123°)		(79°, 7°)		(39°,312°)	
(52.9°,159.2°)		106 km	(5°,307°)		(83°,170°)		(40°,131°)	
(48.3°,153.0°)		120 km	(15°,320°)		(72°,150°)		(59°,334°)	
(49.1°,153.6°)		134 km	(15°,300°)		(73°, 92°)		(59°,311°)	
(48.0°,153.2°)		134 km	(5°,310°)		(80°, 70°)		(39°,122°)	
(49.5°,154.4°)		136 km	(3°,316°)		(80°, 64°)		(41°,127°)	
(49.7°,155.0°)		143 km	(12°,300°)		(72°, 70°)		(32°,108°)	
(43.1°,140.6°)		163 km	(10°,148°)		(70°, 30°)		(34°,326°)	
(44.2°,145.3°)		180 km	(12°,321°)		(74°, 98°)		(32°,132°)	
(51.7°,150.9°)		544 km	(13°,310°)		(58°,198°)		(50°,278°)	
(49.5°,147.0°)		578 km	(10°,155°)		(76°,288°)		(34°,326°)	
(52.4°,151.6°)		645 km	(40°,310°)		(50°,130°)		(85°,310°)	

6-5.   In the course of mapping an uncharted ocean basin, you discover prominent linear features shown as heavy lines in the squares of the figure, but you're uncertain whether the lineations mark transforms, ridges, or zones of convergence. Fortunately you are able to record the earthquakes with epicenters shown by solid circles (shallow earthquakes) or solid squares (intermediate depth earthquakes). These provide you with additional information about the nature of stress release along the plate boundaries. Using the focal mechanisms shown next to each square map, try to decide whether the boundaries are ridges, trenches, or transforms. Don't forget to determine the polarity of the boundary where applicable. As a preliminary, (1) indicate which of the nodal planes is the fault plane, (2) whenever possible plot and label the lower hemisphere slip vector(s) and **P** and **T** axes, and (3) describe the nature of the faulting (for example, normal faulting along a fault trending north-northeast with a small component of right-lateral strike-slip motion).

Earthquakes *g*, *h*, and *i* are earthquakes all on the same boundary, which you know from bathymetry is a trench. Use the focal mechanisms to learn all you can about the subducting slab, including the important question of whether the stress in the slab in the down dip direction is compression or tension for earthquakes *h* and *i*, which are at intermediate depths. It might be helpful to draw a cross section of this boundary.

EPICENTER OF
INTERMEDIATE DEPTH
EARTHQUAKE

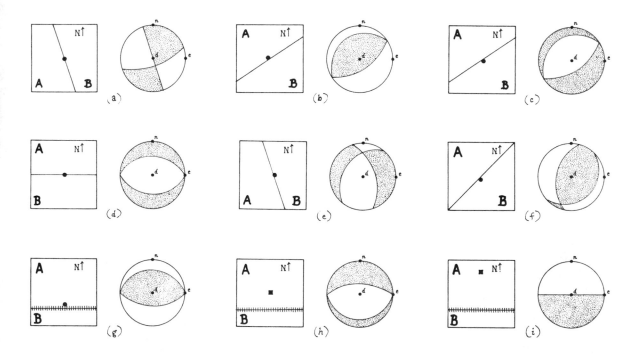

6-6. Four shallow earthquakes along the boundary between the Pacific and North America plates have the focal mechanisms listed in the following table. Your task is to find the Pacific-North America Euler pole from these data. The following steps may help in your analysis.

(1) For each earthquake plot the poles, the corresponding nodal planes, and the **P** and **T** axes.

(2) Shade the quadrants of compression.

(3) On the basis of regional tectonics, decide which nodal plane is the fault plane. Describe the motion of the fault (for example, left-lateral strike-slip).

(4) Label the slip vector **S**.

(5) Read the azimuth *A* of the slip vector.

(6) From the four values of *A* determine the location of the Euler pole using the techniques described in Chapter 4.

(7) Compare your Euler pole with the one found from a global analysis of plate motions, as discussed in Chapter 4.

Epicenter		Nodal Planes				P Axis	
λ	ϕ	I_1	D_1	I_2	D_2	I_p	D_p
(23°,	252°)	(10°,	305°)	(0°,	35°)	(7°,	350°)
(57°,	208°)	(29°,	333°)	(60°,	140°)	(15°,	149°)
(51°,	178°)	(28°,	322°)	(54°,	185°)	(14°,	160°)
(44°,	148°)	(26°,	301°)	(64°,	125°)	(19°,	122°)

Suggested Readings

Mechanism of Earthquakes and Nature of Faulting on the Mid-Ocean Ridges, Lynn Sykes, Journal of Geophysical Research, v. 72, p. 2131–2153, 1967. *Classic article confirming that plate motion along transforms is in the sense predicted by plate tectonics.*

Seismology and the New Global Tectonics, Bryan Isacks, Jack Oliver, and Lynn Sykes, Journal of Geophysical Research, v. 73, p. 5855–5899, 1968. *Classic paper showing that slip vectors along transforms and subductions have the orientations predicted by plate tectonics; shows that many seismological and geophysical observations from island arcs can be explained by plate tectonics.*

Mantle Earthquake Mechanisms and the Sinking of the Lithosphere, Bryan Isacks and Peter Molnar, Nature, v. 223, p. 1121–1124, 1969. *Classic article using focal plane mechanisms to determine the stress in lithospheric slabs.*

Tectonic Implications of the Microearthquake Seismicity and Fault Plane Solutions in Southern Peru, F. Grange, D. Hatzfeld, P. Cunningham, P. Molnar, S. Roecker, G. Suárez, A. Rodrígues, and L. Ocola, Journal of Geophysical Research, v. 89, p. 6139–6152, 1984. *Good example of the use of first-motion analyses of earthquakes to provide insight into regional tectonics.*

Mid-Ocean Ridges: Fine Scale Tectonic, Volcanic and Hydrothermal Processes Within the Plate Boundary Zone, Ken Macdonald, Annual Reviews Earth and Planetary Sciences, v. 10, p. 155–190, 1982. *Review of the geology and geophysics of spreading centers.*

The East Pacific Rise in Cross Section: A Seismic Model, James McClain, John Orcutt, and Mark Burnett, Journal of Geophysical Research, v. 90, p. 8627–8640, 1985. *Attempts to delineate magma chamber beneath rise using seismic ray tracing and other seismic modeling.*

Fault Motion in the Larger Earthquakes of the Kurile-Kamchatka Arc and of the Kurile, Hokkaido Corner, William Stauder and Lalliana Mualchin, Journal of Geophysical Research, v. 81, p. 297–308, 1976. *Source of data for problems 6-3 and 6-4.*

7

Finite Rotations

Finite rotations aren't new to us—they're simply rotations by a given angle about a fixed Euler pole. In most of our constructions on a sphere, we used finite rotations without describing them as such. Because finite rotations are used for solving so many problems in plate tectonics, let's pause for a moment and take a systematic look at how they are done and what they mean.

On the spherical earth finite rotations boil down to the following physical model. Let's say that we want to rotate Africa back to its position relative to South America prior to the opening of the South Atlantic Ocean. Let's call the Africa plate "AF" and the South America plate "SA." Let's start with a small globe showing the outlines of the continents and a set of latitude and longitude circles. Place a plastic shell over the sphere and trace onto it an outline of Africa and the present coordinate grid (Figure 7-1). Now place a pin through the shell at the Euler pole for the opening of the South Atlantic, point **E**. The plastic shell carrying the outline of Africa and the coordinate system is free to rotate about the pin through **E**. Points on the rotating shell move along small circles centered on **E**. When the finite rotation is complete, a point located 90° from **E** will have moved along a great circle an angular distance Ω. The symbol we use for describing this operation is ROT[**E**, Ω], where a positive value of Ω indicates rotation about **E** in a counterclockwise or right-handed sense.

It's very important to keep your wits about you in defining and keeping track of coordinate systems. In fact, before

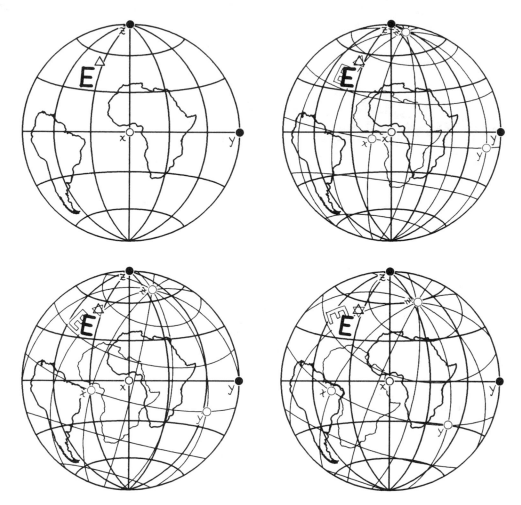

Figure 7-1.
A plastic spherical shell is pivoted at the Euler pole **E** to rotate relative to the underlying globe. Geographical grid lines and the outline of Africa are traced onto the plastic shell. Note that during the rotation, the Euler pole **E** retains the same latitude and longitude in both coordinate systems.

tackling a finite rotation problem it's usually worthwhile to be painfully explicit in describing your coordinate system. An appropriate (if somewhat wordy) statement describing the present problem is: "We will use a fixed coordinate system that remains attached to South America (SA) so that points on South America will have the same latitude and longitude before and after the rotation whereas points on Africa (AF) will have different coordinates." We can summarize this statement by the shorthand notation $_{SA}ROT_{AF}$, meaning "the rotation of plate AF relative to plate SA." The subscript "SA" on the left indicates that SA (South America) is the fixed coordinate system. Figure 7-1 shows the rotation of Africa and its coordinate grid relative to a fixed present coordinate system. Note in Figure 7-1 that the Euler pole **E** is the only point that stays at the same latitude and longitude in both the SA and AF coordinate systems during the rotation.

Turning from an intuitive to a mathematical approach, two main techniques are available for doing finite rotations. In the first technique, which is suitable for programmable pocket calculators and for graphical constructions on a projection, a finite rotation about an Euler pole **E** is broken down into a series of consecutive rotations about the **2** and **3** axes of the projection as follows:

$$\text{ROT}[\mathbf{E},\Omega] = \text{ROT}[\mathbf{3}, -\phi_E] + \text{ROT}[\mathbf{2}, -\theta_E] + \text{ROT}[\mathbf{3},\Omega] + \text{ROT}[\mathbf{2},\theta_E] + \text{ROT}[\mathbf{3},\phi_E]$$

where ϕ_E is the longitude of the Euler pole **E** and $\theta_E = 90°$ $- \lambda_E$ is the colatitude of **E**. An explanation for this technique is given in Boxes 7-1 and 7-2. A second technique, which is more suitable for use on a computer, is to use a matrix transformation of the three Cartesian components of position vectors, as described in Box 7-3. Computer programs using this technique were used to draw Figure 7-1 and many of the other maps in this book.

Jumping Poles

We saw earlier that the Euler pole describing the relative motion between a pair of plates remains fixed for long periods of time. If the plates rotate about a fixed Euler pole **E** between times t_2 and t_1 at a constant rate ω, the resulting finite rotation can be described by the operation $\text{ROT}[\mathbf{E}, \omega(t_1 - t_2)]$. However plates don't go on rotating about the same Euler poles forever. It's quite common, in fact, for an Euler pole to jump to a new position. How can we recognize when this has happened?

Let's imagine that after a long time at position **E** (Figure 7-2), at 30 Ma an Euler pole jumps to position **F**. Now let's observe what happens after the jump. The old pole **E** develops a split personality and becomes two poles, $_A\mathbf{E}_B$ and $_B\mathbf{E}_A$, which begin to diverge. Pole $_B\mathbf{E}_A$ remains fixed in the "B" coordinate system which is attached to plate B. The old transforms on plate B remain concentric about $_B\mathbf{E}_A$, whereas the newer transforms on plate B develop concentrically about the new pole **F**. Now let's watch what happens to plate A. Since **F** is the only point that remains fixed in the "B" coordinate system, it is inevitable that $_A\mathbf{E}_B$ must rotate about **F** in the fixed "B" reference frame. Moreover, transforms on plate

Box 7-1. How to Rotate Graphically.

To reconstruct point $\mathbf{K} = (\lambda_K, \phi_K)$ on the India plate using the Euler pole $\mathbf{E} = (\lambda_E, \phi_E)$ and angle Ω that reconstructs India with respect to Africa, recall that one of our favorite tricks is to align the Euler pole with the **3** axis at the top of the projection. This is done because during rotation about the **3** axis, a point simply moves along one of the latitude circles of the projection. These circles are concentric about the **3** axis.

Using a stereographic or an equal area projection, any finite rotation can be accomplished using five successive rotations about the **2** and **3** axes. The first two rotations align the Euler pole $\mathbf{E}$ with **3**, the third rotation is the desired rotation, and the last two rotations return $\mathbf{E}$ (and the globe) to its original position. The position of India is shown before the rotation (with point $\mathbf{K}$) and after the rotation (with point $\mathbf{K}'$).

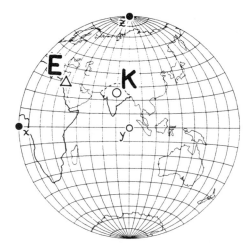

- Plot $\mathbf{E}$, $\mathbf{K}$

Set up projection.

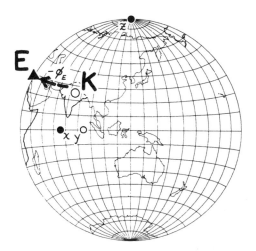

(continued)

Box 7-1.   *(continued)*

- ROT[**3**, $-\phi_E$]

Rotate **E** to the index meridian.

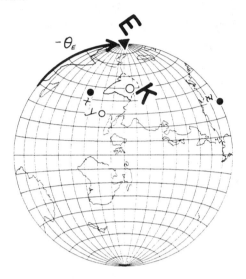

- ROT[**2**, $-\theta_E$]

Rotate **E** to **3**. $\theta_E = 90° - \lambda_E$ is the colatitude of **E**.

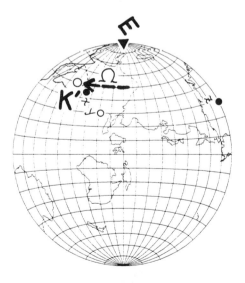

- ROT[**3**, Ω]

This is the desired rotation from **K** to **K'**. Note that in our example Ω is negative.

(continued)

Box 7-1.   *(continued)*

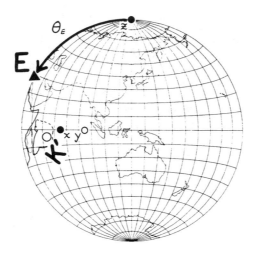

- ROT[**2**, θ_E]

Rotate **E** back to its original latitude.

- ROT[**3**, ϕ_E]

Rotate **E** back to its original longitude.

- Read **K'**

The globe coordinates of **K'** can now be read from the projection.

Note: The first and last rotations can be omitted if you begin and end with the index meridian = ϕ_E.

Box 7-2. How to Rotate Using a Programmable Pocket Calculator.

Most programmable pocket calculators do not have enough memory to store a 3×3 matrix and perform the operations outlined in Box 7-3. Using the functions described in Box 3-9 and the technique described in Box 7-1, however, it is still possible to perform finite rotations on a programmable pocket calculator. Again the actual programming is for the HP series 40 calculator, but transfer onto other calculators is straightforward.

Program	Stack				Registers		
	x	y	z	t	R_{11}	R_{12}	R_{13}
LBL "FINROT"							
LBL a							
"LAT ↑ LON ↑ ANG?"							
PROMPT	Ω	ϕ_E	λ_E	-	-	-	Ω
XEQ "SETROT"	Ω	ϕ_E	λ_E	θ_E	θ_E	ϕ_E	Ω
"LAT ↑ LON?"							
PROMPT	ϕ	λ	-	-	θ_E	ϕ_E	Ω
LBL A	ϕ	λ	-	-	θ_E	ϕ_E	Ω
1	1	ϕ	λ	-	θ_E	ϕ_E	Ω
XEQ "ROTATE"	1	ϕ'	λ'	-	θ_E	ϕ_E	Ω
R ↓	ϕ'	λ'	-	1	θ_E	ϕ_E	Ω
CLA	ϕ'	λ'	-	1	θ_E	ϕ_E	Ω
FIX 2	ϕ'	λ'	-	1	θ_E	ϕ_E	Ω
ARCL Y	ϕ'	λ'	-	1	θ_E	ϕ_E	Ω
← ","	ϕ'	λ'	-	1	θ_E	ϕ_E	Ω
ARCL X	ϕ'	λ'	-	1	θ_E	ϕ_E	Ω
AVIEW	ϕ'	λ'	-	1	θ_E	ϕ_E	Ω
STOP	ϕ'	λ'	-	1	θ_F	ϕ_E	Ω
GTO A							

This program calls two general application subroutines. The first one (SETROT) stores the finite rotation in registers R_{11} through R_{13}:

Function	Stack				Registers		
	x	y	z	t	R_{11}	R_{12}	R_{13}
LBL "SETROT"	Ω	ϕ_E	λ_E	-	-	-	-
90	90	Ω	ϕ_E	λ_E	-	-	-
R ↓	Ω	ϕ_E	λ_E	90	-	-	Ω
STO 13	Ω	ϕ_E	λ_E	90	-	-	Ω
R ↓	ϕ_E	λ_E	90	Ω	-	-	Ω
STO 12	ϕ_E	λ_E	90	Ω	-	ϕ_E	Ω
R ↓	λ_E	90	Ω	ϕ_E	-	ϕ_E	Ω
-	θ_E	Ω	ϕ_E	ϕ_E	-	ϕ_E	Ω
STO 11	θ_E	Ω	ϕ_E	ϕ_E	θ_E	ϕ_E	Ω
LAST X	λ_E	θ_E	Ω	ϕ_E	θ_E	ϕ_E	Ω
R ↓	θ_E	Ω	ϕ_E	λ_E	θ_E	ϕ_E	Ω
R ↓	Ω	ϕ_E	λ_E	θ_E	θ_E	ϕ_E	Ω
RETURN	Ω	ϕ_E	λ_E	θ_E	θ_E	ϕ_E	Ω

(continued)

Box 7-2. *(continued)*

The colatitude θ_E is stored instead of the latitude λ_E because it is θ_E that is actually used (Box 7-1). The second subroutine (ROTATE) performs the rotation:

Function	Stack				Registers		
	x	y	z	t	R_{11}	R_{12}	R_{13}
LBL "ROTATE"	r	ϕ	λ	-	θ_E	ϕ_E	Ω
RCL 12	ϕ_E	r	ϕ	λ	θ_E	ϕ_E	Ω
CHS	$-\phi_E$	r	ϕ	λ	θ_E	ϕ_E	Ω
ST + Z	$-\phi_E$	r	ϕ_1	λ	θ_E	ϕ_E	Ω
CLX	0	r	ϕ_1	λ	θ_E	ϕ_E	Ω
RCL 11	θ_E	r	ϕ_1	λ	θ_E	ϕ_E	Ω
CHS	$-\theta_E$	r	ϕ_1	λ	θ_E	ϕ_E	Ω
XEQ "ROT 2"	r	ϕ_2	λ_1	$-\theta_E$	θ_E	ϕ_E	Ω
RCL 13	Ω	r	ϕ_2	λ_1	θ_E	ϕ_E	Ω
ST + Z	Ω	r	ϕ_3	λ_1	θ_E	ϕ_E	Ω
CLX	0	r	ϕ_3	λ_1	θ_E	ϕ_E	Ω
RCL 11	θ_E	r	ϕ_3	λ_1	θ_E	ϕ_E	Ω
XEQ "ROT 2"	r	ϕ_4	λ'	θ_E	θ_E	ϕ_E	Ω
RCL 12	ϕ_E	r	ϕ_4	λ'	θ_E	ϕ_E	Ω
ST + Z	ϕ_E	r	ϕ'	λ'	θ_E	ϕ_E	Ω
CLX	0	r	ϕ'	λ'	θ_E	ϕ_E	Ω
RCL 13	Ω	r	ϕ'	λ'	θ_E	ϕ_E	Ω
R↓	r	ϕ'	λ'	Ω	θ_E	ϕ_E	Ω
XEQ "S-C"	x'	y'	z'	Ω	θ_E	ϕ_E	Ω
XEQ "C-S"	r	ϕ'	λ'	Ω	θ_E	ϕ_E	Ω
RETURN	r	ϕ'	λ'	Ω	θ_E	ϕ_E	Ω

The "S-C" and "C-S" conversions between spherical and Cartesian coordinates might seem to negate each other, but they perform the important function of converting ϕ' to the range $-180° \leq \phi' \leq 180°$. The operation ST + Z, CLX was used instead of the equivalent operation "ROT 3" because it takes fewer bytes of memory and operates faster.

Box 7-3. How to Rotate Using a Computer.

If you would rather work with algebra than geometry, you can do finite rotations using a 3×3 matrix. If point **A** is a vector with global Cartesian coordinates (A_x, A_y, A_z) prior to rotation, then the components (A_x', A_y', A_z') of **A** after rotation to **A'** may be found from the matrix multiplication

$$\mathbf{A'} = \mathbf{R}\,\mathbf{A}$$

where **R** represents a 3×3 matrix. Writing all of the terms of the vector and matrix we have

(continued)

Box 7-3. *(continued)*

$$\begin{bmatrix} A_x' \\ A_y' \\ A_z' \end{bmatrix} = \begin{bmatrix} R_{11} & R_{12} & R_{13} \\ R_{21} & R_{22} & R_{23} \\ R_{31} & R_{32} & R_{33} \end{bmatrix} \begin{bmatrix} A_x \\ A_y \\ A_z \end{bmatrix}$$

Applying the usual rules for matrix multiplication gives the equations

$$A_x' = R_{11}A_x + R_{12}A_y + R_{13}A_z$$
$$A_y' = R_{21}A_x + R_{22}A_y + R_{23}A_z$$
$$A_z' = R_{31}A_x + R_{32}A_y + R_{33}A_z$$

To define the elements of the rotation matrix **R** we need to know the Cartesian coordinates of the Euler pole $\mathbf{E} = (E_x, E_y, E_z)$ and the angle of rotation Ω. The elements of the matrix are then given by

(first row of matrix)
$$R_{11} = E_x E_x (1 - \cos \Omega) + \cos \Omega$$
$$R_{12} = E_x E_y (1 - \cos \Omega) - E_z \sin \Omega$$
$$R_{13} = E_x E_z (1 - \cos \Omega) + E_y \sin \Omega$$

(second row of matrix)
$$R_{21} = E_y E_x (1 - \cos \Omega) + E_z \sin \Omega$$
$$R_{22} = E_y E_y (1 - \cos \Omega) + \cos \Omega$$
$$R_{23} = E_y E_z (1 - \cos \Omega) - E_x \sin \Omega$$

(third row of matrix)
$$R_{31} = E_z E_x (1 - \cos \Omega) - E_y \sin \Omega$$
$$R_{32} = E_z E_y (1 - \cos \Omega) + E_x \sin \Omega$$
$$R_{33} = E_z E_z (1 - \cos \Omega) + \cos \Omega$$

If you decide to program this operation on your computer, the following numerical example may help in debugging your program. For the rotation

$$\text{ROT}[\mathbf{E}, \Omega] = \text{ROT}[(-37°, 312°), 65°]$$

the values of the elements of the matrix are

$$\mathbf{R} = \begin{bmatrix} 0.588 & 0.362 & -0.724 \\ -0.729 & 0.626 & -0.278 \\ 0.352 & 0.691 & 0.632 \end{bmatrix}$$

The point $\mathbf{A} = (20°, 130°)$ has the Cartesian components $(A_x, A_y, A_z) = (-0.604, 0.720, 0.342)$. Performing the matrix multiplication yields the Cartesian components $(A_x', A_y', A_z') = (-0.342, 0.796, 0.500)$, which when converted to spherical coordinates is $(30.0°, 113.2°)$.

(continued)

Box 7-3. *(continued)*

The rotation ROT[**E**, $-\Omega$] may be regarded as the negative of ROT[**E**, Ω]. These two rotations done in succession leave a point at the coordinates from which it started. (A good check on your computer program is to run the two rotations back-to-back and see if you end up at the point you started with). If the matrix for the operation ROT[**E**, Ω] is the matrix **R** as previously defined, then for the operation ROT[**E**, $-\Omega$] the corresponding matrix is the inverse $\mathbf{R}^{-1}$ of the rotation matrix **R**. If you've taken linear algebra, you'll recognize that the inverse $\mathbf{R}^{-1}$ of a rotation (orthonormal) matrix is the transpose $\mathbf{R}^{T}$ of the matrix:

$$\mathbf{R}^{-1} = \mathbf{R}^{T} = \begin{bmatrix} R_{11} & R_{21} & R_{31} \\ R_{12} & R_{22} & R_{32} \\ R_{13} & R_{23} & R_{33} \end{bmatrix}$$

where the elements R_{11}, R_{12}, R_{13}, etc. are as defined for the matrix **R**. Upon comparing the matrices **R** and $\mathbf{R}^{T}$, we find that the first row of $\mathbf{R}^{T}$ is the first column of **R**, the second row of $\mathbf{R}^{T}$ is the second column of **R**, and the third row of $\mathbf{R}^{T}$ is the third column of **R**. This has the effect of exchanging R_{13} with R_{31}, R_{12} with R_{21}, and R_{23} with R_{32}. If you examine the equations defining R_{13} and R_{31}, you will see that changing the sign of Ω is equivalent to exchanging these two elements of **R**, and likewise for R_{12}, R_{21}, and R_{23}, R_{32}.

A produced before 30 Ma remain concentric about $_A\mathbf{E}_B$ as they and $_A\mathbf{E}_B$ rotate together about the new pole **F**. After rotation about **F** has continued for a while, we can recognize that the pole has jumped to a new position by noting that the older transforms on the two sides of the ridge are not concentric about the same pole. Two Euler poles, $_A\mathbf{E}_B$ and $_B\mathbf{E}_A$, are needed to fit the transforms. Let's say that the total amount of rotation about **F** from 30 Ma to the present can

Figure 7-2.
The Euler pole jumps from **E** to **F** at 30 Ma. Prior to that, rotation about **E** produced the arcuate fracture zone shown to the left with a radius of 30°. After the pole jumps to **F** the new fracture zones have a radius of 90°. In the figure all of the points on plate B remain fixed whereas the pole $_A\mathbf{E}_B$ and all of the other points on plate A rotate 40° about pole **F**. The active spreading center rotates 20° about pole **F** during the time 30 Ma to the present.

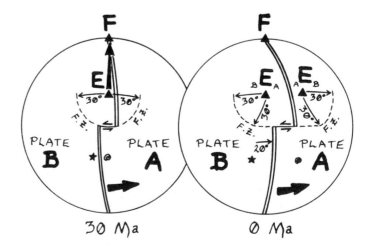

be described by the rotation ROT[**F**, Ω_F] (Figure 7-2), where $\Omega_F = 40°$. Then if we shift the rotation process into reverse and restore the plates into their 30 Ma positions by the rotation ROT[**F**, $-\Omega_F$], we will find that $_A\mathbf{E}_B$ and $_B\mathbf{E}_A$ now coincide and the older transforms are now concentric about the common pole $\mathbf{E} = {}_A\mathbf{E}_B = {}_B\mathbf{E}_A$.

How can you restore to their original position two plates that have moved apart by rotations about a succession of Euler poles? Two techniques are available, one requiring patience and the other mathematical insight. If you're patient, you can first close up the most recent rotation about the appropriate Euler pole, then the second most recent about its Euler pole, and so on until you reach the time for which you desire a reconstruction. The alternative approach is to take a shortcut and combine the entire sequence of rotations into a single rotation about one Euler pole. This is possible because the result of successive rotations about different Euler poles can always be duplicated by a single rotation about one Euler pole. For example, Boxes 7-1 and 7-2 demonstrate that for an Euler pole with colatitude θ_E and longitude ϕ_E),

$$\text{ROT}[\mathbf{3}, -\phi_E] + \text{ROT}[\mathbf{2}, -\theta_E] + \text{ROT}[\mathbf{3}, \Omega] + \text{ROT}[\mathbf{2}, \theta_E] + \text{ROT}[\mathbf{3}, \phi_E] = \text{ROT}[\mathbf{E}, \Omega]$$

This is a very simple example of a very important idea: For consecutive rotations about any two or more Euler poles, ROT[**E**, $\omega_E(t_3 - t_2)$] and ROT[**F**, $\omega_F(t_2 - t_1)$], there is an Euler pole **P** and an angular velocity ω_P such that

$$\text{ROT}[\mathbf{E}, \omega_E\,(t_3 - t_2)] + \text{ROT}[\mathbf{F}, \omega_F\,(t_2 - t_1)] = \text{ROT}[\mathbf{P}, \omega_P\,(t_3 - t_1)] \tag{7.1}$$

The left-hand side of the equation describes the actual history of the rotation. The right-hand side of the equation describes how to get from the initial position (time t_3) to the final position (time t_1) as directly as possible. How to combine finite rotations is explained graphically in Box 7-4 and mathematically in Box 7-5.

In the early days of plate tectonics geophysicists were impressed by the way an individual transform could be fit by a small circle that in some cases extended almost across the entire Atlantic or Pacific Ocean. Currently geophysicists are finding that these long transforms can be fit much better by a series of small circles, which shows that the Euler poles have had many small but real shifts in position during the

Box 7-4. How to Add Rotations Graphically

In plate tectonics we commonly need to make several rotations in succession. Let's assume that we first make the rotation ROT[**E**, Ω] and that we then make the rotation ROT[**F**, Ω']. We want to find the single rotation that would have the same final result as these two successive rotations. Symbolically we describe the equivalent single rotation as ROT[**P**, Ω_T] = ROT[**E**, Ω] + ROT[**F**, Ω']. We will keep our reference coordinates fixed relative to plate A.

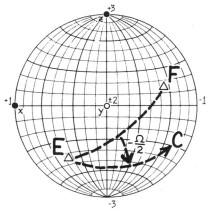

- Plot **E**, **F**

These are the two Euler poles.

- Draw <**E**, **F**>

This is the great circle through **E** and **F**.

- Draw <**E**, **C**>

Construct this great circle so that it passes through **E** and makes an angle $\Omega/2$ with <**E**, **F**>. This great circle can be found by rotating <**E**, **F**> by an angle $-\Omega/2$ around **E**. Remembering our sign convention, if Ω is positive then $-\Omega/2$ is negative and the rotation is clockwise, as shown above. **C** serves to designate or identify the great circle and is not used explicitly in the construction.

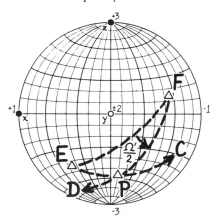

(continued)

Box 7-4.   *(continued)*

● Draw <**F, D**>

Construct this great circle so that it passes through **F** and makes an angle $\Omega'/2$ with <**E, F**>. To generate this great circle, you can rotate <**E, F**> by an angle $\Omega'/2$ around **F**. **D** serves to identify the great circle and is not used in the construction.

● Find **P**

This is the point of intersection of the great circles <**E, C**> and <**F, D**>.

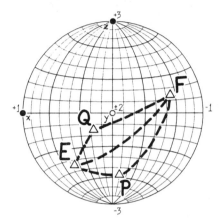

● Draw triangle **EQF**

Make this triangle similar to **EPF** by reflecting the angles $-\Omega/2$ and $\Omega'/2$ across <**E, F**> as shown.

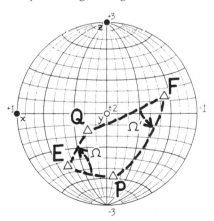

● Read **P**

The first rotation about **E** rotates **P** to **Q**. The second rotation about **F** rotates **Q** back to **P**. Therefore **P** is the desired Euler pole. Note that if we reverse the sequence of the two rotations, then the combined Euler pole is **Q**.

(continued)

Box 7-4. *(continued)*

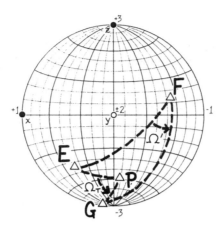

- Find **G**

G can be found by rotating **E** by the angle Ω' around **F**. In this example, Ω' is positive and the rotation is counterclockwise.

- Measure angle Ω_T

Point **E** on plate B does not move during the first rotation, ROT[**E**, Ω]. During the second rotation, ROT[**F**, Ω'], **E** moves to **G**. Since any point on plate B can be used to determine the total rotation and since **E** is on plate B, the angle Ω_T, which is the angle between the great circles <**P**, **E**> and <**P**, **G**>, is the combined angle of rotation. Ω_T is measured in the direction going from the great circle <**P**, **E**> to <**P**, **G**> and is positive if this angle is measured in the counterclockwise direction. (If you've forgotten how to find the angle between two great circles, review Box 3-5).

Note that Ω_T remains the same if you reverse the sequence of rotations, that is

$$\text{if ROT}[\mathbf{E}, \Omega] + \text{ROT}[\mathbf{F}, \Omega'] = \text{ROT}[\mathbf{P}, \Omega_T]$$
$$\text{then ROT}[\mathbf{F}, \Omega'] + \text{ROT}[\mathbf{E}, \Omega] = \text{ROT}[\mathbf{Q}, \Omega_T]$$

Note also that finding **Q** is unnecessary in finding ROT[**P**, Ω_T] = ROT[**E**, Ω] + ROT[**F**, Ω'], and is added only to explain what is going on.

Box 7-5. How to Add Rotations Using Matrix Multiplication.

Let's first rotate the point on the globe described by the vector **A** a new position **A**′ using the rotation matrix **R** so that **A**′ = **R A**. Next let's rotate **A**′ using the rotation matrix **R**′ so that **A**″ = **R**′ **A**′. How can we find a rotation matrix **T** that will rotate **A** directly to **A**″, such that **A**″ = **T A**?

We begin by using the associative property of matrix multiplication. This says that **A**″ = **R**′ **A**′ = **R**′ (**R A**) = (**R**′ **R**)**A** = **T A**, or

(continued)

Box 7-5. *(continued)*

$$\mathbf{T} = \mathbf{R'\, R}$$

the product of the two 3×3 matrices.

Matrix multiplication is defined as follows. In element form it is written

$$\mathbf{T} = \begin{bmatrix} T_{11} & T_{12} & T_{13} \\ T_{21} & T_{22} & T_{23} \\ T_{31} & T_{32} & T_{33} \end{bmatrix} = \begin{bmatrix} R'_{11} & R'_{12} & R'_{13} \\ R'_{21} & R'_{22} & R'_{23} \\ R'_{31} & R'_{32} & R'_{33} \end{bmatrix} \begin{bmatrix} R_{11} & R_{12} & R_{13} \\ R_{21} & R_{22} & R_{23} \\ R_{31} & R_{32} & R_{33} \end{bmatrix}$$

where

$$T_{11} = R'_{11}R_{11} + R'_{12}R_{21} + R'_{13}R_{31}$$
$$T_{12} = R'_{11}R_{12} + R'_{12}R_{22} + R'_{13}R_{32}$$
$$T_{13} = R'_{11}R_{13} + R'_{12}R_{23} + R'_{13}R_{33}$$

$$T_{21} = R'_{21}R_{11} + R'_{22}R_{21} + R'_{23}R_{31}$$
$$T_{22} = R'_{21}R_{12} + R'_{22}R_{22} + R'_{23}R_{32}$$
$$T_{23} = R'_{21}R_{13} + R'_{22}R_{23} + R'_{23}R_{33}$$

$$T_{31} = R'_{31}R_{11} + R'_{32}R_{21} + R'_{33}R_{31}$$
$$T_{32} = R'_{31}R_{12} + R'_{32}R_{22} + R'_{33}R_{32}$$
$$T_{33} = R'_{31}R_{13} + R'_{32}R_{23} + R'_{33}R_{33}$$

In index notation this is written:

$$T_{ij} = RR_{i1}R_{1j} + RR_{i2}R_{2j} + RR_{i3}R_{3j} = \sum_k (R'_{ik}R_{kj})$$

Note that T_{ij}, the element of the ith row and jth column of $\mathbf{T}$, is dot product of the ith *row* of $\mathbf{R'}$ and the jth *column* of $\mathbf{R}$.

Matrix multiplication does not commute. We shouldn't expect it to because it describes adding rotations, which we know doesn't commute. Algebraically, if we define $\mathbf{T'} = \mathbf{RR'}$, we can show that

$$T_{ij} = \sum_k (R'_{ik}R_{kj}) \neq \sum_k (R_{ik}R'_{kj}) = T'_{ij}$$

and therefore

$$\mathbf{T} = \mathbf{R'\, R} \neq \mathbf{R\, R'} = \mathbf{T'}$$

It is very important not to commute when adding rotations.

Now that we have found $\mathbf{T}$ from matrix multiplication, can we find the Euler pole $\mathbf{E}$ and finite angle Ω such that $\mathbf{T} = \text{ROT}[\mathbf{E}, \Omega]$? Yes we can. The trick is to use the definition of the rotation elements listed in Box 7-3 to derive

(continued)

Box 7-5.   *(continued)*

$$T_{32} - T_{23} = 2E_x \sin \Omega$$
$$T_{13} - T_{31} = 2E_y \sin \Omega$$
$$T_{21} - T_{12} = 2E_z \sin \Omega$$
$$T_{11} + T_{22} + T_{33} - 1 = 2 \cos \Omega$$

from which we can derive

$$\phi_E = \tan^{-1} \left(\frac{T_{13} - T_{31}}{T_{32} - T_{23}} \right)$$

$$\lambda_E = \sin^{-1} \left(\frac{T_{21} - T_{12}}{\sqrt{(T_{32} - T_{23})^2 + (T_{13} - T_{31})^2 + (T_{21} - T_{12})^2}} \right)$$

$$\Omega = \tan^{-1} \left(\frac{\sqrt{(T_{32} - T_{23})^2 + (T_{13} - T_{31})^2 + (T_{21} - T_{12})^2}}{T_{11} + T_{22} + T_{33} - 1} \right)$$

with the range of Ω such that $0° \leq \Omega \leq 180°$. It is very important to be sure that Ω is within this range. In most computer languages, the arctangent function returns a value from $-90°$ to $90°$, which must be converted to the desired range.

---

history of the formation of major ocean basins. These shifts correspond to small but real changes in the direction of plate motions. The history of spreading between major plates now appears to consist of a sequence of finite rotations about slightly different Euler poles. The time intervals between successive pole jumps appears to vary from several tens of millions of years to several million years.

In summary, during a sequence of finite rotations about different Euler poles, plates move apart along a series of different small circle arcs, whereas in motion about one Euler pole they move apart along a single arc. Using a single summary rotation is mathematically convenient for making reconstructions, but it does not generally give the true path followed by the plates.

Finite Rotations Versus Angular Velocity Vectors

Both angular velocity vectors and finite rotations can be described by a scalar angle and a unit vector **E**. Both unit vectors are commonly called Euler poles. Because of this

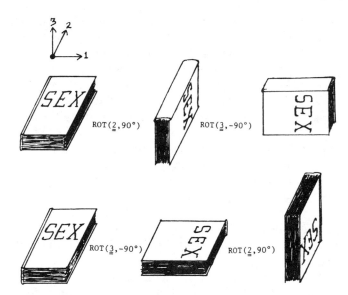

Figure 7-3.
The Evils of Commuting. Changing the order in which you rotate changes the position you end up in.

similarity, it is tempting to think of both angular velocities and finite rotations as vectors. Resist the temptation—finite rotations are *not* vectors. They must be handled with greater care. An angular velocity vector ω can be represented either by an Euler pole $\mathbf{E}$ and a scalar angle ω or by three Cartesian components $(\omega_x, \omega_y, \omega_z)$, whereas the Cartesian representation of a finite rotation is a 3×3 matrix (Box 7-3). Moreover, angular velocity vectors are commutative:

$$\omega_A + \omega_B - \omega_B + \omega_A$$

whereas finite rotations are not. The rotation you obtain depends upon the sequence in which you perform the rotations, as may be seen in Figure 7-3 and Box 7-2. If by "ROT$_A$ + ROT$_B$" we mean "rotation A followed by rotation B," then in general,

$$\text{ROT}_A + \text{ROT}_B \neq \text{ROT}_B + \text{ROT}_A \qquad (7.2)$$

Having emphasized the difference between finite rotations and angular velocity vectors, we must now back off a little and admit that in plate tectonics we are never able to measure the instantaneous velocity of plates, much as we would like to. What we measure is the finite rotation ROT[$\mathbf{E}$, Ω] that has occurred during a short time interval Δt. To determine the present "instantaneous" relative angular velocity vector of two plates separated by a ridge, for example, marine geophysicists first identify the same young isochron of age Δt on both sides of the ridge and then find the finite rotation needed to bring the two isochrons into coincidence. If Δt is short, say on the order of a million years, then for practical purposes we can treat the finite rotation of magnitude Ω as if it were an angular velocity vector $\boldsymbol{\omega}$ that occurred at a constant rate of $\Omega/\Delta t$, the direction of $\boldsymbol{\omega}$ being that of **E**. The basis for doing this is that to a good approximation, very small finite rotations can be added, commuted, and otherwise treated like vectors.

The following worst-case example will give you a feeling for the magnitude of the errors that are introduced when small finite rotations are handled as if they are angular velocities. Suppose we have two Euler poles which are oriented at right angles to each other in the plane of the equator, **E** $= (0°, 45°)$ and **F** $= (0°, 135°)$. Let's further assume that both have unusually high angular velocities of 1°/my, corresponding to a maximum rate of spreading or convergence of 111 mm/yr. If **E** describes the rotation of plate B with respect to plate A and if **F** describes the rotation of plate C with respect to plate B, then two finite rotations of 1° can be added to give the rotation of plate C with respect to plate A after a period of one million years:

$$\text{ROT}[\mathbf{F}, 1°] + \text{ROT}[\mathbf{E}, 1°] = \text{ROT}[\mathbf{G}, \Omega]$$

where **G** $= (0.35°, 90°)$ and $\Omega = 1.414°$. An approximation to this exact finite rotation can be made by converting the two finite rotations to instantaneous angular velocities, each with a scalar amplitude of 1°/my. Vector addition gives an Euler pole **G**$' = (0°, 90°)$ and a scalar amplitude of $\omega' = 1.414°$/my. The corresponding approximation to the finite rotation after a period of one million years is ROT[**G**$'$, ω']. The errors introduced by analyzing these two small finite rotations as if they were vectors is quite small—the error in the angle Ω of the rotation being only 0.0001° and the error in the location of the Euler pole being only 0.35°. Usually "instantaneous" angular velocity vectors are estimated from finite rotations over time intervals of 10 my or less. Repeating this numerical experiment for a time interval of 10 my with corresponding finite rotation angles of $\Omega = 10°$ gives errors of 0.01° in the rotation angle and an error of 3.5° in the location of the Euler pole. Even in this "worst case" for an interval of 10 my, the maximum errors are no larger than the errors commonly encountered in calculating Euler poles.

To determine plate motions in the distant past, we find the finite rotations that bring into coincidence pairs of isochrons of increasingly older ages on the two sides of the ridge. Since these older isochrons are spaced widely apart on either side of the ridge, the rotations needed to bring the anomalies into coincidence cannot be approximated by vectors. Tables listing these finite rotations as a function of the age of the isochrons show, in compact form, the final results of analyzing thousands of kilometers of magnetometer and depth-sounding profiles across a given ridge. These tables are extremely useful because they provide the basis for constructing paleogeographic maps that show the locations of plates in earlier times, and also because the tables provide the basis for determining the angular velocities of plates in the past.

In working with these finite rotations, you will need to keep your wits about you in order to avoid adding rotations in the wrong sequence or using the wrong sign. It helps to stick to a rigid set of conventions and procedures. The following is one that works, although it isn't the only one.

Rules of Finite Rotations

1. We need to keep track of which plate is being moved and which is fixed. We will use the symbol $_A\text{ROT}_B$ to mean "the rotation of plate B with respect to plate A." The first or left-hand subscript describes the fixed reference frame.

2. We also need to keep track of the age of the rotations. Consider two plates A and B separated by a ridge. An isochron on either plate corresponds to age $t = 40$ Ma and the active ridge between the plates to age $t = 0$ Ma. We follow the general convention that time and geologic age move in opposite directions: As time moves from the past toward the present, the age t decreases. We will use the symbol $_A^0\text{ROT}_B^{40}$ to mean "the rotation needed to rotate plate B from its present position to its position at 40 Ma with respect to plate A." This is the rotation needed to bring the 40 Ma isochron on plate B into coincidence with the 40 Ma isochron on plate A (Figure 7-4). A video of this rotation in progress would show motion going backward in time: Plate A would remain fixed as plate B moved toward it, seeming to disappear into the ridge. We will use the symbol $_A^{40}\text{ROT}_B^0$ to mean "the rotation necessary to move plate B from its position at 40 Ma to its present position with respect to plate A." This rotation moves the 40 Ma isochron from where it formed on the ridge at 40 Ma to where it now is located as viewed from

a fixed point on plate A. A video of this rotation in progress would show motion going forward in time: Plate A would remain fixed as plate B moved away from it, the 40 Ma isochron moving with plate B as new plate formed at the ridge.

3. The following relationship can be seen from Figure 7-4.

$$_A^t\text{ROT}_B^0 = {}_B^0\text{ROT}_A^t \tag{7.3}$$

Rotations to and from the present are called **total rotations**. Total rotations are special because they are anchored to the present day, in which all reference frames are equal. Note that both of the rotations in Figure 7-4 are total rotations.

Very often in plate tectonics, you will come across the term **total reconstruction pole**. A total reconstruction pole is a total rotation starting at the present and going backward in time, reconstructing the previous positions of the plates. (Note on usage: We have used "Euler pole" to refer to the unit vector **E** corresponding to the axis of an angular velocity vector or a finite rotation. The term "pole" is often used to refer to the angular velocity vector or finite rotation itself. Here we use "pole" to refer to a finite rotation, and "Euler pole" to refer to the corresponding unit vector.) Rotations going forward in time are called **forward motion poles**. $_A^0\text{ROT}_B^{40}$ is an example of a total reconstruction pole, and $_A^{40}\text{ROT}_B^0$ is a total forward motion pole.

4. When adding rotations, the first one listed is done first. $^0\text{ROT}^{40} + {}^{40}\text{ROT}^{60}$ means "first reconstruct back to 40 Ma, then rotate from 40 Ma back to 60 Ma." Adjacent superscripts cancel when subscripts are the same and the rotations combine to give:

$$_A^0\text{ROT}_B^{40} + {}_A^{40}\text{ROT}_B^{60} = {}_A^0\text{ROT}_B^{60}$$

We can generalize this as

$$_A^{t_1}\text{ROT}_B^{t_2} + {}_A^{t_2}\text{ROT}_B^{t_3} = {}_A^{t_1}\text{ROT}_B^{t_3} \tag{7.4}$$

Superscripts can be cancelled only if the subscripts are the same for all terms and if successive superscripts are the same, indicating that no time elapsed between successive rotations.

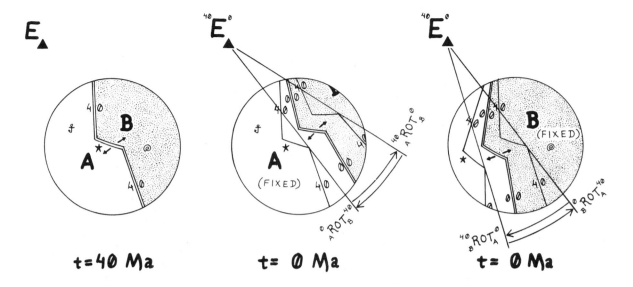

$t=40$ Ma $t= 0$ Ma $t= 0$ Ma

5. The idea of a **negative rotation** is sometimes useful. By $-\mathrm{ROT}[\mathbf{E}, \Omega]$ we mean either $\mathrm{ROT}[-\mathbf{E}, \Omega]$ or $\mathrm{ROT}[\mathbf{E}, -\Omega]$, which are equivalent. In Figure 7-4 it is apparent that

$$_A^t\mathrm{ROT}_B^0 = -_A^0\mathrm{ROT}_B^t \qquad (7.5)$$

The fact that the successive operations $-_A^{40}\mathrm{ROT}_B^0 + {}_A^{40}\mathrm{ROT}_B^0$ and $_A^{40}\mathrm{ROT}_B^0 + (-_A^{40}\mathrm{ROT}_B^0)$ both leave plate B at its starting position can be expressed by the equations

$$-_A^{40}\mathrm{ROT}_B^0 + {}_A^{40}\mathrm{ROT}_B^0 = 0 \qquad (7.6\mathrm{a})$$
$$_A^{40}\mathrm{ROT}_B^0 + (-_A^{40}\mathrm{ROT}_B^0) = 0 \qquad (7.6\mathrm{b})$$

In general,

$$_A^{t_2}\mathrm{ROT}_B^{t_1} = -_A^{t_1}\mathrm{ROT}_B^{t_2} \qquad (7.7)$$

Note from Figure 7-4 that $_B^0\mathrm{ROT}_A^t = -_A^0\mathrm{ROT}_B^t$. This relationship is valid only for rotations to and from the present ($t = 0$), that is, for total rotations.

6. The rotation $_A^{t_2}\mathrm{ROT}_B^{t_1}$ is called a **stage pole**. A stage is simply some interval of time in the history of a pair of plates, and a stage pole is the amount of rotation that occurred during the stage. Generally it is assumed that the Euler pole remained fixed during the stage and jumped between successive stages. Manipulating stage poles is

Figure 7-4.
Finite rotations about an Euler pole $^{40}\mathbf{E}^0$. Symbols such as $_A^{40}\mathrm{ROT}_B^0$ that describe finite rotations are placed at the heads of arrows showing the sense of the motion going either forward in time ($^{40}\mathrm{ROT}^0$) or backward in time ($^0\mathrm{ROT}^{40}$).

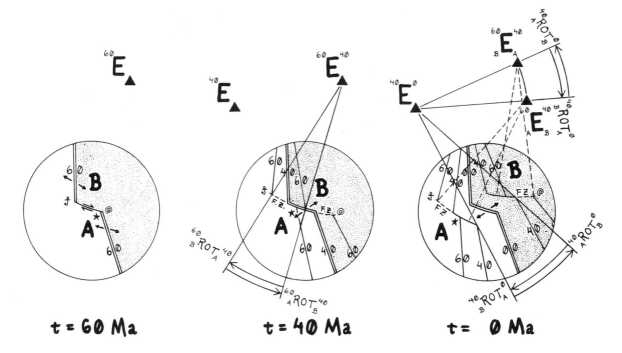

$t = 60$ Ma $t = 40$ Ma $t = 0$ Ma

Figure 7-5.
An earlier stage of rotation from 60 to 40 Ma, $_A^{60}\mathrm{ROT}_B^{40}$, is followed by the rotation from 40 to 0 Ma shown in Figure 7-4. The later rotation "splits" the older Euler pole. Note that $_B^{60}\mathrm{ROT}_A^{40}$ is determined from fracture zones and isochrons on plate B and $_A^{60}\mathrm{ROT}_B^{40}$ from observations on plate A.

tricky, because except for the special case where the Euler pole $_A^{t_2}\mathbf{E}_B^{t_1} = {}_A^{t_1}\mathbf{E}_B^0$,

$$_A^{t_2}\mathrm{ROT}_B^{t_1} \neq -{}_B^{t_2}\mathrm{ROT}_A^{t_1} \qquad (7.8)$$

Many of the errors that arise in manipulating finite rotations can be traced to violations of this rule. From Figures 7-5 and 7-6 it can be seen that the two Euler poles will differ by the rotation $\pm {}_A^{t_1}\mathrm{ROT}_B^0$. Because the Euler pole is the only point that does not change coordinates during a finite rotation, it is only in the special case mentioned above, where $_A^{t_2}\mathbf{E}_B^{t_1} = {}_A^{t_1}\mathbf{E}_B^0$, that $_A^{t_2}\mathrm{ROT}_B^{t_1}$ does indeed equal $-{}_B^{t_2}\mathrm{ROT}_A^{t_1}$.

7. Rotations cannot be commuted or otherwise treated as algebraic variables. For example, if A, B, and C are rotations and if $A = B + C$, then $-A \neq -B - C$. However, you can always add a rotation to both sides of an equation. The rotation must be added either to the left of the terms on both sides of the equation or to the right of both terms; that is, if $A = B$, then $C + A = C + B$ and $A + C = B + C$ but $C + A \neq B + C$. With some ingenuity, you can solve most problems using this rule. Suppose, for example, that you know total reconstruction poles for times t_1 and t_2 and you would like to find the stage pole $_A^{t_1}\mathrm{ROT}_B^{t_2}$.

Start with equation 7.4 and add ${}^{t_3}_A\text{ROT}^{t_2}_B$ to the right of the terms on both sides, using equation 7.7 to obtain

$${}^{t_1}_A\text{ROT}^{t_2}_B = {}^{t_1}_A\text{ROT}^{t_3}_B + {}^{t_3}_A\text{ROT}^{t_2}_B \qquad (7.9)$$

Now set $t_3 = 0$:

$$
\begin{aligned}
{}^{t_1}_A\text{ROT}^{t_2}_B &= {}^{t_1}_A\text{ROT}^{0}_B + {}^{0}_A\text{ROT}^{t_2}_B \qquad (7.10)\\
&= -{}^{0}_A\text{ROT}^{t_1}_B + {}^{0}_A\text{ROT}^{t_2}_B
\end{aligned}
$$

where ${}^{0}_A\text{ROT}^{t_1}_B$ and ${}^{0}_A\text{ROT}^{t_2}_B$ are both total reconstruction poles. (The first and second terms of equation 7.10 are consistent with the rules for the cancellation of superscripts.) In plate tectonics, stage poles are generally found by a two-step process: Total reconstruction poles are found first using observations of fracture zones and isochrons; stage poles are then found using equation 7.10.

8. Probably the most useful of all rules, other than to remain aware of equation 7.8, is to be absolutely certain what kind of pole you are working with. For this reason, we recommend that you answer the following questions about every finite rotation you encounter.

   • *Which plate is fixed?*

   • *Is it a reconstruction pole or a forward motion pole?*

   • *Is it a total pole or a stage pole?*

   With two possible answers for each question, there is only a 12.5 percent chance you will randomly be right.

Analyzing Data

Finding Stage Poles from Total Reconstruction Poles

Now let's see how these ideas can be used to analyze some real data. In Table 7-1 are listed the finite rotations required to bring the isochrons on the eastern side of the northern Mid-Atlantic Ridge into coincidence with those on the western side. The fixed reference frame was arbitrarily assigned to North America, and the total reconstruction poles (TRPs), which are described in Table 7-1 as ${}^{0}_{NA}\text{ROT}^{t}_{EU}$, are those that best bring the Eurasia (EU) isochrons into coincidence with the North America (NA) isochrons. Note that if Eurasia is kept fixed, the corresponding TRPs, ${}^{0}_{EU}\text{ROT}^{t}_{NA}$, are the same

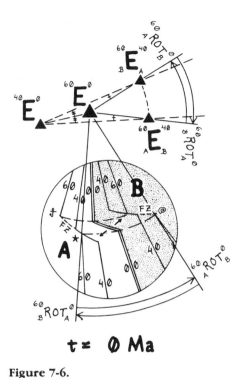

Figure 7-6.
Rotation about the pole ${}^{60}\text{ROT}^{0}$ restores the two plates relative to their position 60 Ma with starfish and snail next to each other. Note that there are no fracture zones or isochrons on either plate corresponding to ${}^{60}\text{ROT}^{0}$ because this is not a real Euler pole in the sense of describing the instantaneous relative motion of plates.

Table 7-1. Poles that rotate the Eurasian mid-Atlantic isochrons of age t back into coincidence with corresponding North American mid-Atlantic isochrons.

| | **Total Reconstruction Poles** | | |
| | ${}_{NA}^{0}\text{ROT}_{EU}^{t} = -{}_{EU}^{0}\text{ROT}_{NA}^{t}$ | | |
$t(\text{Ma})$	λ	ϕ	Ω
37	68.0°	129.9°	−7.8°
48	50.8°	142.8°	−9.8°
53	40.0°	145.0°	−11.4°
83	70.5°	150.1°	−20.3°
90	75.5°	152.9°	−24.2°

as those listed in the table with the sign of Ω changed from negative to positive. From these TRPs, let's see if we can find the stage pole in fixed Eurasia coordinates for the opening of the North Atlantic from 83 to 53 Ma.

Our first step is to answer the questions of rule 8:

● *Which plate is fixed?* Eurasia is fixed.

● *Is it a reconstruction pole or a forward motion pole?* Forward motion.

● *Is it a total pole or a stage pole?* Stage pole.

The rotation we're after is ${}_{EU}^{83}\text{ROT}_{NA}^{53}$. Applying our earlier definitions and rules, we have

$$
\begin{aligned}
{}_{EU}^{83}\text{ROT}_{NA}^{53} &= {}_{EU}^{83}\text{ROT}_{NA}^{0} + {}_{EU}^{0}\text{ROT}_{NA}^{53} \\
&= -{}_{EU}^{0}\text{ROT}_{NA}^{83} + {}_{EU}^{0}\text{ROT}_{NA}^{53} \\
&= {}_{NA}^{0}\text{ROT}_{EU}^{83} + (-{}_{NA}^{0}\text{ROT}_{EU}^{53}) \\
&= \text{ROT}[(70.5°, 150.1°), -20.3°] + \text{ROT}[(40.0°, 145.0°), 11.4°] \\
&= \text{ROT}[(-80.44°, 157.32°), 11.97°]
\end{aligned}
$$

As a rough check, let's apply the right-hand rule to this rotation. The Euler pole is near the South Pole. If you point your right thumb in that direction, your fingers curve around the globe toward the west. Since that is the direction of motion of North America relative to Eurasia, you know you haven't made an error in sign or some other gross error in calculation.

Now let's try to find the motion of Eurasia for this same time interval keeping North America fixed.

Table 7-2. Rotations between North America and Eurasia going forward in time from age t_2 to age t_1.

		Forward Motion Stage Poles		
		$^{t_2}_{NA}\text{ROT}^{t_1}_{EU}$		
t_2**(Ma)**	t_1	λ	ϕ	Ω
90	83	77.93°	283.76°	4.36°
83	53	78.09°	284.06°	11.97°
53	48	−6.19°	146.21°	2.57°
48	37	5.57°	150.35°	3.43°
37	0	68.00°	129.90°	7.80°
		$^{t_2}_{EU}\text{ROT}^{t_1}_{NA}$		
t_2**(Ma)**	t_1	λ	ϕ	Ω
90	83	−80.34°	156.45°	4.36°
83	53	−80.44°	157.32°	11.97°
53	48	5.38°	334.43°	2.57°
48	37	−6.75°	337.29°	3.43°
37	0	−68.00°	309.90°	7.80°

$$
\begin{aligned}
{}^{83}_{NA}\text{ROT}^{53}_{EU} &= {}^{83}_{NA}\text{ROT}^{0}_{EU} + {}^{0}_{NA}\text{ROT}^{53}_{EU} \\
&= -{}^{0}_{NA}\text{ROT}^{83}_{EU} + {}^{0}_{NA}\text{ROT}^{53}_{EU} \\
&= \text{ROT}[(70.5°, 150.1°), 20.3°] + \text{ROT}[(40.0°, 145.0°), -11.4°] \\
&= \text{ROT}[(78.09°, 284.06°), 11.97°]
\end{aligned}
$$

According to the right-hand rule, Eurasia should be moving in an easterly direction, which also checks.

Comparing the two stage poles, note that although the rotation angles are equal, the poles are not exactly antipodal, that is,

$$
{}^{83}_{EU}\text{ROT}^{53}_{NA} \neq -{}^{83}_{NA}\text{ROT}^{53}_{EU}
$$

as predicted by rule 6 and equation 7.6.

Table 7-2 lists stage poles for the opening of the Atlantic going forward in time from 90 Ma, the age of the oldest isochrons used in this analysis, to the present. The first entry, $\text{ROT}[\mathbf{E}, \Omega] = \text{ROT}[(77.93°, 283.76°), 4.36°]$ is a stage pole which describes, in a reference frame fixed with respect to North America, the motion of Eurasia from 90 to 83 Ma. The

Figure 7-7.
Finite rotation poles for restoring North America and Eurasia to their relative positions corresponding to the data in Table 7-1. These were found by bringing together the isochrons with the ages shown on the two sides of the Mid-Atlantic Ridge.

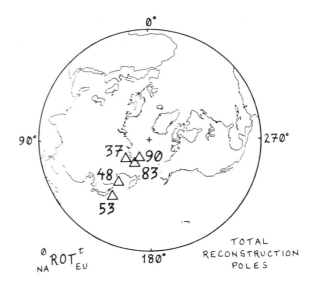

two sets of stage poles relative to a fixed Eurasia and a fixed North America have somewhat different Euler poles, although the rotation angles Ω are equal. In Figures 7-7 and 7-8 the poles can be seen to undergo jumps that are generally small except for the time interval 53 to 37 Ma. Whether this jump is due to observational error or to a real change in the direction of plate motion is still being debated.

Finding Instantaneous Rates

Is it possible to use these data to obtain an estimate of the angular velocity vector $\boldsymbol{\omega}$ of Eurasia relative to North America at some specific time, say at 85 Ma? Note that the first line of Table 7-2 lists the quantity $^{90}_{NA}\text{ROT}^{83}_{EU}$, which gives the motion of Eurasia away from North America during the time interval (or stage) from 90 to 83 Ma. During this time, the stage pole had the direction of a unit vector $\mathbf{E}$ with latitude and longitude of $(77.93°, 283.76°)$. This is the Euler stage pole expressed in a reference frame that is fixed relative to North America. Assuming a constant rate of motion for this interval of time $\Delta t = t_2 - t_1$, the rate of rotation is

$$\omega = \frac{\Omega}{\Delta t} = \frac{4.36°}{(90 - 83)\text{ my}} = 0.623°/\text{my}$$

The desired angular velocity vector for 85 Ma thus has the orientation of $\mathbf{E}$, $(77.93°, 283.76°)$, and the rotation rate $0.623°/\text{my}$.

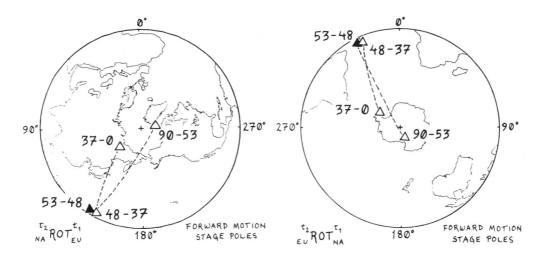

Figure 7-8.
Stage poles for the motion between North America and Eurasia, corresponding to the data in Table 7-2. Note that $_{EU}^{t_1}\mathrm{ROT}_{NA}^{t_2}$ is not the antipole of $_{NA}^{t_1}\mathrm{ROT}_{EU}^{t_2}$.

To find the angular velocity of North America relative to a fixed Eurasia, begin with the second set of entries in Table 7-2, which lists values for $_{EU}^{t_2}\mathrm{ROT}_{NA}^{t_1}$. The average rotation rate of North America relative to Eurasia during this time interval is the same,

$$\omega = \frac{4.36°}{(90-83)\ \mathrm{my}} = 0.623°/\mathrm{my}$$

However, the Euler pole $\mathbf{E} = (-80.34°, 156.45°)$ for this stage of rotation and for the corresponding angular rotation vector $\boldsymbol{\omega}$ in the fixed Eurasia reference frame is different from the earlier one: The antipole of the Euler pole for this same stage in the fixed North America reference frame is $\mathbf{E} = (-77.93°, 103.76°)$. This example illustrates the important general result that the orientations of angular velocity vectors for times other than the present are generally different in different frames of reference.

Finding Intermediate Positions Between Two Total Reconstruction Poles

Several research groups have produced movies showing continents drifting and ocean basins opening and closing during the course of geologic time. In doing this, it is necessary to make reconstructions for times between the ages of the isochrons that define stage boundaries. If the latter ages are t_2 and t_1 and if the age t of the desired reconstruction lies between them, then by linear interpolation the angle of rotation beginning at time t_2 is

$$
{}^{t_2}_A\Omega^t_B = (t_2-t){}^{t_2}_A\omega^{t_1}_B = (t_2-t)\frac{{}^{t_2}_A\Omega^{t_1}_B}{t_2 - t}
$$

Similarly the total reconstruction pole back to time t is given by

$$
\begin{aligned}
{}^0_A\text{ROT}^t_B &= {}^0_A\text{ROT}^{t_2}_B + {}^{t_2}_A\text{ROT}^t_B \\
&= {}^0_A\text{ROT}^{t_2}_B + \delta{}^{t_2}_A\text{ROT}^{t_1}_B
\end{aligned}
$$

where

$$
\delta{}^{t_2}_A\text{ROT}^{t_1}_B = \text{ROT}[{}^{t_2}_A\mathbf{E}^{t_1}_B, \delta{}^{t_2}_A\Omega^{t_1}_B]
$$

and

$$
\delta = \frac{t_2 - t}{t_2 - t_1}
$$

Note that this simple linear interpolation is valid only if the rate of angular motion was constant during the stage.

To find the location of Eurasia at 40 Ma in the fixed North America reference frame we would use

$$
\begin{aligned}
{}^0_{NA}\text{ROT}^{40}_{EU} &= {}^0_{NA}\text{ROT}^{48}_{EU} + \left(\frac{48-40}{48-37}\right){}^{48}_{NA}\text{ROT}^{37}_{EU} \\[2mm]
&= \text{ROT}[(50.8°, 142.8°), -9.8°] + (0.727)\text{ROT}[(5.57°, 150.35°), 3.43°] \\
&= \text{ROT}[(50.8°, 142.8°), -9.8°] + \text{ROT}[(5.57°, 150.35°), 2.49°] \\
&= \text{ROT}[(-62.65°, 315.61°), 8.25°] \\
&= \text{ROT}[(62.65°, 135.61°), -8.25°]
\end{aligned}
$$

To find the location of North America at 40 Ma in the fixed Eurasia reference frame we would use

$$
\begin{aligned}
{}^0_{EU}\text{ROT}^{40}_{NA} &= {}^0_{EU}\text{ROT}^{48}_{NA} + \left(\frac{48-40}{48-37}\right){}^{48}_{EU}\text{ROT}^{37}_{NA} \\[2mm]
&= \text{ROT}[(50.8°, 142.8°), 9.8°] + (0.727)\text{ROT}[(-6.75°, 337.29°), 3.43°] \\
&= \text{ROT}[(50.8°, 142.8°), 9.8°] + \text{ROT}[(-6.75°, 337.29°), 2.49°] \\
&= \text{ROT}[(62.65°, 135.62°), 8.25°]
\end{aligned}
$$

Note that ${}^0_{NA}\text{ROT}^{40}_{EU}$ is the antipole of ${}^0_{EU}\text{ROT}^{40}_{NA}$ as we would expect from equation 7.3.

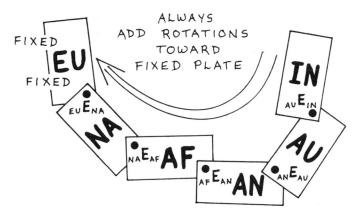

Figure 7-9.
Schematic global circuit of plates connecting Eurasia (EU) to India (IN).

Global Circuits

Sometimes we would like to make a paleogeographic reconstruction showing the ancient position of two plates like the Eurasia and India plates between which there has been a convergent margin for a long time. Lacking the isochrons and fracture zones needed to calculate finite rotations directly, we are then forced to calculate the ancient positions of the two plates using a global circuit or chain in which the links are plates separated by well-defined sets of fracture zones and isochrons. To find the position of the India plate relative to Eurasia at 40 Ma, for example, we can use the 0 to 40 Ma total reconstruction poles of Eurasia-North America, North America-Africa, Africa-Antarctica, Antarctica-Australia, and Australia-India (Figure 7-9). In Chapter 4 we traversed a global circuit using instantaneous angular velocity vectors; we will now do it using finite rotations.

We start as always by deciding which coordinate system to keep fixed. Let's pick Eurasia. Although we know that the rotations were going on simultaneously, in making the reconstructions it is conceptually simpler if we consider one rotation at a time while keeping all of the other Euler poles "locked" (Figure 7-9). In completing the circuit, two possible routes are obvious: route 1 starting at Eurasia (fixed) and ending at the India plate and route 2 starting at the India plate and ending at Eurasia. Both give the correct answer, but a moment's reflection will show you that route 2 involves less computation. With route 1 (Figure 7-9), if we first rotate about Euler pole $_{EU}\mathbf{E}_{NA}$ while keeping all of the other poles locked, we must rotate the coordinates of the other Euler poles $_{NA}\mathbf{E}_{AF}$, $_{AF}\mathbf{E}_{AN}$, $_{AN}\mathbf{E}_{AU}$ and $_{AU}\mathbf{E}_{IN}$. When we next rotate about Euler pole $_{NA}\mathbf{E}_{AF}$ we must rotate the coordinates of $_{AF}\mathbf{E}_{AN}$, $_{AN}\mathbf{E}_{AU}$ and $_{AU}\mathbf{E}_{IN}$. And so on. However if we use route 2 and begin by rotating about Euler pole $_{AU}\mathbf{E}_{IN}$ while keeping the other Euler poles locked, these poles do not rotate during this first rotation. Therefore all we need to do is to rotate the Indian plate first about $_{AU}\mathbf{E}_{IN}$, then about $_{AN}\mathbf{E}_{AU}$, and so on. Since in doing these rotations we do not rotate other Euler poles that will be used in subsequent steps, the expression for the rotation via route 2 is simply

$$_{EU}^{0}\text{ROT}_{IN}^{40} = {}_{AU}^{0}\text{ROT}_{IN}^{40} + {}_{AN}^{0}\text{ROT}_{AU}^{40} + {}_{AF}^{0}\text{ROT}_{AN}^{40} + {}_{NA}^{0}\text{ROT}_{AF}^{40} + {}_{EU}^{0}\text{ROT}_{NA}^{40}$$

Note that in this expression, adjacent subscripts do not cancel in the way that subscripts did earlier.

The first step in calculating this global circuit rotation using the poles given in Table 7-3 is to calculate the individual rotations needed for the above addition, interpolating where necessary to the time 40 Ma. No motion took place between India and Australia after 40 Ma, that is, $_{AU}^{0}\text{ROT}_{IN}^{40} = 0$. The other rotations are

$$\begin{aligned}
_{AN}^{0}\text{ROT}_{AU}^{40} &= {}_{AN}^{0}\text{ROT}_{AU}^{42} + \left(\frac{42-40}{42-37}\right) {}_{AN}^{42}\text{ROT}_{AU}^{37} \\
&= {}_{AN}^{0}\text{ROT}_{AU}^{42} + (0.4)({}_{AN}^{42}\text{ROT}_{AU}^{0} + {}_{AN}^{0}\text{ROT}_{AU}^{37}) \\
&= \text{ROT}[(10.3°, 34.8°), -23.6°] + \\
&\quad (0.4)(\text{ROT}[(10.3°, 34.8°), 23.6°] + \text{ROT}[(11.9°, 34.4°), -20.5°]) \\
&= \text{ROT}[(10.3°, 34.8°), -23.6°] + (0.4)\text{ROT}[(-0.38°, 35.16°), 3.16°] \\
&= \text{ROT}[(10.3°, 34.8°), -23.6°] + \text{ROT}[(-0.38°, 35.16°), 1.26°] \\
&= \text{ROT}[(-10.89°, 214.65°), 22.36°]
\end{aligned}$$

$$_{AF}^{0}\text{ROT}_{AN}^{40} = \text{ROT}[(5.8°, 322.8°), 7.2°]$$

$$\begin{aligned}
_{NA}^{0}\text{ROT}_{AF}^{40} &= {}_{NA}^{0}\text{ROT}_{AF}^{66} + \left(\frac{66-40}{66-37}\right) {}_{NA}^{66}\text{ROT}_{AF}^{37} \\
&= {}_{NA}^{0}\text{ROT}_{AF}^{66} + (0.897)({}_{NA}^{66}\text{ROT}_{AF}^{0} + {}_{NA}^{0}\text{ROT}_{AF}^{37}) \\
&= \text{ROT}[(80.8°, 351.4°), -22.5°] + \\
&\quad (0.897)(\text{ROT}[(80.8°, 351.4°), 22.5°] + \text{ROT}[(70.5°, 341.3°), -10.4°]) \\
&= \text{ROT}[(80.8°, 351.4°), -22.5°] + (0.897)\text{ROT}[(85.36°, 62.26°), 12.42°] \\
&= \text{ROT}[(80.8°, 351.4°), -22.5°] + \text{ROT}[(85.36°, 62.26°), 11.14°] \\
&= \text{ROT}[(-72.57°, 162.38°), 11.62°]
\end{aligned}$$

$$_{EU}^{0}\text{ROT}_{NA}^{40} = \text{ROT}[(62.65°, 135.62°), 8.25°]$$

Table 7-3. Total reconstruction poles for selected fragments of Pangaea.

	$t(\text{Ma})$	λ	ϕ	Ω
$_{AF}^{0}\text{ROT}_{AR}'$	20	36.5	18.0	−6.1
$_{AR}^{0}\text{ROT}_{SM}'$	20	26.5	21.5	7.6
$_{SM}^{0}\text{ROT}_{MA}'$	100	—	—	0
	123	−13.1	319.0	8.1
	138	−16.3	328.6	13.8
$_{AF}^{0}\text{ROT}_{AN}'$	40	5.8	322.8	7.2
	50	12.0	311.4	7.5
	83	19.7	316.2	19.2
	123	−3.9	332.2	45.5
	138	−2.4	327.3	55.4
$_{AN}^{0}\text{ROT}_{AU}'$	37	11.9	34.4	−20.5
	42	10.3	34.8	−23.6
	50	11.9	30.8	−30.9
$_{AU}^{0}\text{ROT}_{IN}'$	50	—	—	0
	63	8.9	333.4	−17.2
	83	5.6	355.3	−38.6
	123	−9.6	9.7	−59.1
	138	−11.1	3.5	−62.1
$_{AF}^{0}\text{ROT}_{SA}'$	40	59.0	325.0	14.9
	50	63.0	324.0	20.1
	83	63.0	324.0	33.8
	138	44.0	329.4	57.0
$_{AF}^{0}\text{ROT}_{NA}'$	37	70.5	341.3	10.4
	66	80.8	351.4	22.5
	71	80.4	347.5	23.9
	75	78.3	341.4	27.1
	83	77.1	340.8	29.4
	119	66.3	340.1	54.3
	127	66.1	341.0	56.4
	145	66.1	341.6	59.8
	169	67.0	346.8	72.1
	180	67.0	348.0	75.6

(continued)

Table 7-3. *(continued)*

	t(Ma)	λ	ϕ	Ω
$_{NA}^{0}\mathrm{ROT}_{GR}'$	37	—	—	0
	48	13.0	302.0	1.1
	53	26.0	301.5	2.8
	83	−69.6	46.6	9.5
	90	−71.5	53.7	11.0
$_{NA}^{0}\mathrm{ROT}_{EU}'$	37	68.0	129.9	−7.8
	48	50.8	142.8	−9.8
	53	40.0	145.0	−11.4
	83	70.6	150.2	−20.3
	90	75.5	153.0	−24.2
$_{EU}^{0}\mathrm{ROT}_{IB}'$	75	—	—	0
	83	45.2	359.4	−31.8

Plate codes: AF-Africa; AR-Arabia; SM-Somalia; MA-Madagascar; AN-Antarctica; AU-Australia; IN-India; SA-South America; NA-North America; GR-Greenland; EU-Europa; IB-Iberia.

Then we start adding the rotations together.

$$_{AN}^{0}\mathrm{ROT}_{IN}^{40} = {}_{AU}^{0}\mathrm{ROT}_{IN}^{40} + {}_{AN}^{0}\mathrm{ROT}_{AU}^{40}$$

$$= {}_{AU}^{0}\mathrm{ROT}_{IN}^{40} + 0$$

$$= \mathrm{ROT}[(-10.89°, 214.65°), 22.36°]$$

$$_{AF}^{0}\mathrm{ROT}_{IN}^{40} = {}_{AN}^{0}\mathrm{ROT}_{IN}^{40} + {}_{AF}^{0}\mathrm{ROT}_{AN}^{40}$$

$$= \mathrm{ROT}[(-10.89°, 214.65°), 22.36°] + \mathrm{ROT}[(5.8°, 322.8°), 7.2°]$$

$$= \mathrm{ROT}[(-13.12°, 233.84°), 21.14°]$$

$$_{NA}^{0}\mathrm{ROT}_{IN}^{40} = {}_{AF}^{0}\mathrm{ROT}_{IN}^{40} + {}_{NA}^{0}\mathrm{ROT}_{AF}^{40}$$

$$= \mathrm{ROT}[(-13.12°, 233.84°), 21.14°] = \mathrm{ROT}[(-72.57°, 162.38°), 11.62°]$$

$$= \mathrm{ROT}[(-34.11°, 220.37°), 27.06°]$$

And finally:

$$_{EU}^{0}\mathrm{ROT}_{IN}^{40} = {}_{NA}^{0}\mathrm{ROT}_{IN}^{40} + {}_{EU}^{0}\mathrm{ROT}_{NA}^{40}$$

$$= \mathrm{ROT}[(-34.11°, 220.37°), 27.06°] + \mathrm{ROT}[(62.65°, 135.62°), 8.25°]$$

$$= \mathrm{ROT}[(-17.21°, 214.89°), 24.34°]$$

$$= \mathrm{ROT}[(17.21°, 34.89°), -24.34°]$$

In making global circuits, hidden errors will result if one of the plates in the circuit contains a hidden boundary that makes the plate, in effect, two plates during all or part of the time interval of interest. The greater the number of plates in the circuit, the more opportunities there are for errors of this type. For example, a particularly long circuit has been used to determine the motion of North America relative to the Farallon plate, which lay to the west of North America during most of the Cretaceous and Tertiary periods. The circuit is North America-Africa-India-Antarctica-Pacific-Farallon. Recent research has raised the possibility that deformation may have occurred either in the Antarctica or the Pacific plate, introducing errors of unknown magnitude into the global circuit.

Finite Rotations in a Hotspot Reference Frame

In Chapter 4 we found that we could simplify the listing of relative rotation vectors between pairs of plates by referring all plate velocities to a common reference frame, which we visualized as either a hypothetical plate M or as a reference frame fixed relative to the lower mantle. We will now discuss how we can follow the same procedure with finite rotations. Before doing this, however, let's reflect for a moment on the question of "absolute" reference frames and how we might set about determining such a reference frame.

Consider for example a hypothetical planet designed by a mathematician in which six plumes of unusually hot magma rise from the deep interior of the planet exactly along the $\pm \mathbf{x}$, $\pm \mathbf{y}$, $\pm \mathbf{z}$ axes of a coordinate system with the origin at the center of the planet. Each plume is hot enough to partially melt its way through to the surface of plates moving over the plume. At any one time there are six major centers of volcanic activity on the planet spaced 90° apart, each leaving a track of dead volcanoes which mark the motion of plates in the (x, y, z) coordinate system. By dating the volcanoes we can determine the angular velocities of the plates in (x, y, z) coordinates. On such a planet we would have no hesitation in deciding what to use as our absolute reference frame: it would obviously be the (x, y, z) coordinates.

There is growing evidence that several dozen such hot plumes rise through the mantle of planet Earth. They produce volcanism at centers like Hawaii, Iceland, and Yellowstone, which in plate tectonic jargon are termed "hotspots." The angular distances between the individual plumes are not 90°, as on the mathematician's planet, but they appear to remain nearly constant over long periods of time. Since the positioning of this system of hotspots appears to be more nearly constant than that of any other feature on the earth's surface, they provide a useful basis for defining the motion of plates in an "absolute" reference frame.

The tracks of hotspots are marked by arcuate features, like the Hawaiian seamount chain, and appear to lie along segments of small circles. A small-circle path corresponds to motion about an Euler pole that is absolute in the sense that it tends to remain fixed in the hotspot reference frame. The history of motion of a plate in the hotspot reference frame can be described by giving the finite rotation required to rotate the plate so that a dead volcano of known age is positioned over the present location of the active hot spot that originally produced the volcano. This process is exactly analogous to finding the finite rotation between two plates needed to superimpose corresponding isochrons on each plate. Therefore the motion of a plate relative to the hotspot reference frame can be analyzed in the same way as the motion of the plate relative to another plate. All of the techniques used for finite rotations between plates can be applied directly to the finite rotation of a plate relative to a hotspot.

In Figure 7-10 the present position of plates EU and IN are shown schematically relative to a hypothetical plate HS, which represents the hotspot reference frame. If you prefer, HS may represent the lower mantle or some other fixed reference frame—more about this in Chapter 10. The rotation needed to rotate plate EU back to its position relative to the hotspots at some time in the past, say 40 Ma, is described by the symbol $_{HS}^{0}\text{ROT}_{EU}^{40}$. Similarly the finite rotation needed to rotate IN back to its position relative to hotspots at 40 Ma is described by the symbol $_{HS}^{0}\text{ROT}_{IN}^{40}$. In Table 7-4 are listed the finite rotations required to move Eurasia and India to their former positions in the hotspot reference frame. From these two sets of data, we can find the finite rotation of IN

Figure 7-10.
The global set of hotspots can be analyzed as if they were one plate labeled "HS". If both the motion of Eurasia (EU) relative to the hotspots and the motion of India (IN) relative to the hotspots are known, then the motion of India relative to Eurasia can be found using methods discussed in the text.

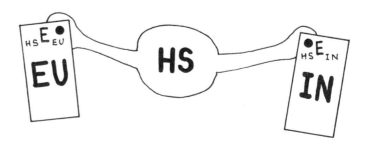

Table 7-4. Total reconstruction poles in the fixed hotspot reference frame.

	t(Ma)	λ	ϕ	Ω
	20	-50.40	143.10	5.40
	40	-48.00	142.20	10.40
	60	-44.20	131.40	14.60
	80	-35.40	121.60	21.70
	100	-29.00	133.40	27.90
$_{HS}^{0}\text{ROT}_{AF}^{t}$	120	-25.90	137.50	34.20
	140	-23.70	134.00	34.00
	160	-17.00	126.00	29.40
	180	-5.50	112.90	23.80
	200	17.60	86.70	22.70
	20	78.85	347.32	2.26
	40	78.96	2.36	5.10
	60	72.90	31.58	11.67
	80	59.72	51.49	17.74
	100	62.09	47.01	20.25
$_{HS}^{0}\text{ROT}_{SA}^{t}$	120	58.63	33.41	22.30
	140	48.93	10.15	29.09
	160	46.03	7.74	35.99
	180	43.58	3.26	45.23
	200	40.73	1.12	58.75
	20	32.29	120.54	2.10
	40	39.67	117.55	5.13
	60	48.28	104.59	12.34
	80	45.45	89.50	20.28
	100	54.75	82.93	28.30
$_{HS}^{0}\text{ROT}_{NA}^{t}$	120	57.70	69.44	36.98
	140	57.27	59.84	42.24
	160	58.27	42.13	55.93
	180	58.31	27.44	71.33
	200	56.85	17.15	83.18
	20	-83.84	75.92	2.81
	40	-75.12	47.45	4.24
	60	-12.42	39.50	5.88
	80	-19.08	69.35	12.75
	100	-3.71	73.01	15.24
$_{HS}^{0}\text{ROT}_{EU}^{t}$	120	19.18	63.98	21.33
	140	24.85	57.51	26.66
	160	35.60	45.05	39.91
	180	41.43	34.46	55.31
	200	42.78	26.60	67.71

(continued)

Table 7-4. *(continued)*

	$t(\text{Ma})$	λ	ϕ	Ω
	20	32.29	120.54	2.10
	40	42.41	117.03	4.95
	60	48.82	94.31	10.61
	80	21.11	86.33	17.59
	100	32.67	84.19	23.18
$_{HS}^{0}\text{ROT}'_{GR}$	120	41.97	73.88	31.09
	140	43.94	65.94	36.32
	160	49.45	50.51	49.52
	180	52.58	36.71	64.56
	200	52.72	26.39	76.35
	20	−87.07	277.97	3.80
	40	−86.61	260.78	7.00
	60	−88.09	251.84	7.14
	80	−64.67	43.72	7.86
	100	−62.90	349.66	13.49
$_{HS}^{0}\text{ROT}'_{AN}$	120	−55.11	335.59	22.85
	140	−35.47	325.92	30.22
	160	−26.61	336.30	32.53
	180	−15.99	343.92	38.31
	200	−5.92	352.01	48.43
	20	−29.11	213.44	12.51
	40	−26.69	211.96	24.90
	60	−23.72	207.56	33.26
	80	−25.72	204.91	30.12
	100	−37.01	211.97	31.71
$_{HS}^{0}\text{ROT}'_{AU}$	120	−50.55	223.90	35.39
	140	−50.00	253.22	37.93
	160	−50.87	268.40	34.00
	180	−46.59	292.92	31.80
	200	−34.05	321.61	32.68
	20	−29.11	213.44	12.51
	40	−26.69	211.96	24.90
	60	−25.48	191.30	42.42
	80	−19.70	184.13	62.12
	100	−17.62	185.85	71.64
$_{HS}^{0}\text{ROT}'_{IN}$	120	−17.42	186.46	77.87
	140	−21.28	187.04	74.07
	160	−20.38	188.04	67.08
	180	−19.94	190.78	57.82
	200	−18.23	194.86	44.49

Plate codes: HS-Hotspots; AF-Africa; SA-South America; NA-North America; EU-Europa; GR-Greenland; AU-Australia; AN-Antarctica; IN-India.

in the EU reference frame as follows. From the geometrical relationships in Figure 7-10 and using the rules for finite rotations we can write,

$$_{EU}^{0}\mathrm{ROT}_{IN}^{40} = {}_{HS}^{0}\mathrm{ROT}_{IN}^{40} + {}_{EU}^{0}\mathrm{ROT}_{HS}^{40}$$
$$= {}_{HS}^{0}\mathrm{ROT}_{IN}^{40} + (- {}_{HS}^{0}\mathrm{ROT}_{EU}^{40})$$
$$= \mathrm{ROT}[(-26.69°, 211.96°), 24.90°] + \mathrm{ROT}[(-75.12°, 47.45°), -4.24°]$$
$$= \mathrm{ROT}[(-17.20°, 214.88°), 24.35°]$$
$$= \mathrm{ROT}[(17.20°, 34.88°), -24.35°]$$

in good agreement with the global circuit.

The Three-Plate Problem

We've already discussed a dilemma that lies close to the heart of plate tectonic theory. Much of the usefulness of plate tectonics derives from the fact that the relative motion between pairs of plates can be described by Euler poles that remained fixed relative to both plates for long periods of time. Yet as we gather more and more information about plate motions, we find that Euler poles may undergo many small shifts in location after short intervals of time. The reason for this is apparent when we consider what is sometimes called the "three-plate problem." Consider three plates A, B, and C as shown schematically in Figure 7-11, with corresponding Euler poles $_A\mathbf{E}_B$, $_B\mathbf{E}_C$, and $_C\mathbf{E}_A$, where each Euler pole describes the instantaneous angular rotation of a pair of plates at one point in time. Is it possible for all of the Euler poles to remain fixed relative to each other after the plates have undergone a finite rotation during the time interval Δt? Let's adopt a fixed reference frame attached to plate A and specify that Euler poles $_A\mathbf{E}_B$ and $_C\mathbf{E}_A$ remain fixed in the "A" reference frame. This is very straightforward because by definition, an Euler pole between plate A and any other plate must keep the same coordinates in the "A" reference frame. Therefore Euler pole $_B\mathbf{E}_C$ must be moving relative to the "A" reference frame and also, therefore, relative to $_A\mathbf{E}_B$ and $_C\mathbf{E}_A$, which are fixed in the "A" reference frame. This demonstrates that on a globe with more than two plates, at least some of the poles must be moving relative to the plates whose motion they describe.

We can gain some additional insight into the ways Euler poles move if we shift our attention from relative Euler poles

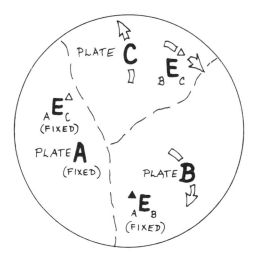

Figure 7-11.
The classic three-plate problem. If plate A and Euler poles $_A\mathbf{E}_C$ and $_A\mathbf{E}_B$ are fixed, the pole $_B\mathbf{E}_C$ must be moving relative to plates B and C.

that describe the relative motion between pairs of plates to absolute Euler poles that describe the motion between one plate and an "absolute" reference frame fixed (at least in our imaginations) to a stable lower mantle. This "mantle" reference frame is not necessarily the same as the hotspot reference frame discussed earlier. Rapidly moving plates appear to move along arcuate paths toward their trench boundaries, where they bend downward and become descending slabs of lithosphere. Since these slabs descend deep into the mantle, it seems unlikely that they undergo rapid lateral motion within the mantle—to do so would require displacing large volumes of asthenosphere. This implies that the trench boundaries of plates generally remain more or less stationary in a fixed mantle reference frame. As a hypothetical plate A moves toward its stationary trench boundary, the simplest path it can follow is one in which all points on the plate move along small circles concentric about a fixed point which we can regard as a fixed Euler pole $_M\mathbf{E}_A$, where the subscript "M" refers to a stable lower mantle reference frame. The above argument does not apply in a strict sense to plates that are not attached to slabs descending into trenches. However, it turns out that these plates have very low velocities in our "mantle" reference frame, so they also have nearly constant (i.e. zero) velocities. A self-consistent set of present angular velocities for all of the major plates in a fixed "mantle" reference frame is listed in Table 4-4. We will now examine the consequences of assuming that absolute angular velocities tend to remain constant for long periods of time.

Consider two plates A and B (Figure 7-11) which are rotating about Euler poles $_M\mathbf{E}_A$ and $_M\mathbf{E}_B$ that remain fixed relative to a stable lower mantle reference frame. In this reference frame the Euler pole $_A\mathbf{E}_B$, describing the instantaneous angular rotation vector for the motion of plate B relative to plate A, is located on the great circle between $_M\mathbf{E}_A$ and $_M\mathbf{E}_B$ and, like the latter, remains fixed in the lower mantle reference frame. The angular rotation vector can be obtained by vector addition: $_A\boldsymbol{\omega}_B = {}_A\boldsymbol{\omega}_M + {}_M\boldsymbol{\omega}_B = -{}_M\boldsymbol{\omega}_A + {}_M\boldsymbol{\omega}_B$. As plates A and B move in the mantle reference frame they carry the fracture zones, isochrons, and associated Euler poles with them away from the instantaneous Euler pole $_A\mathbf{E}_B$, so that the locus of increasingly older Euler poles $_A\mathbf{E}_B$ found from fracture zones and isochrons on the plate A will lie along a small circle centered on $\pm {}_M\mathbf{E}_A$ and passing through $\pm {}_A\mathbf{E}_B$. Similarly in the B reference frame, increasingly older Euler poles will lie along a small circle centered on $\pm {}_M\mathbf{E}_B$. From

Figure 7-12 it is obvious that the pole of relative motion between two plates undergoing constant motion about fixed Euler poles will remain nearly fixed relative to both plates if either of two conditions is met: if the two absolute Euler poles are nearly coincident; or if at least one of the angular velocity vectors is very small. We will now consider two examples of plate pairs, one of which satisfies these criteria and one of which does not.

The North America and Eurasia plates provide an example of plates that satisfy the second condition for nearly stationary Euler poles, since the angular velocity of the Eurasia plate is only 0.038°/my. The relative rotation vector for Eurasia relative to North America is given by

$$_{NA}\boldsymbol{\omega}_{EU} = -_{M}\boldsymbol{\omega}_{NA} + _{M}\boldsymbol{\omega}_{EU}$$

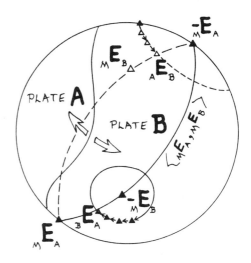

From the values in Table 4-4 we find that $_{NA}\mathbf{E}_{EU} = (65.9°, 132.4°)$ and $\omega = 0.231°/my$. If we assume that both plates will maintain their present absolute motion for the next 20 my, then the position of the Euler pole $_{NA}\mathbf{E}_{EU}$ in a fixed North America reference frame may be found by rotating $_{NA}\mathbf{E}_{EU}$ about the absolute North America Euler pole $_{M}\mathbf{E}_{NA} = (-58.3°, 319.3°)$, as listed in Table 4-4, by an angle $\Omega = (20 \text{ my } 0.247°/\text{my}) = 4.94°$, where 0.247°/my is the absolute angular velocity of North America. Performing the rotation $\text{ROT}[(-58.3°, 319.3°), 4.94°]$ on $_{NA}\mathbf{E}_{EU}$ gives for the future position of the Euler pole the coordinates $(66.2°, 134.0°)$. The great circle distance that the Euler pole $_{NA}\mathbf{E}_{EU}$ will move in 20 my is only 0.72°.

Let's check these calculations by repeating them in the Eurasia reference frame. The Euler pole $_{EU}\mathbf{E}_{NA} = (-65.9°, 312.4°)$ will rotate about $_{M}\mathbf{E}_{EU}$ by the rotation $\text{ROT}[(0.7°, -23.2°), 0.76°]$ to a future position at $E = (-66.2°, 314.0°)$. The great circle distance that $_{EU}\mathbf{E}_{NA}$ will move in 20 my is again only 0.72°, which checks.

Now let's look at the motion between the Pacific and Cocos plates, which satisfies neither of the conditions for nearly stationary Euler poles. The relative rotation vector for the Cocos plate relative to the Pacific plate is given by

$$_{PA}\boldsymbol{\omega}_{CC} = -_{M}\boldsymbol{\omega}_{PA} + _{M}\boldsymbol{\omega}_{CC}$$

From Table 4-4 we find that $_{PA}\mathbf{E}_{CC} = (38.7°, 252.6°)$ and $\omega = 2.21°/my$. If we again assume that both plates will maintain their present absolute motion for the next 20 my, then the

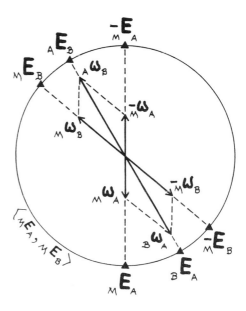

Figure 7-12.
The absolute motion of plates A and B in an upper mantle reference frame is described by fixed Euler poles $_{M}\mathbf{E}_{A}$ and $_{M}\mathbf{E}_{B}$. The Euler poles of relative motion between plates A and B move along small circles as shown.

future position of the Euler pole $_{PA}\mathbf{E}_{CC}$ in a fixed Pacific reference frame may be found by rotating $_{PA}\mathbf{E}_{CC}$ about the absolute Pacific Euler pole $_M\mathbf{E}_{PA} = (-61.7°, 97.2°)$, as listed in Table 4-4, by an angle $\Omega = (20 \text{ my})(0.967°/\text{my}) = 19.34°$, where $0.967°/\text{my}$ is the absolute angular velocity of the Pacific plate. Performing the rotation ROT[($-61.7°$, 97.2°), 19.34°] on $_{PA}\mathbf{E}_{CC}$ gives for the future position of $_{PA}\mathbf{E}_{CC}$ the coordinates (43.2°, 242.5°). The great circle distance that $_{PA}\mathbf{E}_{CC}$ will move in 20 my is 8.9°, which is 12 times larger than the motion we found for $_{NA}\mathbf{E}_{EU}$ during the same time interval.

In the Cocos reference frame, the Euler pole $_{CC}\mathbf{E}_{PA} = (-38.7°, 72.6°)$ will rotate about $_M\mathbf{E}_{CC}$ by the rotation ROT[(21.9°, 244.3°), 28.44°] to a future position at $_{CC}\mathbf{E}_{PA} = (-40.0°, 61.3°)$. The great circle distance that $_{CC}\mathbf{E}_{PA}$ will move in 20 my is 8.8° which checks within the accuracy of our calculations. From the viewpoint of this analysis, it is not surprising that the Cocos-Pacific Euler pole has been changing its position rapidly in recent time, whereas the Eurasia-North America pole has remained relatively fixed.

Problems

7-1. Using the data in Table 7-1 find the coordinates of the following Eurasian cities in fixed North America coordinates during the Cretaceous (at 83 Ma):

(a) Madrid (40°, 356°),

(b) Murmansk (69°, 33°),

(c) Magnitogorsk (53°, 59°),

(d) Naoming (22°, 108°).

(Check: the answer to (c) is (60°, 38°).)

7-2. Using the data in Table 7-1 find the coordinates of the following North American cities in fixed Eurasia coordinates at 83 Ma:

(a) St. Johns (48°, −53°),

(b) Barrow (72°, −156°),

(c) Stanford (38°, −122°),

(d) Miami (26°, −80°).

(Check: the answer to (b) is (78°, −151°).)

7-3. In Figure 7-2, the coordinates of $_B\mathbf{E}_A$ in a reference frame fixed with respect to plate B are (30°, 90°) and those of pole $\mathbf{F}$ are (90°, *). Assume that the rotation about $\mathbf{E}$ took place from 50 to 30 Ma and that about **F** from 30 Ma to the present. As a marine geophysicist living at time $t = 0$ you find the fracture zones and isochrons produced by both episodes of spreading. To analyze your results you find the rotations needed to superimpose the 50 Ma isochrons and the 30 Ma isochrons, which are the finite rotations needed to restore the plates to their relative positions at those times. What values do you find for

(a) $_B^0\text{ROT}_A^{30}$,

(b) $_B^0\text{ROT}_A^{50}$,

(c) $_A^0\text{ROT}_B^{30}$,

(d) $_A^0\text{ROT}_B^{50}$,

(e) $_A^{50}\text{ROT}_B^0$,

(f) $_A^{50}\text{ROT}_B^{30}$, and

(g) $_B^{50}\text{ROT}_A^{30}$?

7-4. In a reference frame fixed with respect to North America, find the 138 Ma coordinates of Casablanca, Morocco, with present coordinates $(33°, -8°)$. How far away from Atlantic City, New Jersey $(39°, -74°)$ was Casablanca at that time?

7-5. The finite rotations shown in Table 7-3 for India, Australia, and Antarctica are based on a model in which (1) India moves relative to Australia from 138 to 50 Ma, and (2) India becomes locked to Australia at 50 Ma and both plates move together away from Antarctica from 50 Ma to the present. (a) Find the rotation needed to restore India to its position relative to Antarctica at 138 Ma, using a reference frame fixed relative to Antarctica. (b) In this reference frame, find the 138 Ma coordinates of Bombay with present coordinates $(18°, 72°)$.

7-6. (a) Find the average angular velocity vector for the motion of India in the hotspot reference frame for the time interval from 80 to 40 Ma.

(b) What was the average linear velocity of Bombay during this interval (in mm/yr)?

7-7. The Deccan Traps are an enormous deposit of basaltic rock 64 my old located in India in a broad area around the coordinates $(19°, 76°)$. If these basalts were produced by a hotspot, and if this hotspot has remained fixed in the hotspot reference frame since 64 Ma, where would you expect to find the active hotspot today?

7-8. Assume that a new hotspot breaks through the lithosphere on Sunset Boulevard in Los Angeles $(34°, -118°)$ and continues to erupt for 60 my as the North America plate continues its present motion in the hotspot reference frame. Find the coordinates of the center of volcanic activity 60 my from now

(a) in the fixed hotspot reference frame and

(b) in the fixed North America reference frame.

7-9. Use the information in Table 7-3 to find the total reconstruction pole of South America relative to a fixed North America for 83 Ma.

7-10. Find the average rate (mm/yr) and total distance (km) of opening during the past 20 million years of the Great Rift Valley (between the Somalian plate and African plate) at Lake Nakuru, Kenya (0°, 36°). The width of the Great Rift Valley at this point is 60 km measured from the Mau Escarpment on the west to the Aberdare Range on the east. How does this compare with the displacement you would predict from your rate of spreading?

7-11. Find the angular velocity vector for the motion of Greenland relative to Eurasia at 60 Ma in a reference frame with Eurasia fixed.

7-12. Find the rotation matrix that describes the following rotations.

(a) ROT[**3**, 90°]

(b) ROT[**3**, 90°] + ROT[**2**, 90°]

(c) ROT[**2**, 90°] + ROT[**3**, 90°]

7-13. Show by matrix multiplication that ROT[**1**, 90°] + ROT[**−1**, 90°] = 0

7-14. Starting with the simple matrices for ROT[**2**, Ω] and ROT[**3**, Ω], derive by matrix multiplication the equation for R_{32} as defined in Box 7-3.

Suggested Readings

Texts

Plate Tectonics, Xavier Le Pichon, Jean Francheteau, and Jean Bonin, Elsevier, New York, 302 pp., 1973. *Excellent description of the mathematics of rotations.*

Sources of Data

Relative Motions between Oceanic and Continental Plates in the Pacific Basin, D. Engebretson, A. Cox, and R. Gordon, Special Paper 206, Geological Society of America, 59 pp., 1985. *Gives rotation poles of Tables 7-1, 7-2, and 7-4, together with ref-*

erences to original sources. Also gives motions of the oceanic plates of the Pacific basin: Farallon, Kula, Phoenix, and Izanagi.

Hotspot Tracks and the Opening of the Atlantic and Indian Oceans, W. Morgan, in *The Sea, 7.* Edited by C. Emiliani, Wiley Interscience, New York, p. 443–487, 1981. *Classic paper about hotspots. Primary source of information about the motion of plates over hotspots.*

A Model for the Evolution of the Indian Ocean and the Breakup of Gondwanaland, I. Norton and J. Sclater, Journal of Geophysical Research, v. 84, p. 6803–6830, 1979. *Source of data listed in Table 7-3.*

8

Magnetism and Isochrons

The fossil magnetism or **paleomagnetism** of lava flows on the seafloor provides the basis for determining the location of the seafloor isochrons discussed in earlier chapters. With magnetic isochrons it became possible to determine not only the age of the ocean floor but also the all-important rates of tectonic processes in ocean basins. Moreover, isochrons can be mapped without collecting samples from the seafloor simply by towing or flying a magnetometer over an ocean basin. This remarkable feat of dating rocks by remote sensing is possible because many rocks are imprinted by the magnetic field of the earth at the time the rocks form, much in the way checks are imprinted with digital magnetic numbers at a bank. This fossil magnetism, although weak in intensity, is locked so tightly into the crystal structure of iron oxide minerals contained in the rock that the magnetism is still measurable after millions and even billions of years. This chapter will discuss how this amazing magnetic recorder works.

Earth's Magnetic Field

As a preliminary, let's take a brief look at the earth's magnetic field, without which there would be no paleomagnetic imprints. Starting from scratch, how could you determine whether a magnetic field exists on the earth or some other planet? If you are an experimentalist, the answer is simple:

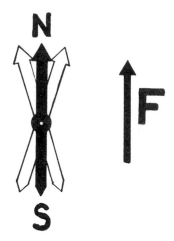

Figure 8-1.
A simple compass needle points in the same direction as the geomagnetic field vector **F**.

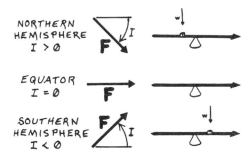

Figure 8-2.
The magnetic field vector, **F**, plunges downward in the northern hemisphere and the corresponding inclination (angle below the horizontal) is positive. In the southern hemisphere **F** plunges upward and the inclination is negative. The weight W balances the needle to keep it horizontal rather than parallel to **F**. No weight is required at the equator, where the inclination is zero.

do an experiment. The simplest is to place a magnetized needle on a pivot so it can rotate—in other words, start with a simple compass. In the absence of a magnetic field, the needle will remain listlessly in its initial position. In the presence of a magnetic field, the compass needle will oscillate for a while and then settle down to a rest position. This rest position is parallel to the direction of the field (Figure 8-1). The intensity of the field can be determined from the rate of oscillation of the needle, larger fields producing faster oscillations.

Because it has a magnitude and direction, the magnetic field is a vector quantity. We will describe this vector by the symbol **F**. Having determined the length and orientation of **F**, the final step in our experiment is to define the head and tail of the vector. By convention, these are the same as the head and tail of a simple compass needle, the head being the north-seeking end of the needle (Figure 8-1). It's that simple: the earth's field is presently directed to the north.

Those who have navigated using compasses will recall that a compass doesn't point exactly north. A compass needle points toward **magnetic north**, which may deviate from **true north** by as much as several tens of degrees.

Is **F** everywhere horizontal? One might conclude that it is from the observation that compass needles assume a horizontal orientation in all parts of the world. To convince yourself that this conclusion is in error, try the following thought experiment if you like to travel. Borrow an Ecuadorian boy scout compass that works perfectly at the equator and take it to Alaska. You will find that the compass needle will get hung up as its head end tries to dip below the horizontal. The reason is that at localities north of the equator, the earth's magnetic field vector plunges downward (Figure 8-2). The farther north the locality, the steeper the downward plunge of the field. At the equator the field vector is horizontal and in the southern hemisphere it points above the horizontal. The **magnetic inclination** is the angle between the magnetic field vector and the horizontal. Field vectors that dip below the horizontal are said to have + inclinations whereas vectors that point upward have − inclinations. Compass makers, well aware of the variation of inclination with latitude, attach small weights fore or aft of the pivot for balance. In order to balance the needles of the compasses carried by geologists into the field, a piece of wire is generally twisted around the needle south of the pivot point if the geologist is working in the northern hemisphere and north of the pivot point if he is working in the southern hemisphere. No balancing weight is needed at the equator.

We turn now from the the earth's magnetic field vector **F** to the earth's magnetic moment vector **M**. The relationship of the magnetic moment **M** to the magnetic field **F** is the same as the relationship of the earth's mass to the earth's gravitational field. In computing gravity the basic element is a **point mass**, which produces a radially symmetrical gravitational field that points toward the **point mass**. In magnetism the analog of the point mass is the **magnetic dipole M**. Unlike a point mass, a magnetic dipole is a vector quantity. As a consequence, the magnetic field produced by a magnetic dipole does not have the simple radial symmetry of the gravitational field produced by a point mass. Instead, it has rotational symmetry about the axis of the dipole.

You can make a map of the magnetic field produced by a dipole by the following simple experiment. Draw a large circle on a piece of paper and place a bar magnet at the center of the circle. The bar magnet's magnetic field closely approximates that of a magnetic dipole. Now place some small compasses around the circle and observe the orientation of the compass needles (Figure 8-3). Note that two compasses are placed in alignment with the axis of the bar magnet. One of these points in a radial direction away from the center of the circle. Mark the end of the bar magnet closest to this compass with a " + ". The other compass points toward the center of the circle. Mark this end of the magnet " − ". The direction of the dipole moment vector **M** is along the axis of the bar going from the − end to the + end. At the two points at the top and bottom of the circle aligned with the axis of the magnet, the orientations of the compasses and therefore of **F** are both parallel to **M**.

If we think of the circle as a cross section of the earth, the point at the top of the circle is the **north magnetic pole**, which is defined as the point on the earth's surface where the magnetic field vector is directed straight down. It is the point where the value of the inclination *I* is 90°. The **south magnetic pole** is defined as the location where **F** points straight up and *I* is − 90°. The magnetic moment **M** points toward the south magnetic pole (Figure 8-3), which today is located in Antarctica.

Along the magnetic equator, which intersects the circle at 3 and 9 o'clock in Figure 8-3, the orientation of the field **F**

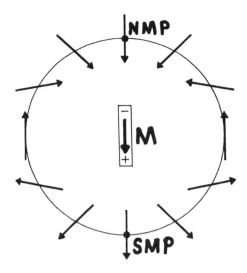

Figure 8-3.
Compass needles (arrows) on the circle are aligned by the field of the bar magnet at the center of the circle. The field of a magnetic dipole **M** is essentially the same. The directions shown are a good first approximation to the field of the earth viewed in a cross section through the north pole, which is at the top of the circle. **NMP**: North magnetic pole. **SMP**: South magnetic pole.

is antiparallel to that of the magnetic moment **M**. At points other than the magnetic equator and poles, **F** is neither parallel nor antiparallel to **M**.

Today, as in the past, the magnetic dipole axis is aligned approximately but not exactly with the earth's rotation axis, the angle between the two axes being 11.5°. The distance between the north magnetic pole and the geographic north pole is therefore also 11.5°. The magnetic pole wobbles around the geographic pole in an irregular manner with variable time constants ranging from centuries to about 10,000 years, producing what is called the **secular variation** of the field.

How Rocks Get Magnetized

Geophysicists who measure the paleomagnetism of rocks are called **paleomagnetists**, the deplorable synonym *paleomagician* no longer being used. Paleomagnetists prowl the earth, looking for rocks bearing decipherable magnetic imprints. Fortunately many such rocks are to be found on all of the plates. For example, the lava flows which compose the ocean floor acquired magnetic imprints of the fields that existed when they cooled. On land a typical volcano continues to erupt lava flows for hundreds of thousands of years. An ancient volcano is the geologic analog of a gigantic floppy disk containing dozens of spot readings of the ancient magnetic field.

Some rocks have good magnetic memories, others don't. Those that do have the ability to record a magnetic image in much the same way that photographic film records a light image. At the time of their formation, these rocks are magnetically sensitized so that they are able to acquire an accurate imprint of the magnetic field, even if the field is weak. Shortly afterwards, this imprint becomes locked into the mineral grains of the rock, much as a photographic image is fixed in film during the process of being developed. Once the magnetic image has been fixed, the ideal rock recorder becomes desensitized to magnetic imprinting. It seems incredible that such a complicated and convenient process should occur in nature without benefit of electronic instrumentation. Yet magnetic imprinting of rocks is going on around us all the time. Let's see how it works.

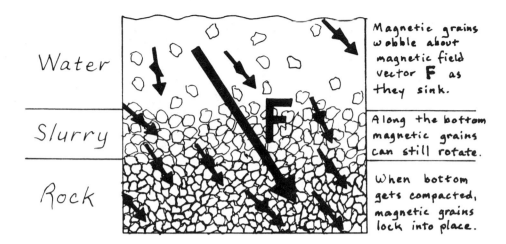

Water

Slurry

Rock

Magnetic grains wobble about magnetic field vector **F** as they sink.

Along the bottom magnetic grains can still rotate.

When bottom gets compacted, magnetic grains lock into place.

Depositional Remanent Magnetization (DRM)

Fine grains of sediment are constantly showering down on the floors of the world's oceans and lakes. These sediments are usually of diverse origin, for example, sand and silt carried to the sea by rivers, ash from volcanoes, and the skeletons of microorganisms. As a consequence, grains of many different types, mostly nonmagnetic, are usually present in sediments. In a typical sediment about one grain in a thousand is a strongly magnetic mineral.

These grains behave like compass needles. As they slowly descend through the water, the grains are subjected to two competing processes, one tending toward order and the other toward disorder. The first is provided by the earth's magnetic field, which exerts a torque on the grains to align them with the field. The second is turbulence in the water, which tends to tumble the grains. Grains of silt size and smaller (less than 1/16 mm in diameter) are dominated by the forces of order and tend to become aligned with their dipole moments parallel to the earth's field. Larger grains have their dipole moments randomized by the turbulence of the water. As a consequence, siltstones generally become magnetized at the time of deposition whereas sandstones do not.

After grains of magnetic iron oxides have touched down on the bottom, they usually find themselves in a loose, water-rich slurry in which they are still free to rotate toward alignment with the magnetic field (Figure 8-4). However as addi-

Figure 8-4.
Process by which a sediment acquires a magnetic memory of the field **F** that existed when magnetic grains (black) and nonmagnetic grains (white) were being deposited.

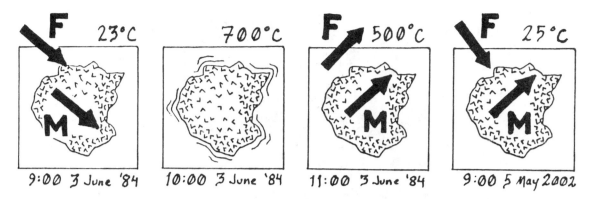

Figure 8-5.
Experiment showing that when a sample of volcanic rock is heated to its Curie temperature, it loses its magnetism. When it is then cooled in a known field **F**, the rock becomes magnetized in the direction of **F** and retains that direction, even after **F** changes.

tional sediments rain down from above, the increased pressure squeezes water from the pore spaces between grains. As the grains touch and interlock with their nearest neighbors, they are unable to reorient themselves when the earth's magnetic field changes. As this happens, the magnetic field becomes locked into the rock. Under ideal conditions, it will stay locked for hundreds of millions of years.

Thermoremanent Magnetization (TRM)

The magnetization acquired by lava flows and other igneous rocks as they cool is termed thermoremanent magnetization. Because igneous rocks start out in a molten state, a reasonable initial hypothesis about TRM would be that it is produced by the magnetic alignment of grains of magnetite in partially molten rock. This hypothesis is wrong. You can convince yourself of this by the following experiment. First fly to Hawaii, collect some oriented samples from a young lava flow, bring them back to your lab, and measure their magnetization. You'll find that they're magnetized to the north with a positive inclination of about 35°. Now place a sample inside a furnace next to a magnetometer and measure its magnetization while heating it (Figure 8-5). During the early stages of heating you will find that the intensity of magnetization decreases by only a few percent; however, above the temperature of 550°C the intensity of magnetization of the sample rapidly decreases to zero. By the time the sample reaches 580°C, the rock has lost all of its magnetization.

The temperature at which a magnetic mineral looses its magnetism is the **Curie temperature** of that mineral. This varies widely with chemical composition and crystal structure. A Curie temperature of 580°C is characteristic of the mineral magnetite, the composition of which is Fe_3O_4. Paleomagnetists have found basalt samples on Hawaii with Curie temperatures ranging from 580°C to 120°C, the latter being characteristic of titanomagnetite, a titanium-rich form of magnetite.

Returning to the hot rock of our experiment, you note that at 580°C it is still completely solid. The magnetic grains in the rock cannot possibly rotate as the sample cools below this temperature. Now cool the rock in a known magnetic field in the laboratory. It is important to know whether, as the solid rock becomes magnetic again, it becomes magnetized in its old direction or in a new direction parallel to the lab field. So let's use the trick of orienting the magnetic field in the laboratory in a direction perpendicular to the original magnetization of the sample. The two can then be easily distinguished. As the sample cools from 580°C down to 550°C, we find that the sample becomes strongly magnetized in a direction exactly parallel to the applied field in the laboratory, not to the old direction of magnetization. Upon cooling the rest of the way down to room temperature, you note that the direction remains parallel to that of the applied field in the laboratory and that the intensity increases by about 10 percent. Most igneous rocks behave in a similar fashion. They acquire most of their TRM while cooling through a narrow temperature range just below their Curie temperatures. After the rock has cooled, the TRM is firmly locked into the rock, where it will remain as a magnetic memory for hundreds of millions of years, provided the rock is not reheated or degraded by chemical processes.

To better understand this remarkable solid-state process, let's focus our attention on the atomic structure of a single grain of magnetite. The crystal lattice of this mineral consists of a regular array of iron and oxygen atoms. Each atom has orbiting electrons which, like the earth, are spinning about their own axes. Because electrons are charged particles, their spins produce magnetic moments. It is the spin moments of electrons that are the cause of the magnetism of magnetite. Above the Curie temperature, the electron spin moments are randomized by the thermal vibrations of the crystal lattice, each magnetic dipole flipping to new orientations independently from its neighbors. As the mineral cools below the Curie temperature, the spinning electrons become influenced by the magnetic moments of their neighbors and lock themselves into alignment either heads-to-heads or heads-to-tails, depending on their quantized preference. As a result, below the Curie temperature the grain as a whole is magnetized with a uniform magnetization termed **ferromagnetism.** The orientation of ferromagnetism relative to the crystal structure of the grain depends on several factors: the shape of the grain, the orientation of its crystallographic axes, and stress applied to the grain by its neighbors in the rock. When they're hot, magnetic minerals are especially sensitive to stress such as that exerted by thermally agitated neighbors.

The magnetism of the grains in a cooling rock is subject to two opposing processes, one tending toward order and the other toward disorder. The ordering process, as in the case of depositional magnetization, is the earth's magnetic field, which tends to bring the magnetic moments of the grains into alignment with its own direction. The disordering process is heat. Distortion and stress resulting from thermal vibrations cause the orientations of the magnetic moments of individual grains to change randomly. This is a solid-state process in which the orientation of each grain remains fixed whereas the orientation of its magnetic moment changes. The earth's magnetic field, although weak relative to fields achievable in the laboratory, is generally able to magnetically align about 1 percent of the magnetic grains in a typical volcanic rock. The magnetic memory of the rock resides in this small fraction of the magnetic mineral present, which in turn comprises only a small fraction of the rock.

The main features of the process by which rocks become thermally magnetized can be modeled by the following simple experiment. Take a dozen small compasses and tape them to a board. Each compass needle represents the magnetic moment of one grain of magnetite. In this analogy the compass needles represent the changing moments of the grains and not the grains themselves, which do not move. The compasses will all quickly become aligned with the earth's field. This state corresponds to that of a rock at low temperature. Now shake the board to simulate heating the rock. With stronger and stronger shaking the needles of the compasses oscillate more and more wildly until, when they are completely random, they have reached a state analogous with being heated above the Curie temperature.

To model the locking-in of ferromagnetism, let's modify our experiment by filling the chamber of each small compass with molten wax. When we shake the board, the compass needles still oscillate. We will want to pay special attention to the time required for one complete swing of a compass needle, the **relaxation time**. When we first add the hot wax, the relaxation time of the needles, although longer than it was in air, is still short. As the wax cools and thickens, the relaxation time increases. Moreover, because of the damping effect of the wax, instead of oscillating the compass needles will approach their final rest position from one direction. In this case, it is useful to redefine the relaxation time as the time required for the angle between the compass needle and the magnetic field **F** to decrease by the factor $1/e$, where e (2.718...) is the natural base of logarithms.

Upon cooling through a fairly narrow range of temperature, wax usually changes from a liquid state to an almost solid state. As this happens in our experiment, the relaxation time of the compasses increases dramatically. The midpoint of this narrow temperature range is called the **blocking temperature**. While the wax was still hot, relaxation times were measured in seconds. After cooling, the relaxation times of the compasses will be measured in years or centuries. The compasses have obviously locked in the direction of the field that existed at the time the wax cooled. Exactly the same thing happens to the ferromagnetic moments of magnetite grains in a cooling rock, except that instead of cooling wax we have a solid-state magnetic process. The result is an exponential increase in the relaxation time during cooling through the blocking temperature. At 580°C the relaxation time of magnetite grains may be less than one second; at 560°C it will have increased to about an hour; and at room temperature it will typically have increased to several hundred million years or more.

Good and Bad Magnetic Memories

Like people, some rocks remember exactly what happened when they were young. In other rocks, these memories are buried beneath the memory of events which happened more recently. Some have completely forgotten events that occurred during their infancy. Let's consider what might make rocks forget.

We'll start with the most extreme case, complete loss of the primary magnetic memory acquired when a rock originally formed. Heat or chemistry is usually responsible. If a rock is reheated above its blocking temperature, by analogy with melting the wax in our experiment the original magnetism will be reoriented and the magnetic memory destroyed. Chemical alteration of a magnetic mineral also destroys its magnetic memory, just as the magnetic moment of a compass needle is destroyed if the needle is dissolved in acid. Even if the magnetic mineral is only oxidized to another magnetic mineral, as in the oxidation of magnetite (Fe_3O_4) to hematite (Fe_2O_3), the original magnetization is usually destroyed. For this reason, rocks that have been oxidized during the course of ordinary weathering at the surface of the ground usually have unreliable magnetic memories.

In **magnetic overprinting**, a rock is magnetized a second time by some process that overprints but does not destroy the primary magnetization acquired when the rock formed. Overprinting may occur in several different ways. To help visualize them, let's return to our experiment in which we modeled the thousands of magnetic grains present in a typical sample using small compasses mounted on a large board. In a new experiment, let's fill the chambers of the compasses with a variety of liquids: hot wax as before, hot tar, various oils and syrups, water, and air. Now instead of having one relaxation time at each temperature we will have a spectrum of relaxation times ranging from seconds (air) to centuries (tar). Instead of a single blocking temperature the ensemble of compasses will have a spectrum of blocking temperatures. Our compass-on-a-board model is beginning to look magnetically more like a real rock.

Now let's put our compasses through a heating and cooling cycle, locking in a magnetic memory of the field. Next, let's change the orientation of the board so that the thermoremanent magnetization of the compasses is perpendicular to the local magnetic field. The undamped compasses with air and the lightly damped compasses with water will quickly align themselves with the field. In the geophysical vernacular, the resulting magnetism is termed **induced magnetization**. Next the compasses with liquids of varying viscosities begin to swing toward the field at varying rates, depending upon their relaxation times. As this process continues, a new component of magnetization forms in a direction parallel to the applied field, which in our experiment is perpendicular to the original thermoremanence. When rocks sit in a magnetic field for a long time at room temperature, they acquire **viscous magnetization** by an analogous process. In volcanic rocks 1 million years old, the viscous magnetization is typically about one-fifth as large as the thermoremanent magnetization. Viscous magnetization grows in proportion to the logarithm of elapsed time. In extremely old rocks it may be larger than the original magnetization acquired when the rocks formed.

A second type of magnetic overprinting is caused by lightning. In a typical lightning discharge, a current of 20,000 amps flows for a fraction of a millisecond, producing a magnetic field thousands of times stronger than that of the earth. A rock within several meters of the location of the discharge will receive a magnetic overprint that is many times larger than the original TRM of the rock. A third type of magnetic overprinting is produced by the growth of new magnetic minerals in rocks. The new minerals acquire the direction of magnetization of the earth's field at the time the new minerals form. If the direction of the field has changed or if the rock has been folded since it originally formed, the direction of the overprint will be different from that of the primary magnetization.

Magnetic Cleaning

It is the job of the paleomagnetist to strip away one or more layers of magnetic overprint in order to be able to measure the primary thermoremanence or depositional remanence of a rock. This stripping, called **magnetic cleaning**, is done by heating the rock or placing it in alternating magnetic fields. The metaphor of magnetic memory might suggest the metaphor of magnetic brainwashing; we prefer the metaphor of the paleomagnetist as a rock psychiatrist, peeling away recent memories in order to discover magnetic memories formed when the rock was in its infancy.

Reversals of the Earth's Magnetic Field

The existence of magnetic isochrons stems not only from the remarkable magnetic memory of rocks but also from an even more remarkable property of the earth's magnetic field: The earth's magnetic field has not always pointed north. It has pointed south during roughly half the earth's history. Changes from a condition of **normal polarity** when the field is pointing north to a condition of **reversed polarity** when the field is pointing south are called **geomagnetic polarity reversals**.

Discovery of Reversals

The discovery that the earth's magnetic field sometimes points south instead of north came as a surprise and was somewhat disconcerting: in a changing environment, scientists share the general human need for fixed points of reference like *up* and *north*. So perhaps it is not surprising that the idea of polarity reversals was forced upon the geophysicist by observations and did not spring from theory.

The first evidence came from paleomagnetism. Working on lava flows several million years old or less in France, Japan, and Iceland, early paleomagnetists found that only half of the flows were magnetized in a northerly direction; the other half were magnetized in exactly the opposite direction. Having the courage to trust their observations, several early paleomagnetists proposed that the earth's magnetic field could switch or reverse its magnetic polarity. During times of normal polarity, the field is that of a magnetic dipole directed toward the geographic south pole, as it is today. During times of reversed polarity, the field is that of a dipole directed toward the geographic north pole. The clear implication of this research is that if the tool users who lived in Olduvai Gorge in East Africa a million years ago had been clever enough to make compasses, these compasses would have pointed south instead of north.

When this startling hypothesis was advanced more than half a century ago, the response of most geophysicists was neither acceptance nor rejection but rather silence. After all, no viable theory had yet been presented to explain why the earth's magnetic field even exists. Since geophysicists did not know why the magnetic field points north, they were not in a position to reject a proposal, based on observations, that it had once pointed south.

Interest in geomagnetic reversals reawakened in the late 1950s because of accelerating interest in the new field of paleomagnetism. With the discovery of reversed remanent magnetization in dozens and then in hundreds of rock strata of different ages, it became ever harder to dismiss reversals as some rare and inexplicable form of geologic noise.

However an alternative explanation, the **self-reversal hypothesis**, was advanced in the late 1950s that did not require a change in the polarity of the earth's magnetic field. Self-reversal refers to the ability of certain minerals to become magnetized backwards—in a magnetic field directed toward the north, these minerals acquire a remanent magnetization directed toward the south. Several synthetic minerals and a few natural minerals were discovered in the 1950s to have this strange self-reversal property. According to the self-reversal hypothesis, these minerals are present in all reversely magnetized rocks.

A Critical Experiment

In the early 1960s paleomagnetists undertook a critical experiment to determine whether the field-reversal theory or the self-reversal theory was correct. If the field-reversal hypothesis is true, then because geomagnetism is a global phenomenon all of the rocks that formed anywhere in the world at any given time should be of the same polarity. On the other hand, if the self-reversal hypothesis is true, then both polarities are to be expected among contemporaneous rocks. The main difficulty in the dating experiment lay in finding a sufficiently precise method of dating rocks. Classical methods for dating rocks such as those based on fossils or on the stratigraphic correlation of glaciations were not precise enough for this particular experiment. Fossil ages, for example, commonly have uncertainties of a million years or more, reflecting the time required for the occurrence of a recognizable evolutionary change. An error of a million years is acceptable in dating rocks 100 million years old but is disastrous in dating rocks only a million years old.

The key to designing a successful experiment was to use the atomic clocks that reside in volcanic rocks. These clocks begin to run synchronously with the magnetic imprinting of lavas at the time of their initial cooling. Because the ages of young rocks can be determined more precisely than the ages of old rocks, the experiment was first attempted with rocks a few million years old or less, which are young by geologic standards. For dating rocks of this age, the atomic clock that works the best is one based on the radioactive decay of potassium to the inert gas argon-40. As a rock ages, argon gas accumulates. Knowing the decay rate of potassium, one can calculate the age of the rock simply by measuring the amounts of potassium and argon present.

The experiment was very straightforward. Several hundred rock samples were collected from lava flows in different parts of the world. Geophysicists then measured the remanent magnetization of the samples and analyzed them chemically to determine their potassium and argon content. The procedure is simple in theory, but at the time the technique of applying potassium-argon dating to young rocks was new and the equipment scarce, so the experiment was performed in only two laboratories, one in the United States and one in Australia. The results were clearcut. The paleomagnetic polarities and the potassium-argon ages fell into neat packets according to the age of the samples. All of the samples with ages falling within certain time intervals were found to be magnetized toward the north; that is, they were normal. All of the samples with ages falling within the intervening time intervals were found to be magnetized toward the south; that is, they were reversed. The experiment demonstrated that self-reversing minerals are rare in nature. It also confirmed unequivocally the hypothesis that the earth's magnetic field has repeatedly changed polarity.

What Causes the Earth's Magnetic Field?

The question of why the earth has a magnetic field was one of the first to be asked by geophysicists and one of the last to be answered. This branch of geophysics got off to an early start in the fifteenth and sixteenth centuries when European explorers, colonists, and buccaneers carried compasses with them for purposes of navigation when they set out to explore the world. An important scientific byproduct of their travels was the rapid growth of what we would now call a geomagnetic data base.

Geophysical data rarely lie around very long without stimulating the imaginations of theoreticians. The early geomagnetic data were no exception. As early as 1600 William Gilbert, a geophysicist who was to become Queen Elizabeth's doctor, proposed an explanation for the global pattern of declination and inclination as recorded in the logs of navigators. In what may be the earliest geophysical modeling experiment, Gilbert cut a sphere from naturally magnetized magnetite and moved a small bar magnet over its surface. He showed that the angles of inclination and declination assumed by the small bar magnet were essentially the same as those of compass needles carried by explorers to the far corners of the world. The field of a uniformly magnetized sphere is that of a dipole moment **M** (Figure 8-3). The close correspondence between observation and model led Gilbert to the reasonable conclusion that the earth itself was a magnetized body of rock and that the origin of the geomagnetic field was the magnetization of the rocks that comprise the earth's interior.

This idea is no longer viable because we have, in the centuries since Gilbert's classic study, measured the intensity of remanent magnetization of millions of rock samples and found that rocks aren't strongly enough magnetized to account for the observed geomagnetic field. A second difficulty with Gilbert's model is that we now know that below depths of about 30 km the interior of the earth is so hot that rocks are above their Curie temperatures and therefore are nonmagnetic. A third difficulty is that secular variation of the field occurs over time scales ranging from a decade to around 10,000 years, whereas geologic processes capable of changing the magnetization of rocks occur much more slowly. Thus magnetic rocks clearly cannot account for the main dipole field of the earth, although they do account for magnetic anomalies which are superimposed on the main field.

Electrical currents flowing in the earth's core provide a much more satisfying explanation. Recall the axial dipole field produced by a bar magnet at the center of a sphere: this is essentially the dipole field of the earth and of Gilbert's sphere of magnetite. However the same axial dipole field is also produced by a loop of current at the center of a sphere when the plane of the loop lies in the plane of the equator. A dipole field produced by such a current loop fits the earth's field just as well as the field of Gilbert's sphere of magnetite and provides a closer physical analogy to the earth's magnetic field.

A lot of current is required to produce a field as large as the earth's. If this current were flowing through the rocks that make up the earth's crust and mantle, which are not very good electrical conductors, the amount of heat generated would be enormous—much larger than is observed. Therefore if the earth's field is produced by current, there must exist some material deep in the earth that is a much better electrical conductor than the rocks of the crust and mantle. Such material is to be found in the earth's core, which is composed of an alloy of molten iron and nickel. Because the temperature of the earth's core is several thousand degrees hotter than the Curie temperature of iron and nickel, the core is not ferromagnetic, but being a metal, it is a good electrical conductor. All modern theories of geomagnetism assume that the field is produced by electrical currents flowing in the earth's metallic core.

When electrical currents flow through good (but not perfect) electrical conductors, part of the electrical energy is transformed into heat. Without input of additional electrical energy, the current will die away. If the earth's field is produced by currents in the core, then a feedback process must exist that transforms some available form of energy in the core into electromagnetic energy.

We're all familiar with machines which do just that in everyday life. The generator in an automobile, for example, transforms mechanical energy to electrical energy. The disk

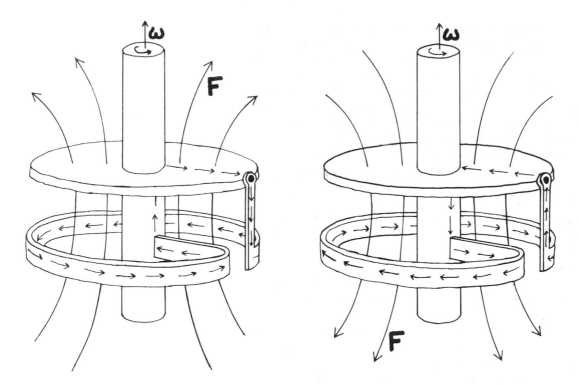

Figure 8-6.
Disk dynamo. The disk rotates through magnetic field lines **F**, generating current (small arrows) flowing through the disk and coil. Feedback is provided by this current, which produces a magnetic field that reinforces the field **F**. The left diagram shows the end state if, when the dynamo started, a weak initial field pointed downward. The right diagram shows the end state for an initial field pointing upward. Although these two antiparallel states are equally stable, switching between them does not occur spontaneously unless another element such as a second coil is added to the system.

dynamo does this with great conceptual simplicity. As a rotating disk of copper moves across magnetic field lines (Figure 8-6), a voltage is induced in the disk which drives current through the system. Mechanical energy is supplied in the form of torque that drives the disk, which otherwise would decelerate because of mechanical friction and the force exerted on the rotating disk by magnetic field lines. The current generated in the disk is led via a sliding contact through a coil, where it generates a magnetic field which reinforces the initial weak field. This provides positive feedback, which is required to maintain all dynamos.

Is the earth's core an electrical dynamo? After all, it's simply a blob of molten metal, not a complicated arrangement of wires, sliding contacts, insulators, and mechanical bearings. Magnetic fields produced by fluid dynamos are, in fact, present almost everywhere in the universe. The sun, for example, has a magnetic field about as strong as the earth's. Many other stars have strong magnetic fields. Since the temperatures of most of these bodies are much too high for permanent magnetism to exist, their magnetic fields must be due to electrical currents. Moreover, for reasons discussed earlier, to compensate for dissipative losses the current systems in these bodies must be sustained by some sort of feedback process. In other words, their magnetic fields are produced by fluid dynamos. So a dynamo in earth's fluid core is not unreasonable.

The exact nature of this magnetic dynamo has intrigued some of the most brilliant theoreticians in geophysics during the past 30 years and continues to be a subject of active research. Theoreticians all agree that it is the movement of fluid in the core that plays the role of the rotating disk in the simple disk dynamo: as the conducting fluid moves through the magnetic field, an electrical current is induced in the fluid. Most theoreticians agree that what keeps the fluid moving is some form of thermal convection: this plays the role of the torque that keeps the disk rotating in the disk dynamo. The ultimate source of energy to drive the dynamo is thus heat. Concerning the source of the heat and the mechanism for converting it to fluid motion, theories differ. One theory is that although most radioactive elements such as potassium, thorium, and uranium reside in the earth's crust and mantle, a small amount is present in the core, producing heat by radioactive decay; this, in turn, causes thermal convection. Another theory is that dense iron-nickel crystals form in the molten outer core and sink to the interface with the solid inner core; as they sink, these dense crystals produce motion in the fluid of the outer core. While many details remain to be worked out, it is now firmly established that the geomagnetic field is produced in the earth's core by a fluid dynamo similar to the many other fluid dynamos known to exist in the universe.

What Causes Reversals?

Why does the earth's magnetic field alternate between north-seeking and south-seeking dipole states? Part of the answer lies with the equations that describe the core dynamo. These equations are symmetrical with respect to the earth's rotation axis in the sense that for every solution to the equations that yields a field of normal polarity there exists another solution that yields a field of reversed polarity. This principle is shown in its simplest form in Figure 8-6 for the disk dynamo: with the disk rotating in the same direction, either of two antiparallel polarities may result, depending upon the orientation of whatever weak magnetic field was present when motion was initiated. So even in the absence of explicit solutions to the dynamo equations, the very form of the equations points to the existence of two polarity states.

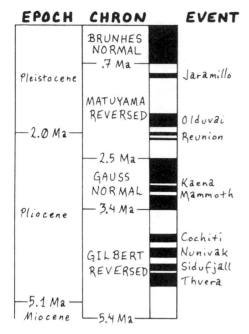

EPOCH   CHRON   EVENT

Figure 8-7.
Reversal time scale determined from potassium-argon dating. On the polarity bar code, the black and white segments are time intervals when the field was normal and reversed, respectively.

Even though all dynamos may exist in either of two symmetrical polarity states, not all dynamos flip back and forth between the two states. The simple disk dynamo shown in Figure 8-6, for example, once established with one polarity, does not flip to the other. If the earth's field were this simple, we would not have geomagnetic reversals. However the disk dynamo can be made to undergo spontaneous reversals by adding other elements such as a second coil. In fact, theoretical dynamos only slightly more complex than the disk dynamo have been studied which spontaneously switch between polarity states in a pattern analogous to that displayed by the geomagnetic field.

Still unanswered are questions related to the timing of reversals. What determines when a particular polarity reversal will occur? What determines the mean time between reversals? The fact that this is extremely variable suggests that the timing of reversals may be a random process analogous to the timing of great storms in the atmosphere, which are also randomly distributed in time. The analogy doesn't seem too far-fetched when one realizes that weather and dynamo processes both occur in geophysical fluids (the atmosphere and the core) within which random turbulence is an important process. Rare atmospheric events like giant storms are produced by certain poorly understood patterns of fluid motion in the earth's atmosphere. Similarly, rare events like reversals may well be due to the random occurrence of certain poorly understood patterns of fluid motion in the molten metal of the core.

Magnetostratigraphy

Geomagnetic Reversal Time Scale from K-Ar Dating

A result that was to prove very important to plate tectonics emerged from the experiment in which potassium-argon ages and paleomagnetic polarities were determined for samples from around the world. This was the development of a **geomagnetic reversal time scale**. This time scale describes the polarity of the geomagnetic field as a function of geologic age for the past 5 million years, which is about the limit of potassium-argon dating (Figure 8-7). On this scale, geologic time is divided into periods called **magnetic polarity intervals** within each of which magnetic polarity is constant.

A major surprise to emerge from early work on the time scale was the discovery that polarity intervals are highly variable in length. During the past 4 million years the shortest known polarity interval is 20,000 years long, the longest 0.7 million years. If the reversing geomagnetic field is seen as the earth's magnetic chronometer, the spacing between the ticks of the clock is very irregular. In fact, it is almost completely random.

Now a geomagnetic clock with regular ticks would provide ideal markers for determining the rates of geologic processes because stratigraphic sequences of alternately normally and reversely magnetized rocks would all have formed in exactly the same length of time. However this geomagnetic clock would not be very useful for determining the age of a sequence of rocks because any paleomagnetic sequence of a half-dozen or so polarity intervals would look like every other sequence, regardless of age. It would be as useless for identification as fingerprints would be if all fingerprints were identical or as tree rings would be if their widths never varied. It turns out that from the viewpoint of the geologist trying to determine the age of rocks, the earth's magnetic field with its randomly spaced reversals provides a clock of optimum design. Because of the random nature of reversals, each segment of the reversal time scale has its own unique magnetic signature, much like the bar code used to identify products in the grocery store. This magnetic bar code, when recognized in the paleomagnetism of a continuous sequence of rocks, gives the geologist the age of the rocks.

Because the magnetic field is produced in the earth's core and not locally near the surface, the onset of a reversal is seen synchronously over the entire globe. This provides the geologist with an unparalleled opportunity to establish the synchroneity of events occurring around the world, especially those occurring near a time when the polarity changed. The reversal horizons themselves are not perfectly sharp; the time required for a complete transition from one polarity to another is about 5000 years. During a transition, local magnetic field vectors follow an irregular path that varies from site to site. Thus there is no basis for global correlations of events occurring within the transition zone. To a historian 5000 years is a long time but to a geologist it is a moment. It is their global synchroneity in a random temporal pattern that makes paleomagnetic reversals so useful in determining the ages of rocks.

Polarity Intervals

Because polarity intervals vary so greatly in length, it has proved useful to introduce a hierarchy of names to describe polarity intervals and groups of polarity intervals of different durations. The basic unit is the **polarity chron,** which is a geochronologic term to describe subdivisions of time based on geomagnetic polarity. The terms used to describe different levels in the hierarchy are as follows:

Name	Duration
Polarity subchron	10^4 - 10^5 yrs
Polarity chron	10^5 - 10^6 yrs
Polarity superchron	10^6 - 10^7 yrs

Subchrons are intervals of constant polarity. Chrons, which are the main subdivisions of time based on polarity, may be either of constant polarity or of mixed polarity. Superchrons, which differ from the smaller units both in length and in the physics of the underlying process, will be described later.

Two types of names were introduced during the development of the original reversal time scale, one for chrons and one for subchrons. Chrons were named for famous geomagnetists and paleomagnetists. Subchrons were named for the localities where they were discovered (Figure 8-7). These names, which describe polarity intervals spanning the interval from 5 Ma to the present, are widely used for the global correlation of Pliocene and Pleistocene rocks.

Reversal Time Scale from Marine Magnetic Anomalies

The reversal time scale was soon extended from 5 to 165 Ma using marine magnetic anomalies—the famous seafloor magnetic stripes, alternating parallel bands where the magnetic field is greater and smaller than the mean regional field by 2 percent or so. These stripes vary in width from 1 km to 100 km. They are many hundreds of kilometers long and are offset along fracture zones. The explanation of these small but consistent anomalies, which were mapped prior to the advent of the theory of plate tectonics and are now known to be present over all ocean basins, is one of the triumphs of plate tectonics. The explanation combines two metaphors. In the first, oceanic plates move away from spreading centers as if on dual conveyor belts. In the second,

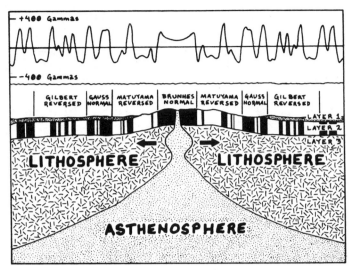

Figure 8-8.
Formation of magnetic anomalies at a ridge. Black and white parts of Layer 2 represent normally (black) and reversely (white) magnetized rocks, which were formed during the Brunhes normal, Matuyama reversed, Gauss normal, and Gilbert reversed polarity chrons. Layer 1 consists of sediments (nonmagnetic). Layer 2 consists of sheet flows and layers of pillow basalt (highly magnetic) with feeder dikes (moderately magnetic). Layer 3 consists of gabbroic oceanic crust (weakly magnetic).

plates become magnetized as if by a magnetic tape recorder with dual tape drives and a single recording head located at the ridge. The recording mechanism is the acquisition of thermoremanent magnetization by basalt as it cools and is accreted to the plates at the spreading center, as was discussed earlier.

To determine a reversal time scale from marine magnetic anomalies, two steps are required. The first is to transform the peaks and lows observed on a given profile into a set of time intervals ΔT_1, ΔT_2, ΔT_3, and so forth, describing the lengths of adjacent polarity intervals. Having thus determined a time scale, the second step is to calibrate the age of the time scale.

The first step has a geophysical part and a mathematical part. The geophysicist must first determine the location of the magnetic rocks that produce the anomalies. From drilling and dredging, the oceanic crust is known to consist of several layers. The uppermost, layer 1, is made up of weakly magnetic sediments that vary in thickness from zero at the ridges to several hundred meters in the deep oceans. The underlying layer 2 is about 0.5 km thick and consists of basalt flows that were cooled quickly (Figure 8-8). These are very fine grained and are highly magnetic. Layer 2A is made up of vertical dikes of basalt that are former conduits to the flows of layer 2. The dikes are coarser grained, cooled more slowly, and are much less magnetic. Layer 3, the deepest layer, is very coarse grained gabbro that crystallized in the magma chamber and is even less magnetic than layer 2A. The anomalies are produced mostly by the highly magnetic basalt flows of layer 2 with much smaller contributions from the deeper layers.

The mathematical model most commonly used to interpret the anomalies is a set of horizontal prisms extending to ± infinity along one axis. In some models layer 2 is a flat sheet, in which case the prisms are rectangles in cross section. In other models layer 2 follows the bathymetry of the bottom, in which case the cross sections of the prisms are irregular polygons. The magnetic polarity of the prisms is alternately normal and reversed with the intensity remaining constant except near the ridge, where higher intensities of magnetization are sometimes needed to model the ridge anomaly. Given the boundaries of the prisms, their intensity of magnetization, the orientation of the regional field, and the azimuth of the traverse along which the magnetic profile was recorded, a model magnetic anomaly can be calculated by a straightforward mathematical procedure. The geophysicist then adjusts the boundaries between the prisms and the intensity of magnetization until an acceptable fit is obtained to the observed profile. The output of this procedure is a set of prism widths W_i. These are related to the lengths ΔT_i of polarity intervals by

$$\Delta T_i = W_i/V$$

where V is the half-velocity of seafloor spreading at the time the oceanic crust formed.

Fidelity and Resolution The length of the shortest polarity interval that can be detected depends upon three factors: the continuity and regularity of the geologic processes that add new ocean floor to the two diverging plates; the width of the shortest prism on the ocean floor that can be detected by a ship at the surface; and the general level of noise on the profile. Noise is produced by mid-plate volcanism, irregular ocean bottom topography, and variations in the intensity of magnetization not accounted for in the model. The length scale of the irregularities in geologic processes at ridges varies from about 1 km at fast spreading centers to several kilometers at slow spreading centers—numbers that appear unbelievably small to the geologist familiar with the irregularities of volcanic processes on land. The narrowest prism that can be detected magnetically at water depths of 3 km is about 1 km. For very rapid spreading at a half-velocity of 50 km/my, a width of 1 km corresponds to a minimum detectable polarity interval of 0.02 my. Several subchrons as short as this have, in fact, been detected where conditions were ideal. Elsewhere the cut-off of observable subchrons is several times greater.

The geophysicist's first impression on viewing magnetic profiles is amazement at the high fidelity of the seafloor magnetic tape recorder, especially over fast spreading centers. However, closer scrutiny of even the best profiles usually reveals spurious anomaly peaks due to mid-plate volcanoes and gaps or duplications due to ridge jumps. Although no one profile is known to contain a perfect record of polarity changes, comparison of many profiles from different spreading centers has revealed a pattern of consistency that provides the basis for a polarity history extending back to 165 Ma.

Calibration   Early reversal time scales were based on the assumption that seafloor spreading had occurred at a constant rate. Several approaches are being used to calibrate and correct these time scales. One is to drill the ocean floor beneath a known anomaly and determine the age of layer 2 either by paleontologic dating of the overlying sediments, which gives a lower bound to the age of the basalt, or by radiometric dating of the basalt itself. A second less direct approach is to look for records of reversals in datable stratigraphic sequences that are not found in direct association with magnetic anomalies. If deposition is at a constant rate, then the thicknesses of normal and reversed strata will be proportional to the widths of the magnetic stripes seen at sea. The trick is to find a recognizable paleomagnetic bar code in a sequence of limestone or other sedimentary strata containing fossils of known age. The fossils can be used to date the set of chrons which compose the bar code. These dates are applicable to the corresponding chrons in the time scale derived from marine magnetic anomalies. Both of these approaches were used in calibrating the reversal time scale shown in Figure 8-9 and listed in Table 8-1.

Superchrons

Superchrons are defined by two related phenomena: reversal frequency and polarity bias. If the reversal time scale is

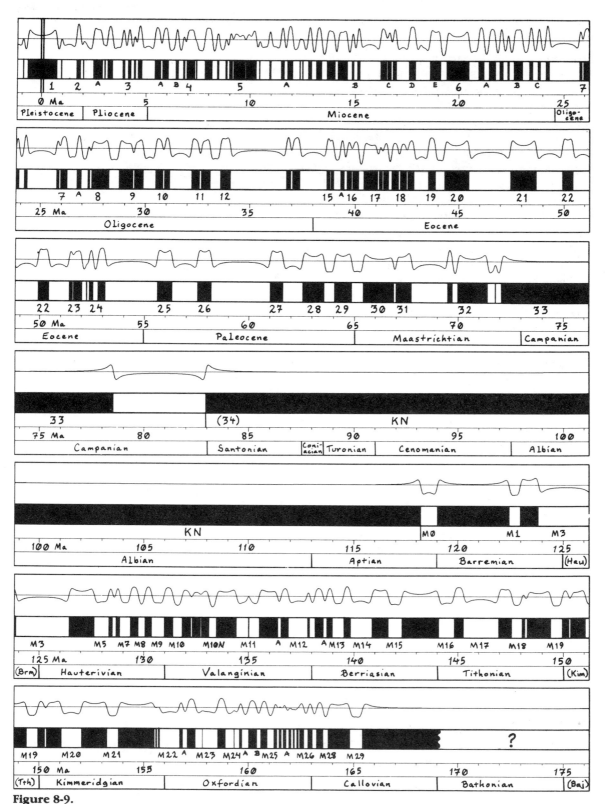

Figure 8-9.

Geomagnetic reversal time scale and corresponding anomalies. A vertical inclination was used for both the field and magnetization, producing symmetrical anomalies for isolated crustal blocks.

Table 8-1. Polarity intervals of the KTQ-M Superchron.

Normal				Reversed		
Chrons	Subchrons	Interval (Ma)		Chrons	Subchrons	Interval (Ma)
1		0.00- 0.73		1r		0.73- 0.92
	1r-1	0.92- 0.97		1r		0.97- 1.67
2		1.67- 1.87		2r		1.87- 2.01
	2r-1	2.01- 2.04		2r		2.04- 2.12
	2r-2	2.12- 2.14		2r		2.14- 2.48
2A		2.48- 2.92			2A-1	2.92- 3.01
2A		3.01- 3.05			2A-2	3.05- 3.15
2A		3.15- 3.40		2Ar		3.40- 3.86
3.1		3.86- 3.98		3.1r		3.98- 4.12
3.2		4.12- 4.26		3.2r		4.26- 4.41
	3.2r-1	4.41- 4.49		3.2r		4.49- 4.59
3.3		4.59- 4.79		3r		4.79- 5.41
3A		5.41- 5.70		3A-1		5.70- 5.78
3A		5.78- 6.07		3Ar		6.07- 6.42
3B		6.42- 6.55		3Br		6.55- 6.77
4		6.77- 6.86			4-1	6.86- 6.94
4		6.94- 7.34			4-2	7.34- 7.39
4		7.39- 7.44		4r		7.44- 7.81
4A		7.81- 8.18		4Ar		8.18- 8.40
	4Ar-1	8.40- 8.49		4Ar		8.49- 8.80
	4Ar-2	8.80- 8.87		4Ar		8.87- 8.98
5		8.98- 9.13			5-1	9.13- 9.17
5		9.17- 9.47			5-2	9.47- 9.48
5		9.48- 9.75			5-3	9.75- 9.78
5		9.78- 10.03			5-4	10.03- 10.05
5		10.05- 10.30		5r		10.30- 10.43
	5r-1	10.43- 10.48		5r		10.48- 10.91
	5r-2	10.91- 10.99		5r		10.99- 11.47
5A.1		11.47- 11.63		5A.1r		11.63- 11.77
5A.2		11.77- 12.03		5Ar		12.03- 12.36
	5Ar-1	12.36- 12.41		5Ar		12.41- 12.49
	5Ar-2	12.49- 12.54		5Ar		12.54- 12.76
5AA		12.76- 12.94		5AAr		12.94- 13.15
5AB		13.15- 13.41		5ABr		13.41- 13.65

(continued)

Table 8-1. *(continued)*

Normal				Reversed		
Chrons	Subchrons	Interval (Ma)		Chrons	Subchrons	Interval (Ma)
5AC		13.65- 14.04		5ACr		14.04- 14.16
5AD		14.16- 14.63		5ADr		14.63- 14.82
5B.1		14.82- 14.93		5B.1r		14.93- 15.09
5B.2		15.09- 15.23		5Br		15.23- 16.20
5C		16.20- 16.50			5C-1	16.50- 16.54
5C		16.54- 16.72			5C-2	16.72- 16.79
5C		16.79- 16.98		5Cr		16.98- 17.58
5D		17.58- 17.91		5Dr		17.91- 18.13
	5Dr-1	18.13- 18.15		5Dr		18.15- 18.59
5E		18.59- 19.12		5Er		19.12- 19.41
6		19.41- 20.50		6r		20.50- 20.95
6A.1		20.95- 21.22		6A.1r		21.22- 21.45
6A.2		21.45- 21.78		6Ar		21.78- 21.97
6AA		21.97- 22.14		6AAr		22.14- 22.34
	6AAr-1	22.34- 22.43		6AAr		22.43- 22.65
6B		22.65- 23.06		6Br		23.06- 23.37
6C.1		23.37- 23.54		6C.1r		23.54- 23.76
6C.2		23.76- 23.90		6C.2r		23.90- 24.15
6C.3		24.15- 24.32		6Cr		24.32- 25.75
7		25.75- 25.88			7-1	25.88- 25.94
7		25.94- 26.27		7r		26.27- 26.74
7A		26.74- 26.95		7Ar		26.95- 27.27
8		27.27- 27.36			8-1	27.36- 27.44
8		27.44- 28.27		8r		28.27- 28.73
9		28.73- 29.39			9-1	29.39- 29.45
9		29.45- 29.91		9r		29.91- 30.48
10		30.48- 30.84			10-1	30.84- 30.90
10		30.90- 31.17		10r		31.17- 32.19
11		32.19- 32.58			11-1	32.58- 32.65
11		32.65- 33.11		11r		33.11- 33.57
12		33.57- 34.06		12r		34.06- 36.73
13		36.73- 36.95			13-1	36.95- 37.02
13		37.02- 37.40		13r		37.40- 38.64

(continued)

Table 8-1. *(continued)*

Normal				Reversed		
Chrons	Subchrons	Interval (Ma)		Chrons	Subchrons	Interval (Ma)
15		38.64- 38.80			15-1	38.80- 38.83
15		38.83- 38.98		15r		38.98- 39.30
15A		39.30- 39.48		15Ar		39.48- 39.60
16		39.60- 39.82			16-1	39.82- 39.86
16		39.86- 40.17		16r		40.17- 40.39
17		40.39- 41.07			17-1	41.07- 41.13
17		41.13- 41.29			17-2	41.29- 41.34
17		41.34- 41.60		17r		41.60- 41.74
18		41.74- 42.08			18-1	42.08- 42.14
18		42.14- 42.47			18-2	42.47- 42.51
18		42.51- 42.84		18r		42.84- 43.52
19		43.52- 43.87		19r		43.87- 44.31
20		44.31- 45.49		20r		45.49- 47.46
21		47.46- 48.69		21r		48.69- 49.91
22		49.91- 50.43		22r		50.43- 51.39
23		51.39- 51.57			23-1	51.57- 51.60
23		51.60- 52.02		23r		52.02- 52.22
	23r-1	52.22- 52.26		23r		52.26- 52.36
24.1		52.36- 52.55		24.1r		52.55- 52.77
24.2		52.77- 53.13		24r		53.13- 55.60
25		55.60- 56.33		25r		56.33- 57.52
26		57.52- 58.19		26r		58.19- 61.00
27		61.00- 61.62		27r		61.62- 62.55
28		62.55- 63.57		28r		63.57- 64.03
29		64.03- 64.86		29r		64.86- 65.39
30		65.39- 66.88		30r		66.88- 66.97
31		66.97- 67.74		31r		67.74- 69.48
32.1		69.48- 69.72		32.1r		69.72- 69.96
32.2		69.96- 71.40		32r		71.40- 71.76
	32r-1	71.76- 71.81		32r		71.81- 72.06
33		72.06- 78.53		33r		78.53- 82.93

Table 8-2. Polarity intervals of the JK-M Superchron.

Normal			Reversed		
Chrons	Subchrons	Interval (Ma)	Chrons	Subchrons	Interval (Ma)
			M0		118.21-119.00
M1n		119.00-122.46	M1		122.46-122.96
M2		122.96-123.83	M3		123.83-126.42
M4		126.42-127.64	M5		127.64-128.31
M6n		128.31-128.49	M6		128.49-128.63
M7n		128.63-128.83	M7		128.83-129.33
M8n		129.33-129.73	M8		129.73-130.03
M9n		130.03-130.38	M9		130.38-130.96
M10n		130.96-131.39	M10		131.39-131.80
M10Nn		131.80-132.24		M10Nn-1	132.24-132.29
M10Nn		132.29-132.71		M10Nn-2	132.71-132.73
M10Nn		132.73-133.11	M10N		133.11-133.43
M11n		133.43-134.42	M11		134.42-134.98
	M11-1	134.98-135.03	M11		135.03-135.49
M11An		135.49-136.39	M11A		136.39-136.52
M12n		136.52-136.89	M12.1		136.89-137.79
M12.2n		137.79-137.91	M12.2		137.91-138.15
M12An		138.15-138.55	M12A		138.55-138.69
M13n		138.69-138.99	M13		138.99-139.51
M14n		139.51-139.84	M14		139.84-140.85
M15n		140.85-141.64	M15		141.64-142.28
M16n		142.28-144.08	M16		144.08-144.81
M17n		144.81-145.27	M17		145.27-146.94
M18n		146.94-147.57	M18		147.57-148.04
M19n		148.04-148.18		M19n-1	148.18-148.27

(continued)

viewed through a sliding window 20 or 30 my wide, the character of the pattern seen in the window can be characterized by two numbers: the **reversal frequency**, which is the number of polarity intervals per unit time, and the **polarity bias**, which is the fraction of the time when the polarity was normal. Both reversal frequency and polarity

Table 8-2. *(continued)*

Normal				Reversed		
Chrons	Subchrons	Interval (Ma)		Chrons	Subchrons	Interval (Ma)
M19n		148.27-149.41		M19		149.41-149.93
M20n		149.93-150.27			M20n-1	150.27-150.34
M20n		150.34-151.07		M20		151.07-152.03
M21n		152.03-153.24		M21		153.24-153.77
M22n		153.77-155.51			M22n-1	155.51-155.57
M22n		155.57-155.63			M22n-2	155.63-155.69
M22n		155.69-155.77		M22		155.77-156.71
M22An		156.71-156.86		M22A		156.86-157.06
M23n		157.06-157.47		M23		157.47-157.79
	M23-1	157.79-157.81		M23		157.81-158.51
M24n		158.51-158.89		M24		158.89-159.30
	M24-1	159.30-159.33		M24		159.33-159.56
M24An		159.56-159.70		M24A		159.70-160.00
M24Bn		160.00-160.40		M24B		160.40-160.58
M25n		160.58-160.90		M25		160.90-161.19
M25An		161.19-161.36			M25An-1	161.36-161.45
M25An		161.45-161.56			M25An-2	161.56-161.65
M25An		161.65-161.80		M25A		161.80-161.92
M26n		161.92-162.03			M26n-1	162.03-162.12
M26n		162.12-162.20			M26n-2	162.20-162.29
M26n		162.29-162.38			M26n-3	162.38-162.43
M26n		162.43-162.65		M26		162.65-162.82
M27n		162.82-163.05		M27		163.05-163.22
M28n		163.22-163.55		M28		163.55-163.78
M29n		163.78-164.82		M29		164.82-165.41

bias have changed repeatedly throughout the history of the earth. Time intervals during which the polarity bias is constant or nearly so are called **polarity bias superchrons** or superchrons for short. The three types of superchrons are normal, reversed, and mixed, the latter describing times when the polarity was unbiased. The following superchrons are recorded on the seafloor.

Name of Superchron	Symbol	Time
		0 Ma
Cretaceous-Tertiary-Quaternary Mixed Polarity Superchron	KTQ-M	
		83 Ma
Cretaceous Normal Polarity Superchron	K-N	
		119 Ma
Jurassic-Cretaceous Mixed Polarity Superchron	JK-M	
		165 Ma

?

Ocean floor that formed during the Cretaceous normal superchron is sometimes described as the **Cretaceous quiet zone** because, in the absence of magnetic stripes due to reversals, magnetic profiles do not show the characteristic high and low anomalies characteristic of other parts of the ocean floor.

Problems

8.1. The following magnetic anomalies were recorded at high latitudes, so they have the symmetrical form shown in Figure 8-9. For each profile, describe in words a plate tectonic scenario that could have produced the profile and determine the full spreading velocity as a function of the age of the crust and the age of the oceanic crust at the ends of the profile.

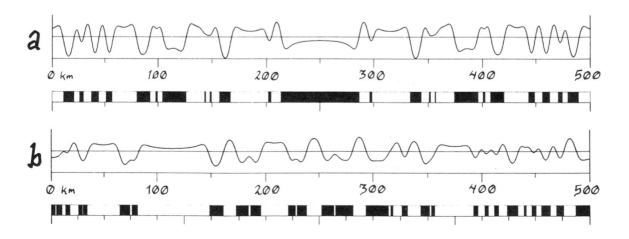

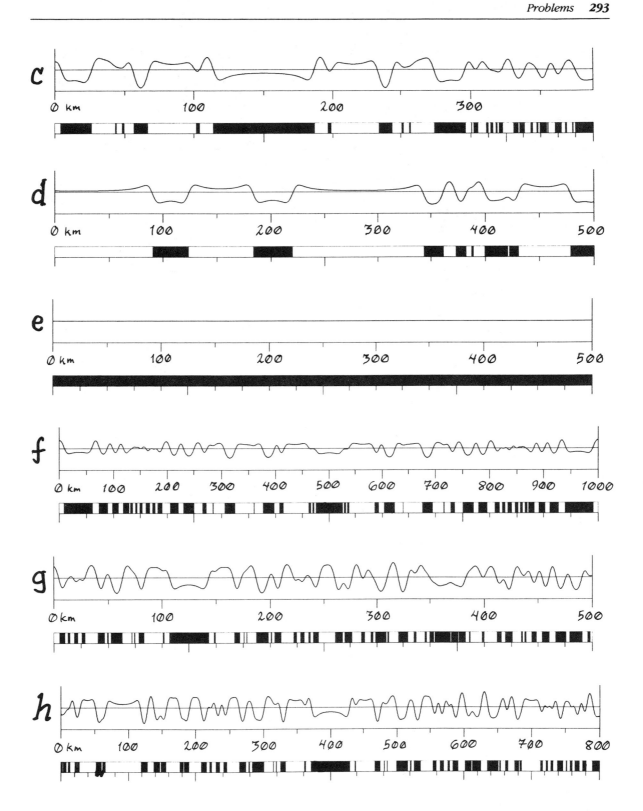

8.2. Devise a plate tectonic scenario complete with spreading velocities and reconstructions showing ridges, trenches, transforms, and isochrons (at all crucial times) to explain the following set of magnetic profiles.

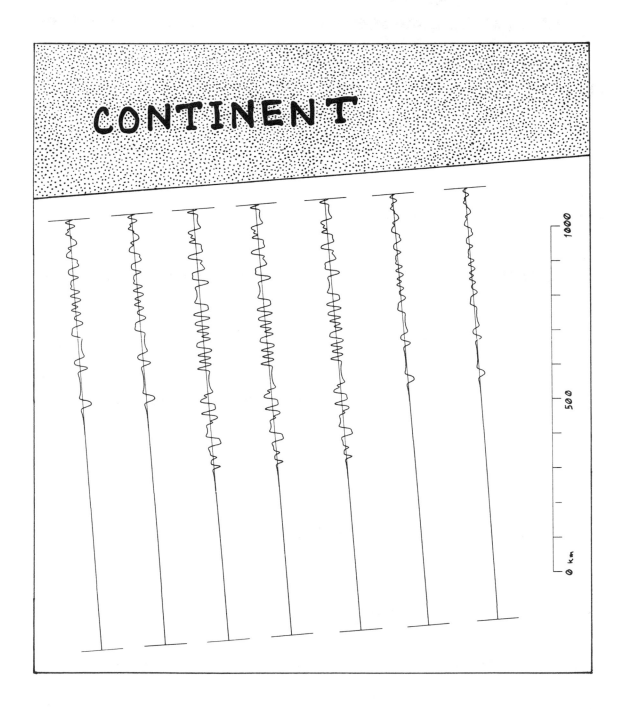

Suggested Readings

Classic Papers on Reversal Time Scale

Geomagnetic Polarity Epochs and Pleistocene Geochronometry, Allan Cox, Richard Doell, and G. Brent Dalrymple, Nature, v. 198, p. 1049–1051, 1963. *Early paper that attempts to define reversal time scale on the basis of six polarities and potassium-argon ages.*

Dating of Polarity Zones in the Hawaiian Islands, Ian McDougall and Don Tarling, Nature, v. 200, p. 54–56, 1963. *Early paper that attempts to define reversal time scale on the basis of 11 additional polarities and ages.*

Reversals of the Earth's Magnetic Field, Allan Cox, Richard Doell, and G. Brent Dalrymple, Science, v. 144, p. 1537–1543, 1964. *Early time scale paper announcing discovery of subchrons on the basis of 64 polarities and potassium-argon ages; polarity chrons and subchrons named.*

Geomagnetic Reversals, Allan Cox, Science, v. 163, p. 237–245, 1969. *Reversal time scale based on 150 polarities and potassium-argon ages; first proposal that the timing of reversals is random.*

The Road to Jaramillo, William Glen, Stanford University Press, Stanford, Calif., 459 pp., 1982. *History of the discovery of the reversal time scale.*

Current Papers on Magnetic Stratigraphy

A Geologic Time Scale, W. Harland, A. Cox, P. Llewellyn, C. Pickton, A. Smith, and R. Walters, Cambridge University Press, Cambridge, 131 pp., 1982. *Magnetic time scale used for Figure 8-9.*

Paleogene Geochronology and Chronostratigraphy, W. Berggren, D. Kent, and J. Flynn, *In* N. Snelling, Ed., Geochronology and the Geological Record, Geological Society of London, Special Paper, in press. *Magnetic time scale for Paleogene.*

Neogene Geochronology and Chronostratigraphy, W. Berggren, D. Kent, and J. Flynn, *In* N. Snelling, Ed., Geochronology and the Geological Record, Geological Society of London, Special Paper, in press. *Magnetic time scale for Neogene.*

Classic Papers on Magnetic Stripes

Magnetic Anomalies over Oceanic Ridges, Fred Vine and Drummond Matthews, Nature, v. 199, p. 947–949, 1963. *First published explanation of the origin of magnetic stripes.*

Spreading of the Ocean Floor: New Evidence, Fred Vine, Science, v. 154, p. 1405–1415, 1966. *Explains origin of marine magnetic anomalies and calculates rates of seafloor spreading.*

Marine Magnetic Anomalies, Geomagnetic Field Reversals, and Motions of the Ocean Floor and Continents, James Heirtzler, G. Dickson, Ellen Herron, Walter Pitman III, and Xavier Le Pichon, Journal of Geophysical Research, v. 73, p. 2119–2136, 1968. *Reversal time scale from 76 Ma to present as determined from marine magnetic anomalies; set the standard for subsequent research.*

9

Paleomagnetic Poles

In earlier chapters, data from the seafloor provided the basis for determining the displacements of both oceanic plates and continental plates. We saw that the motion of the mainly continental Africa and South America plates could be completely determined using plate tectonic data from the South Atlantic ocean floor, without requiring observations from the continents. We will now consider a way to use independent data from the continents in order to test the validity of this and similar reconstructions based on ocean-floor data.

Paleomagnetism provides the basis for this independent test. In essence, the paleomagnetism of a rock provides a north arrow inscribed on the rock. As a plate carrying the rock rotates about some Euler pole, the north vector embedded magnetically in the rock rotates to a new direction that is no longer north. For example, in Permian time when South America and Africa were part of the same plate, the magnetism acquired by rocks as they formed on both continents all pointed toward the same magnetic north pole. After the opening of the Atlantic Ocean, we would expect the Permian paleomagnetic vectors on the two continents to point to two different north poles and, in fact, they do. When the two continents are restored to the positions they had in the Permian by closing the Atlantic about the correct Euler pole, then the Permian paleomagnetic vectors will again point toward the same north pole.

In addition to providing a set of north-seeking vectors, the paleomagnetic imprint carried by the rock tells us the latitude at which the rock formed. In effect, paleomagnetism provides plates with a set of inscribed latitude circles, each neatly labeled with its paleolatitude. If a pair of plates has been correctly reconstructed, the latitude circles will match across the boundary between the plates. The double requirement that circles of latitude and magnetic north arrows must both match after plates have been restored provides a powerful constraint on plate reconstructions.

To appreciate the power of this constraint, imagine a hypothetical planet visited every 20 my or so by astronauts who paint red north-south longitude lines and blue latitude circles on all of the rocks forming at the time of their visit. These colored grid lines would serve the same function as the figures painted on pieces of a jigsaw puzzle: only if the pieces are assembled correctly does a consistent overall pattern emerge. The paleomagnetic imprints carried by plates provides such a pattern in the form of a grid of magnetic latitude and longitude lines.

Obtaining Geographic Coordinates from Paleomagnetic Data

Magnetic Latitude and Colatitude

The distance from a point of observation to the north magnetic pole is the **magnetic colatitude** θ. The **magnetic equator** is the great circle midway between the two magnetic poles, where the magnetic colatitude $\theta = 90°$. The magnetic equator is only one of a set of **magnetic latitude circles** that are concentric around the north and south magnetic poles, just as the geographic equator is but one of a set of geographic latitude circles that are concentric around the north and south geographic poles. The **magnetic latitude** λ is the angular distance of a point of observation from the magnetic equator in the direction of the north magnetic pole. The magnetic latitude λ is related to the magnetic colatitude θ by $\lambda = 90° - \theta$. For a dipole field, at all points along the magnetic equator where $\lambda = 0$, $\mathbf{F}$ is horizontal and the magnetic inclination is zero. At all points along each magnetic latitude circle the magnetic inclination is constant. The relation between the magnetic colatitude θ and magnetic inclination I is the fundamental equation of paleomagnetism:

$$I = \tan^{-1}(2 \cot \theta) \qquad (9.1)$$

The relation between the magnetic latitude λ and magnetic inclination I is

$$I = \tan^{-1}(2\tan\lambda) \qquad (9.2)$$

In the northern magnetic hemisphere where λ is positive, the magnetic inclination I is also positive and **F** is inclined below the horizontal (Figure 9-1). In the southern hemisphere where λ and I are both negative, **F** is inclined above the horizontal.

Now let's try an experiment to see if you can determine latitude from inclination. Draw a circle on a piece of paper and place a bar magnet with dipole moment **M** under the paper at the center of the circle in some unknown orientation. Given the inclination at just one point on the circle, can you find the orientation of the magnet? Yes, you can. The key is to invert the previous equations to give latitude and colatitude as functions of inclination:

$$\theta = \cot^{-1}[(\tan I)/2] \qquad (9.3)$$

$$\lambda = \tan^{-1}[(\tan I)/2] \qquad (9.4)$$

The north magnetic pole is located at an angular distance of θ from the point on the circle where the field was measured. Standing at this point on the circle, you have two options as to which way to go to the pole, a long route and a short route. The component of the magnetic field vector **F** tangent to the circle points toward the north magnetic pole via the shortest route and therefore tells you which way to go.

Dipole Field Observed on the Surface of a Sphere

In an experiment in Chapter 8, a bar magnet **M** placed on a flat piece of paper gave us a two-dimensional cross section of the dipole field (Figure 8-3). We now want to move out of the plane and consider the shape of the field of a dipole viewed in three dimensions. To do this, let's place the dipole **M** at the center of a sphere and take a walk with compass in hand over the surface of the sphere. Any plane passing through the axis of **M** passes through the north and south magnetic poles and intersects the surface of the sphere along a great circle. Let's stand at some point of observation **P** located on this great circle and therefore in the plane through **M**. The following three vectors lie within this plane: the position vector **P** from the center of the sphere to a point of obser-

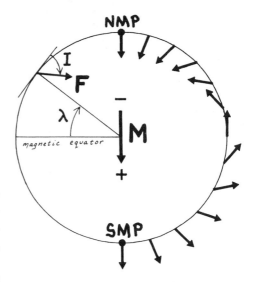

Figure 9-1.
Magnetic dipole **M** at the center of circle produces fields **F** shown along the circle. I: magnetic inclination or plunge beneath horizontal. λ: magnetic latitude. **NMP**: north magnetic pole. **SMP**: south magnetic pole.

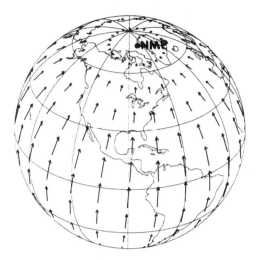

Figure 9-2.
Geomagnetic dipole field. Arrows show horizontal components of magnetic fields, **F**, which lie along great circles converging on the north magnetic pole, **NMP**.

vation on the circle; the orientation of the magnetic field **F** produced by **M**; and of course the vector **M** itself. The declination of **F**, which is the azimuth of the component of **F** tangent to the sphere, is given by the local azimuth of the great circle upon which we are standing. This relationship can be generalized by noting that at any point **P** on the surface of the sphere, a plane can be found within which lie the position vector **P** and vector **M**. At that point the declination is the local azimuth of the great circle from **P** to the north magnetic pole.

Just as a set of planes through the geographic pole defines a set of longitude circles directed north-south, a set of planes through the magnetic dipole defines a set of **magnetic circles of longitude** that pass through the magnetic pole. A set of **F** vectors measured along one of these great circles and projected onto the surface of the sphere has horizontal components directed along the trend of the great circle. If measurements of **F** are available over the surface of the sphere or of a large plate, the projections of **F** onto the surface of the sphere comprise a set of arrows all pointing toward the north magnetic pole (Figure 9-2). These can be used to create a grid of longitude lines through the pole.

An important difference between a grid of magnetic latitude circles and magnetic longitude circles arises because of the symmetry of the dipole field. It is possible to give a unique numerical label to all magnetic latitude circles because the magnetic inclination determines a unique magnetic latitude and therefore a unique distance from the magnetic equator. However because the field of a dipole is symmetrical about the axis of the dipole, the field vectors along any circle of longitude can be superimposed upon those along any other circle by rotation about the dipole axis. In short, all magnetic latitude circles are different but all magnetic longitude circles are the same. As a consequence, paleomagnetic data are able to determine latitude uniquely but not longitude.

Secular Variation

The dipole field that best fits the earth's magnetic field is inclined 11.5° from the rotation axis. The coordinates of the corresponding north magnetic pole are at 78.5° north latitude and 290° east longitude. This so-called **inclined dipole field** accounts for about 80 percent of the observed field.

The remaining 20 percent is an irregularly shaped field called the **nondipole field**.

The earth's magnetic field fluctuates both in direction and magnetic intensity. At some localities the orientation of the geomagnetic vector has changed by 20° in a century, at others less than 2°. Secular variation is produced by very slow fluctuations in the strength of the axial dipole field, by wobble of the inclined dipole about the earth's rotation axis, and by variations in the nondipole field. Components of the field contributing to secular variation are shown in Figure 9-3. These fluctuations, which compose **geomagnetic secular variation**, have a broad spectrum with energy present over the range of periods from 10 to 10^4 yrs.

Of particular importance in tectonics is the field produced by a magnetic dipole at the earth's center aligned with the earth's rotation axis. This is called an **axial dipole field**. On a planet with an axial dipole field, magnetic north is geographic north and the magnetic poles, equator, and latitudes all correspond exactly to the geographic poles, equator, and latitudes. This simplifies plate tectonics: when a plate shifts northward, this is recorded precisely as a shift in magnetic latitude.

Is the earth such a planet? Clearly not. At the present time magnetic north is different from geographic north and the geomagnetic poles are different from the geographic poles. The earth's field is clearly not that of an axial dipole today nor has it likely ever been at any instant in the past. However, over long periods of time the magnetic field wobbles about the true north direction. We know this from paleomagnetic studies. Similarly, the instantaneous inclination varies about a mean inclination which is close to that of an axial dipole. So in contrast with the **instantaneous** magnetic field, the **mean** magnetic field appears to be essentially that of an axial dipole. So the mean magnetic pole and latitude circles of planet Earth are like those of the ideal planet with an axial dipole field.

There are two reasons why the time-averaged field is much more regular and symmetrical than the instantaneous field. The first is that the inclined geomagnetic dipole wobbles symmetrically about the earth's rotation axis. The result is a time-averaged dipole field that is axial. The second is that the nondipole component of the field is transient. When contoured, it resembles a weather map with its changing patterns of highs and lows that drift over the earth's surface. When summed over a long period of time, most (but prob-

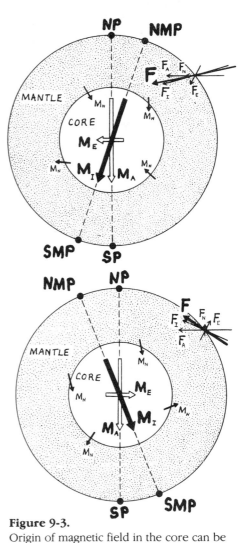

Figure 9-3.
Origin of magnetic field in the core can be represented by the following magnetic moments. M_A: axial dipole moment. M_E: equatorial dipole moment. M_I: inclined dipole moment, the vector sum of M_A and M_E. M_N: nondipole sources, which can be modeled by small dipoles in the outer part of the fluid core. At right are shown the three components of the field, **F**, produced at a site on the equator by the three sources in the core. Two snapshots of the core are shown taken 10^4 years apart. Time average of all components sum to zero except axial dipole component.

ably not all) of the nondipole field cancels. Its net effect on the long-term mean value of the field is at most several degrees.

In summary, if paleomagnetic data are averaged from rock strata with ages that span the full range of periods present in secular variation, about 10^4 yrs, all field components will be canceled by vectorial averaging except those of the axial dipole field. The bottom line for the paleomagnetist is that secular variation superimposes lots of noise on the axial dipole field but very little bias. Were it not for this property, paleomagnetism would be of little use in plate tectonics.

Nuts and Bolts of Paleomagnetism

Having reviewed the theoretical side of geomagnetism and paleomagnetism, we are now ready to look at the practical question of how this knowledge can be used to test a tectonic hypothesis. We will do this by developing a case history from the viewpoint of a geologist using paleomagnetism for the first time.

Has Spain Rotated?

Imagine that you were the first geologist to notice that the geology of Spain and Portugal (the Iberian peninsula) matches up better with that of France if you close up the Bay of Biscay. This can be done by rotating Iberia clockwise by 35° about a local Euler pole. The timing of the rotation appears to be in the range 150 to 100 Ma. You're naturally excited about this idea and write a short article for *Nature,* your first publication in an international journal. To your delight, the article is promptly accepted.

One would expect that when the article is published, you would be acclaimed for having recognized an important relationship that others had missed. To your surprise and chagrin, geologists familiar with the geology of southern Europe tell you that they had always been aware of the possibility that the Iberian peninsula might have rotated. However they concluded long ago that if one looks at all of the relevant geologic data, there's not much of a case for rotation. Their response, although polite, clearly indicates that they think you've gone off the deep end.

So you are on the spot. As a young scientist, you need to maintain your credibility, especially now that you've come under fire. Since your critics clearly don't believe your interpretation of existing data, you need to gather some convincing new data that shows that Spain has rotated. With your career on the line, you turn to paleomagnetism. How do you get started?

Experimental Strategy

You first need to plan your attack. Your basic strategy is simple: collect samples in Spain and France and compare their paleomagnetic declinations. If your theory is right and Spain has rotated 35° relative to France, then the declination of the rocks from Spain will be 35° less than the declination of rocks from France. Several experimental considerations are important. (1) Rocks with the same age from Spain and France are needed for the test. (2) In determining an unrotated reference declination from France to compare with that from Spain, your task is simplified by the fact that France is unrotated relative to Europe. Therefore data from Europe can be combined with data from France. (3) The age of the rocks studied paleomagnetically in Spain and Europe should be greater than the age when the rotation began. (4) Enough rocks with good paleomagnetic memories must be available to permit averaging out scatter in directions due to secular variation and local tectonic disturbances.

Selection of Formations to be Sampled

Your next step is to look for formations of the right age that are suitable for paleomagnetic analysis. Paleomagnetists have learned that certain rock types usually yield useful paleomagnetic results whereas others generally prove to be a waste of time. From conversations with a few old hands you come up with the following helpful advice about sampling.

Volcanics   Extrusive rocks are superb magnetic recorders provided they were still hot when they were extruded or deposited. If they underwent jostling or other deformation after cooling below their Curie temperatures, the directions were probably scrambled. For example, lava flows and hot ash flows are excellent magnetic recorders, whereas cold ash falls are not.

The time required for most volcanic formations to cool is so short that the geomagnetic field doesn't change measurably during cooling. Therefore the paleomagnetic vector obtained from one volcanic cooling unit yields an instantaneous paleofield measurement. Moreover because of geomagnetic secular variation, the paleofield measured from one cooling unit may differ in direction by several tens of degrees from the expected axial dipole field at the sampling site. If you were to sample only one lava flow in Spain and one in France, both flows being Quaternary in age, their declinations might well differ by 30° even in the absence of any rotation simply because of secular variation. Fortunately, as noted earlier, the nonaxial dipole components of the field can be canceled by averaging paleomagnetic results from several dozen lava flows of slightly different ages.

Sediments Depositional remanent magnetization (DRM) is acquired by and preserved in sediments only under rather special circumstances. As noted in Chapter 8, grain size is important: DRM is acquired only by sediment grains that are silt size or smaller. Moreover the accuracy with which sediments record a paleofield direction depends upon the nature of the depositional process. In laboratory experiments, sediments deposited under quiet conditions in which individual grains settle gently to the bottom are found to acquire a paleomagnetic direction somewhat shallower than that of the ambient magnetic field. The size of this inclination error may be as great as 10°. On the other hand, sediments deposited under more turbulent conditions (for example, by pouring a slurry into a beaker) generally do not have an inclination error. In ancient sediments, it is often difficult to know whether inclination errors are present or absent. Therefore caution is advisable in interpreting paleolatitudes determined from sediments—an undetected inclination error of 10° might lead you to the conclusion that a sampling site had moved 1000 km northward when, in fact, it had remained stationary.

Like other types of remanent magnetization, DRM is destroyed if the magnetic minerals in the sediment are chemically altered and replaced by other minerals. Chemical alteration may occur near the surface during weathering or at depth by reaction of the magnetic minerals with circulating water. Sediments with high permeability to fluid flow are especially vulnerable to postdepositional chemical alteration. Sediments adjacent to plutons that have established large-scale groundwater convection systems during cooling commonly are overprinted magnetically by minerals formed in the sediments by circulating fluids.

Sediments offer the paleomagnetist the major advantage of providing close sampling in time of the geomagnetic field.

This is because sedimentation, unlike volcanism, is a nearly continuous process. Consider, for example, a stratigraphic section 100 meters thick that was deposited at a rate of 1 cm/1000 yrs. Because a paleomagnetic sample 2 cm long was deposited over an interval of 2000 yrs, each paleomagnetic vector averages the field over a time of 2000 yrs. The time required for deposition of the 100 meters of the section is 10 my, more than enough time to average out secular variation and record numerous geomagnetic reversals. Sediments almost always provide enough coverage in time to permit removal by averaging of nonaxial dipole components of the geomagnetic field.

Red beds A large fraction of the paleomagnetic results cited in the literature are from red beds, a special class of sediments. In a few red beds the remanence is DRM acquired by hematite grains as they were deposited. In most red beds, the magnetization is chemical in origin and resides in the hematite that gives the beds their red color. The hematite forms in an arid environment shortly after the beds are deposited. The age of the magnetization then is only slightly younger than the age of the rock. Other red beds become magnetized long after deposition.

From a practical viewpoint, red beds are attractive to the paleomagnetist for three reasons. (1) Hematite (Fe_2O_3) is a good carrier of magnetic memory. When grown in the earth's field, hematite becomes strongly magnetized and this magnetization tends to be stable over long periods of time. (2) Hematite is chemically stable in the oxidizing environments in which most rocks spend the longest part of their lives. (3) The magnetic remanence of red beds, whether depositional or chemical in origin, is generally acquired over a long interval of time. Therefore, sampling red beds usually averages out the effects of geomagnetic secular variation. The main experimental difficulty in working with red beds is that fossils are rarely preserved in the arid environment in which red beds form, so that the age of these deposits is commonly uncertain. An additional uncertainty stems from the difficulty of determining whether red beds became magnetized during, shortly after, or long after deposition.

Limestones Fine-grained pelagic limestone often carries a very weak but stable depositional remanent magnetization. These deposits offer the paleomagnetist the advantage of commonly being fossiliferous and therefore accurately datable. The main problem in working with limestones is that they are vulnerable to recrystallization, which destroys their DRM.

Intrusives   Plutons are best regarded as rock formations of last resort. This is because it is rarely possible to determine how much a pluton has been tilted subsequent to emplacement. You can reduce the probability that undetected tilts are present by demonstrating the internal consistency of paleomagnetic results over a broad sampling area. However you will never be as confident in results from intrusives as in results from bedded deposits which carry their own paleohorizontal surfaces.

Having been given these tips by a veteran paleomagnetist, your challenge is to apply them to the problem in which you're interested. You quickly discover from the literature that there are many geologic formations of volcanic rocks and red beds in Spain and France with ages of around 250 Ma. You decide to concentrate your sampling in these formations. Armed with local topographic and geologic maps and, if you're lucky, with the assistance of a geologist who knows the area, you're ready to head for the field.

Collecting Samples

Samples collected for paleomagnetic analysis must be oriented at the time they're collected so that at the end of the study the orientations of the paleomagnetic vectors can be specified relative to geographic coordinates. Most paleomagnetists collect oriented samples using a water-cooled diamond drill driven by a gasoline engine. The drill bit is a metal tube several centimeters in diameter with diamond grit mounted in the cutting edge. As the drill bit rotates, it cuts into the rock face like a rotating cookie cutter, leaving a specimen of rock in the form of a cylinder still attached to the rock outcrop. After the drill has been removed, the cylindrical specimen is oriented by scribing an orientation line parallel to the axis of the cylinder along the top of the cylinder. The orientation of this line in geographic coordinates is then recorded by specifying its azimuth or declination relative to geographic north and its plunge or inclination below the horizontal. Six to eight oriented specimens spread over a lateral distance of several hundred meters are usually collected from each stratum, the purpose of multiple sampling being to assess the internal consistency of the magnetization and to reduce by averaging local scatter in directions of magnetization. If a lava flow or sedimentary layer is tilted, the direction and amount of tilt is carefully recorded at the outcrop for later use in making a **tectonic correction** to the paleomagnetic vector.

Measurement and Magnetic Cleaning

Back in the paleomagnetic laboratory, after cutting the cylindrical cores into specimens about 2 cm long you place your specimens into a magnetometer which measures the paleomagnetic vector of the specimen. Like most modern magnetometers, yours is interfaced with a small computer which queries you for the direction of the azimuth and plunge of the line you inscribed on the cylinder. After a few minutes your magnetometer prints out three numbers which describe the magnetic moment **m** of your specimen: the inclination I and declination D which describe the orientation of **m**, and the scalar number m, which gives the length of the vector. Recall that magnetic moment m is the analog of mass in gravity. The analog of density (mass per unit volume) is the magnetization per unit volume $J = m/V$, where V is the volume of the specimen.

Your next step is magnetic cleaning—the stripping away of magnetic overprints or secondary components of magnetization that may have overprinted the primary magnetization acquired when your rocks originally formed. Lightning bolts that struck near some of your volcanic samples, for example, may have overprinted their original thermoremanent magnetization. In some of your red beds, deep burial and heating may have left a thermal overprint on an initial DRM. Magnetic cleaning is sometimes successful in removing both types of overprint.

In **thermal cleaning**, the specimen is heated to a temperature of 100°C, then cooled in a corner of the lab where the earth's field has been carefully cancelled, and finally measured in the magnetometer to determine its **partial thermal remanent magnetization** (PTRM). This process is repeated at heating steps of about 50°.

AC cleaning or **AC demagnetization** also uses a space where the magnetic field has been cancelled. The specimen is placed in a coil of wire carrying current which produces an alternating magnetic field. The alternating magnetic field is brought to some peak value and smoothly ramped down to zero, after which the magnetization of the specimen is measured. The process is then repeated using a higher peak magnetic field.

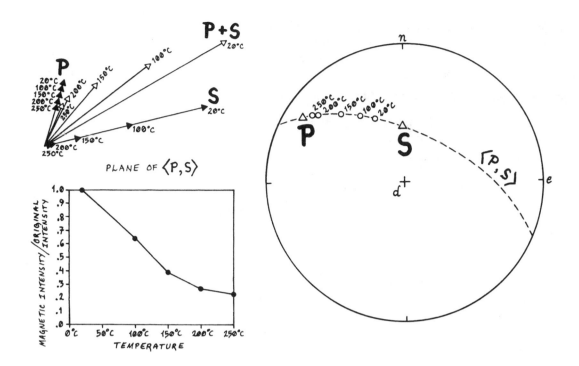

Figure 9-4.

Thermal cleaning. Primary component **P** retains most of its magnetism whereas secondary component **S** is almost completely destroyed on heating to 250°. Magnetic intensity decreases rapidly as secondary component is being removed. The resultant primary + secondary magnetization changes orientation on heating, moving closer and closer to the primary component.

In both types of cleaning, at a given temperature or peak field, the "softer" remanent magnetization—that which is locked into the rock least tightly—is erased by randomizing the magnetic moments of the grains bearing it. This is done by agitating the magnetic moments thermally or magnetically. If two components of magnetization are present, a magnetically hard primary thermal magnetization, for example, and a soft secondary component due to lightning, the secondary component will begin to be destroyed at lower temperatures and smaller alternating fields than will the primary component.

One of your volcanic samples from Spain (Figure 9-4) was close to the location of an old lightning strike. As a result, the secondary component due to the lightning is much stronger than the primary thermoremanent magnetization. Therefore the initial direction of magnetization, which is the vector sum of the two components, is close to that of the secondary component (Figure 9-4). During successive cleaning steps, the secondary component is destroyed more rapidly than is the primary (Figure 9-4). As a result, the direction you measure swings around toward the direction of the primary component. Since only two components are present in your sample, the vector you measure remains in the plane containing the primary and secondary components of magnetization. You can see this happen when you use a projection to plot the vectors after successive demagnetization steps (Figure 9-4): the plane containing the primary and secondary components cuts the unit sphere along the great circle passing through the unit vectors parallel to the primary and secondary components (Figure 9-4). In most of your magnetic cleaning experiments you notice that the magnetic vector is moving along a great circle path on a projection. From this it's safe to conclude that two components of magnetization are present and that one is being destroyed more rapidly than the other. When the secondary component has been effectively destroyed by cleaning, the measured direction of magnetization will stop changing direction on successive demagnetization steps. This indicates that you have succeeded in recovering the primary direction.

Magnetic cleaning is usually effective in removing secondary viscous magnetization and secondary magnetization due to lightning. Thermal cleaning is usually effective in removing secondary magnetization due to post-natal heating events and is sometimes effective in removing secondary magnetization due to a limited amount of chemical alteration.

Statistical Analysis

After you have obtained a number of magnetically cleaned vectors from one site, your next step will be to calculate a site average. The most straightforward way to do this might seem to be to calculate the average value of the declinations and inclinations separately. However, if you do this you will find that you sometimes get absurd answers. Consider, for example, two vectors with I and D values of (60°, 0°) and (89.98°, 180°). Intuition suggests that the mean vector should lie in the plane of the two vectors midway between them. Since the second vector points almost straight down, by intuition the mean should be approximately at (75°, 0°). However, averaging the inclinations and declinations gives a mean vector with I and D of (74.99°, 90°), which is not in the plane of the two vectors. If the second vector is represented by the equivalent expression (89.98°, −180°), the algebraic mean of the values of I and D becomes (74.99°, −90°). Box 9-1 describes an alternate averaging method called **Fisher analysis** that gives more sensible averages. It was invented by the great statistician, Sir Ronald Fisher, explicitly for use

Box 9-1.   Finding the Mean of a Set of Magnetic Moments or Poles.

The following is the standard procedure used in paleomagnetism to find the mean direction of a set of vectors. This analysis was devised by Sir Ronald Fisher, and is respectfully called **Fisher analysis**. The important idea in Fisher analysis is to give each vector equal weight by averaging the unit vectors $\hat{\mathbf{V}}_i$ in the direction of each vector $\mathbf{V}_i$.

The only difference between Fisher analysis in local coordinates and Fisher analysis in global coordinates is the name of each component. If the vectors are slip vectors, magnetic moments, or other **directional vectors** they are described by **local** spherical coordinates (I, D). If the vectors are virtual geomagnetic poles (VGPs) or other **positional vectors** they are described by **global** spherical coordinates (λ, ϕ). The following equations are written in local (n, e, d) coordinates. To rewrite the equations in global coordinates, substitute λ for I, ϕ for D, x for n, y for e, and z for d.

1. Find the Cartesian components (n_i, e_i, d_i) of the unit vector $\hat{\mathbf{V}}_i = (I_i, D_i)$.

$$n_i = \cos I_i \cos D_i \,(\text{northward component})$$
$$e_i = \cos I_i \sin D_i \,(\text{eastward component})$$
$$d_i = \sin I_i \,(\text{downward component})$$

2. Find the vector sum $\mathbf{R} = (R_n, R_e, R_d) = \Sigma \hat{\mathbf{V}}_i$ in Cartesian coordinates:

$$\mathbf{R}_n = \Sigma n_i \,;\, \mathbf{R}_e = \Sigma e_i \,;\, \mathbf{R}_d = \Sigma d_i$$

The best estimate of the mean direction (I_R, D_R) is simply the inclination and declination of $\mathbf{R}$, which are found by converting back from Cartesian to spherical coordinates.

$$I_R = \tan^{-1}\left(\frac{R_d}{\sqrt{R_n{}^2 + R_e{}^2}}\right) \,;\, -90° \leq I_R \leq 90°$$

$$D_R = \tan^{-1}\left[\frac{R_e}{R_n}\right] \,;\, -180° \leq D_R \leq 180°$$

3. The best estimate of precision parameter K is

$$K = \frac{N-1}{N-R}$$

where N is the number of vectors being averaged, and R is the length of $\mathbf{R}$.

4. The angular standard deviation S is

$$S = \frac{81°}{\sqrt{K}}$$

Both K and S provide measures of the angular dispersion of a population of vectors. As the number N of samples collected at a given site is increased, the value of K and S will tend to remain nearly constant.

(continued)

Box 9-1. *(continued)*

5. The 95 percent cone of confidence about the mean vector is usually approximated by

$$\alpha_{95} \approx \frac{140°}{\sqrt{K\,N}}$$

A slightly more accurate and somewhat more complicated equation is

$$\alpha_{95} = \cos^{-1}\left[\, 1 - \frac{N-R}{R}\,(20^{1/(N-1)} - 1)\,\right]$$

The value of α^{95} usually drops as the number of sites is increased.

Programming these equations for a pocket calculator or larger computer is very straightforward. (See Box 3-9 for an efficient way to convert between spherical and Cartesian coordinates using the polar-rectangular function on most pocket calculators.) The following example can be used to check your program. For the following vectors:

(70°, 10°), (88°, 190°), (75°, 350°), (85°, 22°), (80°, 349°), (89°, 159°), (82°, 19°), (74°, 358°)

the analysis yields:

$$R_n = 1.19938\;;\; R_e = 0.04985\;;\; R_d = 7.83739$$
$$N = 8\;;\; R = 7.92879$$
$$I_R = 81.29°\;;\; D_R = 2.38°$$
$$K = 98.30\;;\; S = 8.2°\;;\; \alpha_{95} = 5.61°$$

in paleomagnetism and is now widely used to average all sorts of vectors, including the geographic position of points on a globe.

In Fisher analysis (Box 9-1), all vectors are given unit weight so that they contribute equally to the final result. In addition to giving a reasonable average value of I and D in the case of magnetic directions, or of λ and ϕ in the case of magnetic poles, Fisher analysis also yields several other useful statistical parameters. These include the **precision parameter** K, which is a measure of how tightly grouped the vectors are, and the 95 percent **circle of confidence** or **cone of confidence** α_{95}. The significance of α_{95} can be visualized as follows. Imagine the unit vector describing the mean magnetic moment or the mean paleomagnetic pole with its tail at the center of a unit sphere. Around this vector is a circular cone with its apex at the center of the sphere and with its axis along the mean vector. The size of the opening of the cone, as measured by the angle at its apex, is $2\alpha_{95}$. The intersection of the cone with the surface of the unit sphere is a small circle with a radius of α_{95}. Fisher analysis tells us that with a probability of 0.95, the true mean vector lies inside this cone or, equivalently, within the small circle on the unit sphere.

Tectonic Corrections

If the stratum you sampled has been tilted tectonically, you will need to make what is called a **tectonic correction**. For example, if an initially horizontal bed now dips 60° toward the east, the bed can be restored to its original horizontal position by rotating it 60° about a north-south horizontal axis so that the bed is again horizontal. As the bed is rotated conceptually, the magnetic vector is rotated with it using techniques described in Chapters 3 and 7. In the literature, the expression **stratigraphic coordinates** is sometimes used to describe magnetic vectors after tectonic correction and the term **geographic coordinates** to describe the vectors before tectonic correction.

If you are a structural geologist, you will realize that rotating a bed about a horizontal strike line will restore the bed to its original position only if the axis of the fold that produced the tilt is horizontal. For folds with plunging axes, a more complex sequence of finite rotations is required to bring the bed back to its original orientation and to tectonically correct the paleomagnetic vector.

Having obtained mean values of *I* and *D* at a dozen or so sites in Iberia, let's now suppose that you plot them without tectonic correction and find that the vectors are highly scattered. After making the tectonic corrections you find that the paleomagnetic vectors are tightly grouped (Figure 9-5). What new information does this give you? It tells you that the magnetization was acquired prior to the tectonic event that produced the tilt. This is very important because it rules out the possibility of post-tectonic overprinting.

On the other hand, if you find that the paleomagnetic vectors are tightly grouped prior to unfolding and scattered afterwards, then you know that the magnetization was acquired after the tectonic event that folded the rocks. Interpretation of data in this way is termed a **fold test**. Fold tests provide the paleomagnetist with a powerful tool for assessing whether magnetic overprinting has occurred and for setting limits to the time when the paleomagnetism of the rock was acquired.

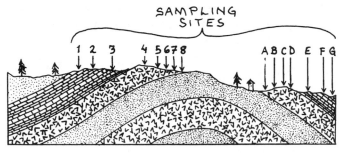

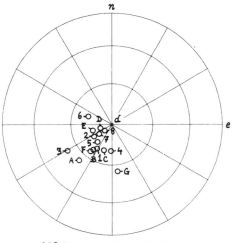

Figure 9-5.
Fold test. Correcting for the folding of this anticline brings the initially scattered paleomagnetic directions into a tight group. Therefore magnetization was acquired before folding.

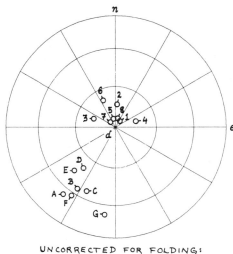

UNCORRECTED FOR FOLDING:
GEOGRAPHIC COORDINATES

CORRECTED FOR FOLDING:
STRATIGRAPHIC COORDINATES

Virtual Geomagnetic Poles and Paleomagnetic Poles

Having completed the analysis of your paleomagnetic data from Spain, you now want to compare your result with the paleofield expected from an unrotated Spain. The difference between the observed and expected declination will be the amount of rotation undergone by Spain. Let's suppose that paleomagnetic data from France are rather sparse. Therefore you want to include in your calculation of a reference paleofield all of the reliable data reported in the literature for sites scattered throughout Europe. Can you simply calculate the mean of all of the available paleofield vectors? This won't work because across a region as large as Europe the latitude and inclination vary by several tens of degrees. Before averaging, you need to convert inclination and declination into a pair of parameters that do not vary with latitude. Here's how to do it.

1. For the mean values of I and D at each site, calculate the pole of the axial dipole field which would produce the observed paleomagnetic direction (Box 9-2).

2. Regarding each of these poles as a unit vector, find their Fisher mean and α_{95} cone of confidence using Box 9-1 with λ in place of I and ϕ in place of D.

3. Finally, use the mean paleomagnetic pole to calculate the expected paleofield direction at some central sampling site in Spain (Box 9-2).

You will find that no matter how good the data are from Europe and Spain, the poles for individual sites will be scattered about the mean pole with an angular standard deviation that typically ranges from 5° to 20°. What could be its cause? Let's look a little more closely at what we're doing when we calculate paleomagnetic poles. A given paleomagnetic direction can be described by two equivalent pairs of numbers. One pair is (I, D), the inclination and declination of the paleomagnetic vector. The other is $(\lambda, \phi)_M$, the latitude and longitude of the north magnetic pole of the geocentric dipole **M** that would produce the observed values of I and D at the sampling site if the geomagnetic field were entirely dipolar. (I, D) and $(\lambda, \phi)_M$ can both be used to describe all possible positions of a vector on a unit sphere or, equivalently, all possible orientations of a unit vector in space. The values of I and λ both range from $-90°$ to $+90°$. The values of D and ϕ both range from 0° to 360°. Box 9-2 shows how to map any point on the (I, D) unit sphere onto an equivalent point on the $(\lambda, \phi)_M$ unit sphere. The mapping function depends upon the coordinates $(\lambda, \phi)_S$ of the site where the field vector (I, D) is measured or calculated.

If the measurement of (I, D) is based on the sampling of a single lava flow or some other spot reading of the geomagnetic field, the assumption that the field was dipolar is not justified because we know that a nondipole field component is always present. However it may still be useful to map field directions into poles as a preliminary to averaging the poles. The name **virtual geomagnetic pole** or VGP is given to such a pole, the word "virtual" signaling that a pole was calculated even though a nondipole component was probably present. The scatter in your VGPs from Spain was due to presence of the nondipole field, wobble of the dipole field about the earth's rotation axis, imperfect tectonic corrections, and other experimental errors. The name **paleomagnetic pole** is used to refer to an average VGP and

Box 9-2.   Finding Paleomagnetic Poles from Paleomagnetic Directions and Vice Versa.

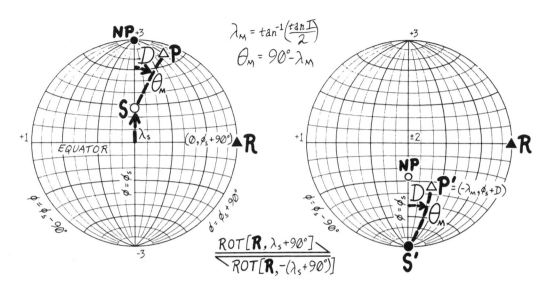

$$\lambda_M = \tan^{-1}\left(\frac{\tan I}{2}\right)$$
$$\theta_M = 90° - \lambda_M$$

$$\underrightarrow{ROT[\mathbf{R}, \lambda_s + 90°]}$$
$$\overleftarrow{ROT[\mathbf{R}, -(\lambda_s + 90°)]}$$

Mapping a Paleomagnetic Direction (I, D) at a Site (λ_S, ϕ_S) to a Paleomagnetic Pole (λ_P, ϕ_P)

The paleomagnetic pole **P** lies at a distance θ_M from the sampling site **S** along the great circle having a local azimuth D, where $\theta_M = 90° - \lambda_M = 90° - \tan^{-1}[(\tan I)/2]$. Following the logic of Box 3-8, start by plotting **S** on the axis of the projection labeled "ϕ_S". Plot a rotation axis **R** on the right half-circle of the projection, which is the meridian $\phi_S + 90°$. Next, rotate **S** about **R** to the position **S'** at the south pole of the projection:

$$\mathbf{S'} = ROT [\mathbf{R}, (\lambda_S + 90°)] \mathbf{S}$$

In practice you can skip this step by simply plotting **S'** as the south pole. Now plot **P'** at projection coordinates ($-\lambda_M$, $\phi_S + D$), which puts it on a great circle with the desired values of D and θ_M. A single rotation about **R** brings the site **S'** back to **S** and pole **P'** to position **P**:

$$\mathbf{S} = ROT[\mathbf{R}, -(\lambda_S + 90°)] \mathbf{S'}$$
$$\mathbf{P} = ROT[\mathbf{R}, -(\lambda_S + 90°)] \mathbf{P'}$$
$$= ROT[(0°, \phi_S + 90°), -(\lambda_S + 90°)] (-\lambda_M, \phi_S + D)$$
$$= (\lambda_P, \phi_P)$$

Mapping a Paleomagnetic Pole (λ_P, ϕ_P) at a site (λ_s, ϕ_s) to a Paleomagnetic Direction (I, D)

To invert the previous procedure, plot **S** and **P** as shown on the left. Again a rotation about **R** brings **S** to **S'** at the South Pole and **P** to **P'**:

(continued)

Box 9-2. *(continued)*

$$\begin{aligned}
\mathbf{P}' &= \text{ROT } [\mathbf{R}, \lambda_S + 90°] \; \mathbf{P} \\
&= \text{ROT } [(0°, \phi_S + 90°), \lambda_S + 90°] \, (\lambda_P, \phi_P) \\
&= (\lambda_{P'}, \phi_{P'})
\end{aligned}$$

The figure on the right shows that the values of I and D are given by the coordinates $(\lambda_{P'}, \phi_{P'})$ of $\mathbf{P}'$:

$$\lambda_M = 90° - \theta_M = -\lambda_{P'}$$
$$I = \tan^{-1}(2 \tan \lambda_M) = \tan^{-1}(2 \tan - \lambda_{P'})$$
$$D = \phi_{P'} - \phi_S$$

---

usually carries the implication that sampling has been adequate to remove the effects of geomagnetic secular variation.

Confidence Limits

How many individual virtual geomagnetic poles must you average to obtain a reasonably small α_{95} circle for your mean paleomagnetic pole for Spain? This depends upon the amount of scatter between site VGPs, as described by the precision parameter K. If the scatter is due entirely to the secular variation, then values of K can be expected to vary from 40 at the equator to 16 at the pole. The confidence circle for VGPs from N sites is given by

$$\alpha_{95} = \frac{140°}{\sqrt{K N}} \tag{9.5}$$

For a site where $K = 36$, the following numbers of sites give the following values for the confidence circles:

$$
\begin{aligned}
N &= 2 & \alpha_{95} &= 16.5° \\
N &= 4 & \alpha_{95} &= 11.7° \\
N &= 9 & \alpha_{95} &= 7.8° \\
N &= 16 & \alpha_{95} &= 5.8° \\
N &= 25 & \alpha_{95} &= 4.7° \\
N &= 49 & \alpha_{95} &= 3.3°
\end{aligned}
$$

These values will be somewhat larger if, as is usually true, the rocks have not been perfect magnetic recorders or if tectonic corrections were imperfect.

With a fairly modest amount of sampling, you can achieve paleomagnetic poles with an accuracy of 5° or better. However to achieve this accuracy it is important that the rocks sampled at your N different sites all have formed at distinctly different times. If you were to sample 49 different lava flows in a given region that had all been erupted in the same year or decade, your virtual geomagnetic poles would be tightly grouped and you might be lulled into believing that you had obtained a highly accurate result. In actuality your pole is based on little more information than if you had sampled only one flow. Assuming $N = 49$ gives you a value of α_{95} of 3.3° whereas a more realistic value would be about 20°. On the other hand, if the strata formed at intervals of 10,000 yrs or more, then your paleomagnetic vectors will be statistically independent and your value of α_{95} realistic.

Vindication

Now back to Spain. Two years have passed since your article appeared in *Nature* and your idea came under fire. In your first field season you conducted reconnaissance sampling of many different formations in Spain and Portugal and found which ones were most promising. Most of these turned out to be Permian in age. In your next field season you sampled the hell out of those formations. After six months of lab work, you now have a beautiful set of paleomagnetic poles which you can show are reliable. So it's time to write another article for *Nature*.

Your first step is to analyze your paleomagnetic data in a way that brings information to bear on the rotation problem as effectively as possible. The best way to do this is to first combine all your data from Permian rocks to obtain a mean Permian paleomagnetic pole for the Iberian peninsula, P_{SP}. This can then be compared with the mean Permian paleomagnetic pole from Europe P_{EU}. The results are as follows.

	λ	ϕ	α_{95}
Spain	45°	212°	10°
Europe	45°	160°	4°

It comes as something of a shock that the result of two years of work can be expressed as three numbers!

Box 9-3. How to Use Paleomagnetic Data to Determine the Amount of Rotation of a Terrane Relative to a Reference Continent.

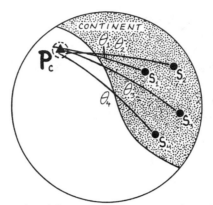

1. Use paleomagnetic data from rocks of the same age at sites on the reference continent to determine a paleomagnetic pole for each site (Box 9-2).

2. Average these poles (Box 9-1) to obtain a mean paleomagnetic pole (P_C) for the continent.

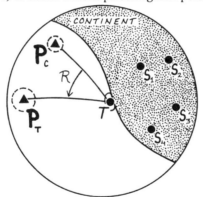

3. Find a paleomagnetic pole (P_T) from rocks on the terrane of the same age as those on the continent.

4. If the terrane has not rotated relative to the continent, the measured paleomagnetic declination of the terrane will be that expected at the sampling site if the paleofield was produced by pole P_C for the continent.

   If the terrane has rotated, the amount of rotation R is equal to the difference between the expected and observed paleomagnetic declinations. R is also equal to the angle between the two great circles from the terrane to the two paleomagnetic poles P_C and P_T.

Before calculating how much rotation is implicit in these data, let's think for a moment about the relationship between a paleomagnetic pole and the plate carrying the rocks from which the pole was determined. Recall the graphical technique used for finding a paleomagnetic pole by drawing a great circle from the sampling site to the pole: The local azimuth of the great circle is determined by the paleomagnetic declination; the length of the great circle from the site to the pole, by definition the paleocolatitude θ, is determined by the paleomagnetic inclination. Because the paleomagnetic declination and inclination are frozen into the rock, the great circle between site and pole can be regarded as a rigid rod attached to and moving with the plate (Figure 9-6). If the plate rotates about a local Euler pole through the sample site, then the paleomagnetic pole will move around a small circle of radius θ centered on the sampling site. If a plate undergoes a finite rotation about any Euler pole, then the paleomagnetic pole will undergo the same finite rotation about the same Euler pole.

You are finally in a position to determine paleomagnetically how much Spain has rotated since the Permian. Your poles and their confidence circles are plotted in Figure 9-7. Deducing the amount of rotation from these paleomagnetic results is straightforward (Box 9-3). First use Box 9-2 to obtain a mean declination D_{obs} at some central point in Spain from your paleomagnetic pole for Spain P_{SP}. Next use Box 9-2 to obtain the declination D_{ex} expected at that same central point from the paleomagnetic pole for Europe P_{EU} in the absence of any rotation. The amount of counterclockwise rotation R is given by

$$R = D_{ex} - D_{obs} = D_{EU} - D_{SP} \qquad (9.6)$$

If α_{SP} and α_{EU} are the confidence limits for the paleomagnetic poles and θ_{SP} and θ_{EU} the corresponding distances from the poles to the central point in Spain, then the corresponding confidence limits for the two declinations are

$$\Delta D_{ex} = \frac{1}{\sqrt{2}} \sin^{-1}\left(\frac{\sin \alpha_{ex}}{\sin \theta_{ex}}\right) \qquad (9.7)$$

$$\Delta D_{obs} = \frac{1}{\sqrt{2}} \sin^{-1}\left(\frac{\sin \alpha_{obs}}{\sin \theta_{obs}}\right) \qquad (9.8)$$

where, as before, the observed declination is that from Spain and the expected that from Europe. These equations describe

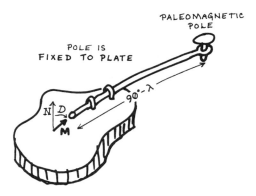

Figure 9-6.
A paleomagnetic pole moves with a plate as if the pole were attached to the plate with a rigid rod.

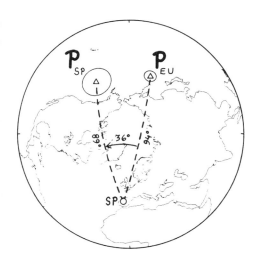

Figure 9-7.
Paleomagnetic demonstration that Spain (SP) has rotated counterclockwise by an angle R of 36° with respect to Europe. $\mathbf{P}_{EU}$: Permian paleomagnetic pole and α_{95} circle based on data from northern Europe. $\mathbf{P}_{SP}$: Permian paleomagnetic pole based on data from Spain. Colatitudes of Spain (89°) and Europe (94°) are not significantly different.

how, for a fixed value of α, the uncertainty in D becomes larger for sites closer to the pole. The confidence limit for the rotation is

$$\Delta R = \sqrt{\Delta D_{ex}^2 + \Delta D_{obs}^2} \qquad (9.9)$$

For a central sampling site in Spain with coordinates (40°, 355°), the values of these quantities are as follows.

$$D_{ex} = D_{EU} = 10.6°$$
$$D_{obs} = D_{SP} = -25.2°$$
$$R = D_{ex} - D_{obs} = 10.6° - (-25.2°) = 35.8°$$

Since R is the amount of counterclockwise rotation, the $+$ sign indicates that the rotation is counterclockwise, as predicted by your hypothesis. The 95 percent confidence limits for this rotation are found as follows.

$$\Delta D_{ex} = 2.8°$$
$$\Delta D_{obs} = 7.1°$$
$$\Delta R = \sqrt{2.8^2 + 7.1^2} = 7.6°$$

The bottom line after two years of paleomagnetic work is that

$$R = 36° \pm 8°$$

Triumph is yours! In your *Nature* article you are able to conclude with due modesty that your original hypothesis has been confirmed by a new and completely independent line of evidence. (Spain *has*, in fact, rotated. The data used in the preceding narrative are approximately correct.)

Polar Wander and Plate Motion

Put yourself in the shoes of a geophysics student who has just read about Wegener's theory of continental drift. The year is 1954 and the text used in your class is *Principles of Geology* by Giluly, Waters, and Woodford, the most advanced text at that time. Your are fascinated by Wegener's idea that Eurasia, Africa, India, Australia, and Antarctica were once part of a supercontinent, Pangaea, which began to break up at the end of the Jurassic, approximately 160 Ma. Your textbook describes the pros and cons of Wegener's hypothesis. Wegener showed that many large-scale geological features such as tectonic belts and glaciated regions are abruptly cut off at today's coastlines, yet when the continents are reassembled in their pre-Jurassic position, these features form continuous patterns like those of an assembled jigsaw puzzle. Being pretty well convinced by this, you are surprised to find that the authors of your text conclude with the following statement about continental drift: "Though the theory is a brilliant tour-de-force, its support does not seem substantial."

You tell your geology professor that you don't agree with this conclusion and ask him what he thinks about continental drift. "A lot of arm waving," he replies. "Ever since Wegener published *The Origin of Continents and Oceans* in 1915, geologists have been arguing about it. I can give you some fascinating references. The bottom line is that the theory *does* seem to explain the observations described in your text and a few others, besides. But neither Wegener nor anyone after him has been able to explain *how* continents move through the mantle. What's the driving force? Most of us have ended up regarding continental drift as one more flaky tectonic idea that *could* be right but probably isn't. We'll remain skeptical until someone comes up with an experiment or observation that can prove or disprove it."

You read some papers your professor recommends and find yourself more and more convinced that Wegener was right. Then something happens that is to change the course of your career. While reading about continental drift, you happen to hear about paleomagnetism from one of your friends who is building a paleomagnetic lab to study geomagnetic reversals. Suddenly a light dawns. Why not use paleomagnetism to test Wegener's theory?

Your first task is to decide which continent to work on. You pick India for three reasons. First, India moved a long ways if continental drift is correct. Second, you learn that paleomagnetic poles are being determined for Eurasia, so you will have a basis for comparison if you can get some poles from India. The clincher is that your university has a special fund for students who want to study or do research in India.

Your second task is to find a faculty sponsor. This proves difficult. Most of your professors think your project would be a waste of time and money. You argue that it's more exciting to try to prove an idea that everyone thinks is wrong than to confirm an idea that everyone agrees is right. Finally a maverick faculty member agrees with you and decides to back your project, even though it looks like a long shot.

Your general research strategy is obvious. Wegener had proposed that India drifted northward away from Africa and Antarctica some time after the Jurassic. By sampling post-Jurassic rocks that formed while India was moving, you should be able to chart the course of India's displacement.

You have beginner's luck. Indian geologists help you find suitable rocks in the field and with the help of your paleomagnetist friend, the work in the laboratory goes well. Within a year you have a beautiful set of paleomagnetic data from India. Now comes the interesting part. What conceptual framework can you use to test Wegener's hypothesis?

Using Paleomagnetic Poles to Validate Plate Reconstructions

Your first step is obviously to combine all of your data from India for a given age. To do this you calculate a virtual geomagnetic pole for each of your sites, then combine these poles using Fisher analysis to obtain a mean Indian paleomagnetic pole. To assess the motion of India relative to Eurasia, you will need to compare the Indian pole with a reference pole for Eurasia. Fortunately the first reliable paleomagnetic poles for Eurasia are just being published as you complete your lab work late in 1954. So you have paleomagnetic poles and confidence circles for rocks of several different ages from India and Eurasia. You are ready to test Wegener's hypothesis. If Wegener was wrong and India was always part of Eurasia, then the confidence circles of the Indian paleomagnetic pole and the Eurasian paleomagnetic pole will coincide. You make this test and find that all Indian paleomagnetic poles older than middle Tertiary are significantly different from the poles of the same age for Eurasia.

So your paleomagnetic result proves Wegener's hypothesis. Or does it? Does the observation that the Indian and Eurasian poles are displaced from each other necessarily prove that Wegener's hypothesis is correct? Generalizing this question, does the coincidence of poles from two plates prove that the plates have not moved relative to each other? Conversely, does discordance of two poles prove displacement? In short, what constraints do paleomagnetic poles place on the relative displacements of two plates?

Answers to these questions will be obvious when you recall (Figure 9-6) that a paleomagnetic pole moves as if it were attached with a rigid rod to the plate from which it was derived. If a small plate undergoes rotation about an Euler pole located within the boundary of the plate, the paleomagnetic pole will sweep out a circle centered on the Euler pole while the plate undergoes negligible translation. As a result, the latitude of the paleomagnetic pole may change by a large angle while the latitude of the plate remains nearly constant. If the Euler pole and the paleomagnetic pole happen to coincide, the paleomagnetic pole will remain fixed while the plate undergoes a large translation. If both the plate and the paleomagnetic pole are distant from the Euler pole, both will move along circles centered on the Euler pole and both will undergo large changes in latitude.

Armed with this intuitive way of thinking about paleomagnetic and Euler poles, you will probably now regard as trivial the earlier questions and those which follow.

Question. If Triassic poles for India and Eurasia are displaced, does this prove that India has undergone a post-Triassic change in latitude relative to Eurasia?

Answer. No. Displacement of the paleomagnetic pole could have been produced by rotation about a local Euler pole without a large shift in latitude.

Question. If the Triassic paleomagnetic poles of India and Eurasia are coincident, does this demonstrate that India has not moved relative to Eurasia since the Triassic?

Answer. Yes and no. Theoretically the two paleomagnetic poles would still be coincident even if India had undergone a large post-Triassic rotation about an Euler pole coincident with the Triassic paleomagnetic pole. However from a practical viewpoint, such a coincidence is unlikely. Most paleomagnetists would interpret this result as demonstrating that India had not moved relative to Eurasia.

Question. Assume that you are in possession of just two pieces of data: the Triassic pole, $\mathbf{P}_{AF}$, for Africa when Africa and India were joined, and the post-Triassic rotation of India relative to Africa, $^{200}_{AF}\text{ROT}^{0}_{IN}$. What are the expected present coordinates of the Triassic pole, $\mathbf{P}_{IN}$, for India?

Answer. As India rotates away from Africa, $\mathbf{P}_{AF}$ rotates with it. Therefore $\mathbf{P}_{IN} = {}^{200}_{AF}\text{ROT}^{0}_{IN} \times \mathbf{P}_{AF}$.

Question. Assume that you know the Triassic paleomagnetic poles, $\mathbf{P}_{IN}$ and $\mathbf{P}_{EU}$, for India and Eurasia. What can you say about the movement of India relative to Eurasia during the Triassic?

Answer. The movement is equal to the difference between two paleocolatitudes. The first is the paleocolatitude, θ_{EU}, expected if India has remained fixed to Eurasia. This is the distance from a central site in India to $\mathbf{P}_{EU}$. The second is the mean paleocolatitude, θ_{IN}, determined from the paleomagnetism of rocks in India. This is the distance from the same central site in India to $\mathbf{P}_{IN}$. The difference between these two paleocolatitudes,

$$d = \theta_{IN} - \theta_{EU} \qquad (9.10)$$

is a measure of the distance India has moved toward or away from $\mathbf{P}_{EU}$. Positive values of d indicate that India has moved in a direction toward $\mathbf{P}_{EU}$. If the 95 percent confidence circles of the two poles are α_{IN} and α_{EU}, then the 95 percent confidence interval for d is given by

$$\Delta d = \frac{1}{\sqrt{2}} \sqrt{\alpha_{EU}^2 + \alpha_{IN}^2} \qquad (9.11)$$

Question. Given the Triassic paleomagnetic poles, $\mathbf{P}_{IN}$ and $\mathbf{P}_{EU}$, for India and Eurasia, can you make a map showing the position of India relative to Eurasia during the Triassic?

Answer. Recall that paleomagnetic poles can be thought of as being attached by rigid great circles to their plates. Start by drawing Eurasia and its Triassic pole, $\mathbf{P}_{EU}$, on a globe. Next draw India and its Triassic pole, $\mathbf{P}_{IN}$, on a clear shell over the globe. This shell ensures that India's pole moves with India. Move the shell until $\mathbf{P}_{IN}$ coincides with $\mathbf{P}_{EU}$. The new locations of India and its paleomagnetic pole now satisfy the condition that India and Eurasia had the same pole during the Triassic. Is your reconstruction unique? Clearly not. Pin the shell to the globe through $\mathbf{P}_{IN}$ and $\mathbf{P}_{EU}$ and rotate the shell about this pivot point. The two Triassic paleomagnetic poles will remain coincident while India moves to a family of different positions, all coincident with the paleomagnetic poles.

This thought experiment brings out the nature of the non-uniqueness inherent in plate reconstructions based solely on paleomagnetic data: they are able to determine paleolatitude but not paleolongitude. The same limitation was inherent to your analysis of the two paleocolatitudes: their difference gave you the amount of poleward motion, which is defined as motion along a great circle from the sampling site to the paleomagnetic pole for Eurasia. However, the analysis told you nothing about the amount of motion transverse to that circle.

Enough thought experiments! You've spent a hard year in the field and lab acquiring some reliable paleomagnetic data. Now it's time for the fun part: analyzing the data to determine whether Wegener was right about India. You weren't able to get any results you trust from rocks that formed in the Triassic prior to the beginning of continental drift. Your best pole positions are from rocks that are about 60 million years old, a time, according to Wegener, when India was en route between Antarctica and Eurasia. Your results are listed below and shown in Figure 9-8.

(*a*)

	λ	φ	α$_{95}$
India	35°	280°	6°
Eurasia	70°	150°	6°

The first line gives the coordinates of your 60 Ma pole for India and its 95 percent confidence circle. The second line gives a pole from rocks of the same age from Eurasia.

Your first approach is to analyze the paleocolatitudes. As measured by the distance from a site in central India (18°, 78°) to the two poles, these are as follows.

$$\theta_{EU} = 67°$$
$$\theta_{IN} = 123°$$

Since the distance to the 60 Ma pole seen from India ($\mathbf{P}_{IN}$) is greater than the distance to the pole ($\mathbf{P}_{EU}$) expected if India had been part of Eurasia at 60 Ma, India clearly has moved toward the Eurasian pole since 60 Ma. The amount of poleward displacement is given by

$$d = 123° - 67° = 56°$$

The 95 percent confidence interval for this poleward displacement is

$$\Delta d = \frac{1}{\sqrt{2}} \sqrt{\alpha_{EU}^2 + \alpha_{IN}^2} = 6°$$

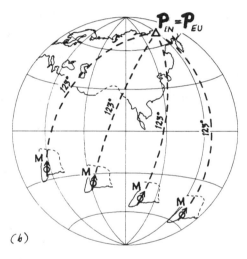

(*b*)

Figure 9-8.
(*a*) Paleomagnetic poles of India $\mathbf{P}_{IN}$ (far hemisphere) and Eurasia $\mathbf{P}_{EU}$ (near hemisphere) for the early Tertiary (60–65 Ma). (*b*) Possible early Tertiary positions of India relative to Eurasia obtained by superimposing $\mathbf{P}_{IN}$ on $\mathbf{P}_{EU}$ while keeping India and $\mathbf{P}_{IN}$ rigidly linked at an angular distance of 123°.

Box 9-4. How to Use Paleomagnetic Data to Determine Possible Positions of a Terrane or Continent Relative to a Reference Continent.

1. Use paleomagnetic data from rocks of the same age at sites on the reference continent to determine a paleomagnetic pole for each site (Box 9-2).

2. Average these poles (Box 9-1) to obtain a mean paleomagnetic pole (P_C) for the continent (see first figure of Box 9-3).

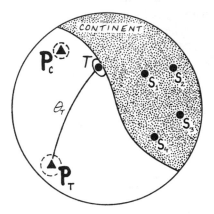

3. Find a paleomagnetic pole (P_T) from rocks on the terrane of the same age as those on the continent.

4. In this figure the paleocolatitude of the terrane θ_T is much larger than the expected paleocolatitude, which is the distance from the terrane to P_C. The difference between the colatitudes is the amount of poleward motion.

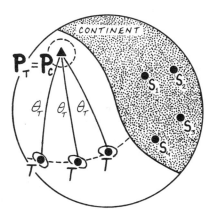

5. To find possible positions of the terrane at the time corresponding to the age of the paleomagnetic poles, superimpose P_T upon P_C keeping the great circle segment θ_T attached to the terrane as if with a rigid rod. Different positions of the terrane consistent with the paleomagnetic data can be found by rotating the terrane about the paleomagnetic poles.

Your colatitude analysis clearly demonstrates that India moved more than 6,000 kilometers toward the Eurasian pole (and by inference, toward Eurasia) during the past 60 Ma.

Let's turn now to the geometrical analysis required to produce a map showing possible positions of India at 60 Ma. Recall that you need to superimpose the Indian pole upon the Eurasian pole while keeping India and its pole rigidly attached to each other. Your first step is to find a finite rotation that will bring $\mathbf{P}_{IN}$ into coincidence with $\mathbf{P}_{EU}$. One way to do this is by trial and error using a projection. Another is to take the cross product $\mathbf{C}$ of the unit vectors corresponding to the two poles. The required rotation is about the direction of $\mathbf{C}$ by the angle $\sin^{-1} C$, where C is the scalar length of $\mathbf{C}$. Numerically the direction is $(-13.3°, 199.5°)$ and the rotation is 69°. Having rotated India and $\mathbf{P}_{IN}$ by this amount, additional rotations by arbitrary angles of India about $\mathbf{P}_{EU}$ will leave the two poles coincident and will move India to new positions consistent with both poles. The final product of this process is shown in Figure 9-8. You're exultant! Your intuition was correct and your textbook was wrong. After 40 years of debate, your experiment has finally settled the controversy about continental drift.

Displaced Terranes

Although India is usually described as a microcontinent that has drifted, it can also be described as a terrane which is clearly out of place because of the obvious mismatch between rocks of similar age on India and Eurasia. **Terranes** are defined and recognized by two criteria. The first is that the geologic history recorded in the rocks of the terrane must be markedly different from that recorded in adjacent rocks, so different, in fact, as to indicate that the terrane was at some distance from the adjacent rocks when the rocks were being formed. An example would be the juxtaposition of Permian glacial tillite on a terrane located next to rocks containing Permian fossils that formed in tropical waters. India satisfies this criterion. The second criterion is that the contacts between the terrane and the continent must be major faults. The Himalayan suture between India and Eurasia is a classic example of such a fault, although many terranes are bounded by faults much less conspicuous than the Himalayas. Some are so inconspicuous, in fact, that geologists are still undecided as to whether certain regions formed as for-

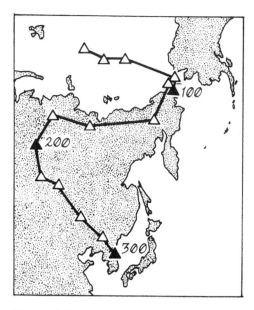

Figure 9-9.
Apparent polar wander (APW) path for North America. Each pole is a Fisher vector sum of the paleomagnetic poles based on rocks with ages in a 30 my window centered on the age shown.

eign terranes or as indigenous parts of the continent. Geologists call these regions **suspect terranes**.

Paleomagnetism has played an important role in validating the suspicions of geologists about suspect terranes. It can be used to measure rotations and translations using the same techniques you used in Spain and India. A classic case history is that of the Wrangellia terrane, made up of similar stratigraphic sequences of rocks in the Wrangell Mountains in Alaska, Queen Charlotte and Vancouver Islands in coastal Canada, and at the bottom of Hell's Canyon in Idaho. Paleomagnetism has shown that all parts of this terrane formed near the equator during the Triassic.

Paleolatitude plots are also useful for checking the amount of displacement across major transform faults like the San Andreas. The amount of paleolatitude mismatch across the suspected fault is a direct measure of the amount of displacement. A mismatch of 5° corresponds to 555 km of displacement, which is close to the lower limit of displacement detectable paleomagnetically, given the usual uncertainty in paleomagnetic data.

Although the term **microplate** is sometimes used to describe crustal blocks that have rotated, many of these are clearly not microplates in the sense of being bounded by ridges, transforms, and trenches that cut through the entire lithosphere. In some cases the tectonic domains that have rotated as a unit appear to be only 10 km long or less, suggesting that structural blocks in the upper part of the crust are being rotated in response to a broad shear zone at greater depth. Rotations on this scale, although extremely interesting to the structural geologist, make it difficult to relate paleomagnetic results to the large-scale plate tectonic picture.

Apparent Polar Wander Paths

The apparent polar wander (APW) path for a plate is a time sequence of paleomagnetic poles. Each of these poles is an average of the poles derived from rocks of a given age sampled at sites over the surface of the plate. Each pole represents the location of the geomagnetic (and presumably geographic) pole as seen by an observer on the plate at the time of formation of the rocks used to find the pole. The APW path for North America is shown in Figure 9-9, and lists of paleomagnetic poles that can be used to construct APW paths for several plates are given in Table 9-1.

Table 9-1. Mean paleomagnetic poles. Each pole is a Fisher vector sum of the paleomagnetic poles based on rocks with ages within a 30-my window centered on the age shown.

t (Ma)	λ	φ	α$_{95}$	t (Ma)	λ	φ	α$_{95}$
NORTH AMERICA (NA)				SOUTH AMERICA (SA)			
20	88	121	3	20	85	178	3
40	84	168	4	40	?	?	
60	81	188	5	60	88	153	17
80	68	191	7	80	88	153	17
100	67	184	10	100	89	225	12
120	68	189	19	120	86	222	9
140	66	163	16	140	81	210	15
160	70	123	19	160	88	140	13
180	71	94	8	180	79	84	14
200	63	92	4	200	74	91	11
220	56	100	5	220	83	78	5
240	55	108	6	240	86	77	6
260	48	118	4	260	82	124	9
280	43	126	4	280	68	159	12
300	39	129	3	300	58	172	9
EURASIA (EU)				AUSTRALIA (AU)			
20	86	196	3	20	77	267	12
40	78	159	6	40	69	329	11
60	75	153	6	60	70	306	?
80	74	145	24	80	57	327	17
100	74	165	14	100	56	330	12
120	77	170	7	120	50	338	?
140	71	145	14	140	?	?	
160	66	123	8	160	47	1	12
180	62	132	31	180	47	0	6
200	49	149	30	200	42	354	14
220	49	148	4	220	?	?	
240	48	153	4	240	?	?	
260	42	167	3	260	45	307	16
280	41	167	3	280	48	316	10
300	37	169	5	300	49	320	12

(continued)

Table 9-1.*(continued)*

t (Ma)	λ	φ	α$_{95}$	*t* (Ma)	λ	φ	α$_{95}$
AFRICA (AF)							
20	84	164	6				
40	80	95	23				
60	?	?					
80	62	226	9				
100	63	234	10				
120	62	258	15				
140	45	248	?				
160	57	259	14				
180	68	249	6				
200	67	242	7				
220	68	255	9				
240	62	265	8				
260	45	233	15				
280	36	244	9				
300	?	?					

On a boring planet with stationary plates, all APW "paths" would, in fact, be fixed points coincident with the geographic pole. On our own interesting planet with its moving plates, APW paths are curves that record the motion of plates relative to the earth's rotation axis. If you stand at one spot on a plate, the pole appears to move along the APW path. If you were able to hover above the geographic and magnetic pole, the plate would appear to move.

One might expect plates to move more or less randomly relative to the earth's rotation axis. However APW paths don't look at all like random walks. A typical APW path is made up of a sequence of smooth, slightly curved segments called **tracks** which approximate small circles. Where these segments join, the local azimuth of the APW path changes markedly and the paths are tightly curved. These sharp turning-points are called **cusps**.

Why do APW paths have this characteristic form? The answer involves all of the three dynamic components of the earth: the fluid core, where the magnetic field originates; the mantle, which undergoes slow deformation; and the rigid lithosphere. The mantle and the core appear to be coupled by viscous and electromagnetic forces. As a consequence, the angular velocity vectors of the core and the mantle are parallel to each other and these rotation axes define, in effect, the earth's geographic poles. Since the axis of symmetry of the geomagnetic dynamo is parallel to the rotation axis of the core, the average position of the geomagnetic pole is coincident with the geographic pole. A paleomagnetic pole path thus marks the motion of a plate not only relative to the magnetic pole but also to the geographic pole and to the earth's mantle.

Problems

9-1. From the direction (I_{obs}, D_{obs}) of the observed remanent magnetic moment $\mathbf{m}_{obs}$ at the following sites, find the paleomagnetic colatitude θ_m and calculate the location of the expected paleomagnetic pole $\mathbf{P}_{ex}$ assuming that the magnetic field is axially dipolar. (Solutions to problems a-l are trivial and should require only a moment's reflection. Solutions to problems m-t are fairly trivial and require a pocket calculator but not a projection. Only the last four problems require a projection or computer.)

	Site λ_S	ϕ_S	$\mathbf{m}_{obs}$ I_{obs}	D_{obs}	θ_M	$\mathbf{P}_{ex}$ λ_P	ϕ_P
a.	(90°,	—)	(90°,	—)	___	(___,	___)
b.	(0°,	0°)	(90°,	—)	___	(___,	___)
c.	(45°,	180°)	(90°,	—)	___	(___,	___)
d.	(0°,	0°)	(0°,	0°)	___	(___,	___)
e.	(0°,	180°)	(0°,	0°)	___	(___,	___)
f.	(0°,	180°)	(0°,	180°)	___	(___,	___)
g.	(0°,	90°)	(0°,	90°)	___	(___,	___)
h.	(0°,	270°)	(0°,	270°)	___	(___,	___)
i.	(−40°,	180°)	(0°,	0°)	___	(___,	___)
j.	(40°,	0°)	(0°,	0°)	___	(___,	___)
k.	(40°,	0°)	(0°,	180°)	___	(___,	___)
l.	(−40°,	180°)	(0°,	180°)	___	(___,	___)
m.	(0°,	240°)	(80°,	0°)	___	(___,	___)
n.	(0°,	240°)	(80°,	90°)	___	(___,	___)
o.	(0°,	240°)	(80°,	180°)	___	(___,	___)
p.	(0°,	240°)	(80°,	270°)	___	(___,	___)
q.	(0°,	240°)	(−80°,	0°)	___	(___,	___)
r.	(0°,	240°)	(−80°,	90°)	___	(___,	___)
s.	(0°,	240°)	(−80°,	180°)	___	(___,	___)
t.	(0°,	240°)	(−80°,	270°)	___	(___,	___)
u.	(−40°,	220°)	(−60°,	160°)	___	(___,	___)
v.	(60°,	100°)	(60°,	60°)	___	(___,	___)
w.	(80°,	350°)	(−70°,	240°)	___	(___,	___)
x.	(−60°,	100°)	(−30°,	200°)	___	(___,	___)
y.	(80°,	330°)	(50°,	300°)	___	(___,	___)

9-2. From the observed paleomagnetic poles $\mathbf{P}_{obs}$ find the paleo-colatitude θ_m and the direction (I_{ex}, D_{ex}) of the expected magnetic moment $\mathbf{m}_{ex}$ for the following sites.

	Site		$\mathbf{P}_{obs}$		θ_M	$\mathbf{m}_{ex}$	
	λ_S	ϕ_S	λ_P	ϕ_P		I_{ex}	D_{ex}
a.	(90°,	—)	(90°,	—)	——	(——,	——)
b.	(0°,	0°)	(90°,	—)	——	(——,	——)
c.	(0°,	0°)	(0°,	0°)	——	(——,	——)
d.	(0°,	180°)	(0°,	0°)	——	(——,	——)
e.	(0°,	90°)	(0°,	0°)	——	(——,	——)
f.	(0°,	0°)	(0°,	270°)	——	(——,	——)
g.	(45°,	180°)	(90°,	—)	——	(——,	——)
h.	(−45°,	180°)	(90°,	—)	——	(——,	——)
i.	(−40°,	180°)	(0°,	0°)	——	(——,	——)
j.	(40°,	0°)	(0°,	0°)	——	(——,	——)
k.	(40°,	0°)	(0°,	180°)	——	(——,	——)
l.	(−40°,	180°)	(0°,	180°)	——	(——,	——)
m.	(0°,	0°)	(80°,	0°)	——	(——,	——)
n.	(0°,	0°)	(80°,	90°)	——	(——,	——)
o.	(0°,	0°)	(80°,	180°)	——	(——,	——)
p.	(0°,	0°)	(80°,	270°)	——	(——,	——)
q.	(0°,	0°)	(−80°,	0°)	——	(——,	——)
r.	(0°,	0°)	(−80°,	90°)	——	(——,	——)
s.	(0°,	0°)	(−80°,	180°)	——	(——,	——)
t.	(0°,	0°)	(−80°,	270°)	——	(——,	——)
u.	(−40°,	220°)	(−60°,	160°)	——	(——,	——)
v.	(60°,	100°)	(60°,	60°)	——	(——,	——)
w.	(80°,	350°)	(−70°,	240°)	——	(——,	——)
x.	(−60°,	100°)	(−30°,	200°)	——	(——,	——)
y.	(10°,	30°)	(80°,	330°)	——	(——,	——)

9-3. If you were to sample Triassic (about 220 Ma) red beds at the localities listed below, what normal and reversed paleomagnetic directions would you expect to find in the absence of experimental errors and local tectonic effects?

Locality	λ	ϕ
South Africa	−25°	20°
Alaska	62°	217°
Japan	35°	138°
Australia	−37°	145°

9-4. Fossils with tropical affinities found in the Wrangellia terrane in Alaska suggest that the terrane formed in the tropics. The first paleomagnetist to study this terrane sampled Triassic volcanic rocks at the location listed in Problem 9-3 and found the following paleomagnetic direction: $I = -20°, D = 74°$, $\alpha_{95} = 6°$. Whether the polarity of the field was normal or reversed during magnetization is uncertain. Imagine that you are this paleomagnetist and are writing an article about the origin of Wrangellia. What can you say about (a) the amount of rotation of the terrane, (b) its paleolatitude in the Triassic, and (c) its displacement relative to North America since the Triassic?

9-5. In the Appalachian Mountains you sample sediments of Carboniferous age (approximately 330 Ma) at a locality with coordinates (35° N, 275° E). The bed you sample is known to have been folded in Late Permian time. To provide a fold test you sample at five sites on the limbs of a syncline. Your results before tectonic correction are listed below, together with the strike and dip of the bed sampled. (Recall that the strike is the azimuth of a horizontal line in the plane of the bed and the dip is the inclination of the bed below the horizontal.)

Uncorrected Moment		Orientation of Bed	
I	*D*	**strike**	**dip**
−59°	18°	30°	60° SE
−45°	348°	35°	25° SE
−17°	327°	120°	5° SW
19°	326°	210°	45° NW
30°	335°	205°	60° NW

Perform the following operations, which are the standard ones used in paleomagnetic analysis.

(a) Find the Fisher mean paleomagnetic direction and the values of K and α_{95} of the uncorrected paleomagnetic directions.

(b) Make the tectonic correction for each site and recalculate the Fisher mean and K and α_{95}.

(c) Plot the tectonically uncorrected data and the corrected data on projections. Comparing these plots and your values of K and α_{95}, what can you conclude about the time of magnetization of the formation you sampled?

(d) Use your results to calculate a paleomagnetic pole and compare this with the expected pole for this time from Table 9-1.

(e) What conclusions can you draw about the rotation and translation of the sampling site?

9-6. Use the paleomagnetic poles in Table 9-1 for the Permo-Carboniferous (approximately 280 Ma) to test Wegener's hypothesis that the continental plates Africa, South America, India, Australia, and Antarctica were once joined to form the supercontinent Gondwanaland. Reconstruction poles that are modern updates of the reconstructions proposed by Wegener are listed in Chapter 7. The timing of the breakup of Gondwanaland is implicit in these tables. In your analysis, you will want to show a projection of the available poles for these plates before and after the plates are reassembled to form Gondwanaland. For the latter, state clearly the reference frame used for the reconstruction (for example, "Poles are shown in a reference frame that has remained fixed relative to Africa").

9-7. (Optional). Wegener also proposed that the supercontinent Gondwanaland (Africa, South America, India, Australia, Antarctica) and the supercontinent Laurasia (North America, Greenland, Eurasia) were once joined to form the hypercontinent Pangaea. To test this idea, use the paleomagnetic poles in Table 9-1 for the Jurassic (180 Ma) and in addition the following paleomagnetic pole from India for approximately 190 Ma : 20° N, 310° E. Reconstruction poles that are modern updates of the reconstructions proposed by Wegener are listed in the tables of Chapter 7. As in the previous problem, you will want to show a projection of the available poles for these plates before and after the plates are reassembled to form Pangaea. For the latter, state clearly the reference frame used for the reconstruction (for example, "Poles are shown in a reference frame that has remained fixed relative to Africa").

9-8. Prior to the discovery of magnetic isochrons in the Atlantic basin, the timing of the opening of the Atlantic was less well known than it is today. Imagine that in those days you had reason to suspect that the Atlantic had opened entirely during the Permian. Describe a paleomagnetic experiment that could provide a definitive test of this hypothesis. Discuss carefully the logic behind this experiment. Conclude by using Table 9-1 to provide the data needed to test your hypothesis.

Suggested Readings

Standard Texts

Palaeomagnetism and Plate Tectonics, M. W. McElhinny, Cambridge University Press, Cambridge, 358 pp., 1973.

Introduction to Geomagnetism, W. D. Parkinson, Scottish Academic Press, Edinburgh, 433 pp., 1983.

Palaeomagnetism: Principles and Applications in Geology, Geophysics, and Archeology, Donald Tarling, Chapman and Hall, London, 379 pp., 1983.

Methods in Rock Magnetism and Paleomagnetism: Techniques and Instrumentation, David Collinson, Chapman and Hall, London, 503 pp., 1983.

Articles

Palaeomagnetic and Palaeoclimatological Aspects of Polar Wandering, Geofisica Pura Applicata, v. 33, p. 23–41, 1956. *Classic paper using paleomagnetism to confirm the drift of India and other continents.*

Apparent Polar Wander Paths Carboniferous Through Cenozoic and the Assembly of Gondwana, E. Irving and G. A. Irving, Geophysical Surveys, v. 5, p. 141–188, 1982. *Lists of paleomagnetic poles, APW paths, and discussion of continental reconstructions.*

Renewed Paleomagnetic Study of the Lisbon Volcanics and Implications for the Rotation of the Iberian Peninsula, R. Van der Voo and J. Zijderveld, Journal of Geophysical Research, v. 76, p. 3913–3921, 1971. *Classic use of paleomagnetism to measure rotation.*

Paleoreconstructions of the Continents, Ed. by M. W. McElhinny and D. A. Valencio, Geodynamics Series v. 2, American Geophysical Union, 194 pp., 1981. *Continental reconstructions based in part on paleomagnetic data.*

10

Putting It All Together

The goal of this book was to show you how to do the basic operations of plate tectonics. However it would be amazing (and disappointing) if, at this stage, your curiosity hadn't carried you beyond the techniques of plate tectonics to deeper scientific questions. In this last chapter we will address some of these, beginning with the question of what drives the plates and going on to a few other topics of current interest in plate tectonics.

What Drives the Plates?

Although almost everyone agrees that the process responsible for plate motion is some sort of thermal convection, there are two schools of thought about the nature of the convection. One school holds that the plates ride as passive passengers on mantle convection cells with dimensions of several thousand kilometers. The location and life cycle of these cells is determined by processes in the mantle as a whole and not by the geometry of thin plates riding on the surfaces of the cells. The other school of thought holds that the plates are not passive passengers but are themselves an active part of the convection process. We will consider each hypothesis in turn and then show why most geophysicists favor the second one.

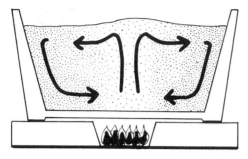

Figure 10-1.
Single convection cell: The kitchen stove experiment.

The general nature of thermal convection itself is well understood. Heat a pan of water over a burner. As the temperature at the bottom rises, a temperature gradient develops and heat flows upward through the layer of water. Initially the physical mechanism of heat transfer is simple thermal conduction like that which occurs in a solid. On further heating, as the water at the bottom of the pan becomes less and less dense, blobs or plumes of hot, light water rise to the top and simultaneously, cold, dense water from the top sinks to the bottom of the pan. This is convection. If the heat source is a single flame at the center of the pan, a plume of hot water will rise at the center and cold water will sink along the sides of the pan, constituting a single convection cell (Figure 10-1). If the pan is shallow and wide and if the bottom is heated uniformly, a regular pattern of convection cells will develop (Figure 10-2). Less regular patterns of convection are commonly observed in nature and in the kitchen.

In some ways convection in the earth's mantle is similar to that in a pan of water and in other ways it is different. Like a pan of water heated over a flame, the mantle is heated from below by heat flowing into it from the earth's core. An additional source of heat in the mantle is the decay of radioactive elements distributed throughout the mantle. A second difference lies in the rheological properties of the two materials. The viscosity of water is low and it is also fairly uniform throughout a pan of water. The viscosity of the mantle is relatively high and it is also highly variable, the viscosity of the lithosphere being much higher than that of the asthenosphere. It is this variability in viscosity that imparts the plate-like character to mantle convection.

Passive Versus Active Plates

Let's first consider the convection model which views mantle convection cells as the prime movers upon which plates ride passively as incidental passengers. In this model plates are analogous to particles of dust carried up and down by convection cells in the earth's atmosphere: They mark the location and velocity of convection cells, but they have virtually no effect on the convection process.

Let's start with an earth in which convection cells carry material from the core-mantle boundary to the surface and back down again. Now add a layer of lithosphere. Along the rising limb of a cell, say the East Pacific Rise or the Mid-Atlantic Ridge, the lithosphere splits and is carried away

Figure 10-2.
Multiple convection cells over distributed heat source.

laterally by mantle flow (Figure 10-3). Along the descending limb of a cell, say along the Japan Trench, the lithosphere is dragged down by mantle currents, causing subduction. This was the basic convection model advanced prior to plate tectonics by leading scientists like Arthur Holmes, David Griggs, and Harry Hess.

The situation is not changed very much if the flow of material through the whole mantle is blocked by density layering, as some geophysicists have suggested. You can model this type of convection by pouring some water and oil into a pan, allowing them to separate into two layers, and then heating the pan with a flame at the center of the pan. A plume of hot water moves up through the center of the water layer and then laterally outward along the oil-water interface. A plume of rising hot oil forms above the rising plume of hot water and moves laterally out along the top of the oil layer. In the mantle, thermal plumes may similarly rise through the entire mantle without the exchange of material between the upper and lower mantle.

In both of these passive models, the location of ridges and trenches is determined by the size and location of deep mantle convection cells. Ridges form along the up-going limbs of convection cells, subduction along descending limbs. In other words, convection cells are causes and ridges and trenches are effects. The distance between adjacent ridges and trenches is determined by the characteristic length scale of the convection cell. This distance would be expected to remain nearly constant during the lifetime of a cell.

Now let's consider a second convection theory in which the plate is an intrinsic part of the convection cell and not a passive passenger. Recall that the lower boundary of a plate is simply an isothermal surface. Three properties distinguish the lithosphere above this isotherm from the asthenosphere beneath: The lithosphere is colder, it is stiffer and more viscous, and it is denser than the asthenosphere. Because of its greater density, the lithosphere tends to sink like the cold, upper layer of any convecting fluid. Subduction occurs not because a slab is pulled down by the dense, sinking limb of a convection cell but because the slab *is* the dense, sinking limb of the cell.

Because the lithosphere is so much more viscous than the asthenosphere, convection in plate tectonics resembles the motion of plates of ice over a pond more than it does convection in a uniform fluid like water. In fact, if ice were denser than water, the analogy would be almost perfect.

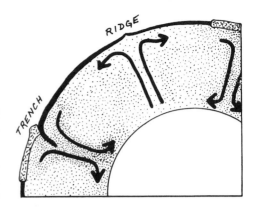

Figure 10-3.
Mantle-wide convection producing ridges and trenches.

Whereas the surface of convecting water undergoes active deformation, plates of lithosphere, like plates of ice, move with almost no deformation. Whereas in ordinary fluids, ascending convection currents originate at great depth, in plate tectonics, ridges are simply cracks between diverging plates filled from local sources of magma in the asthenosphere. As two plates of lithosphere or ice move apart, nearby fluid moves into the crack and solidifies along the edges of the two diverging plates.

First Test: Ridge Offsets   Let's see if we can devise ways to test these two models by comparing their predictions with observed patterns of seafloor spreading. Since the two models make similar predictions about subduction but quite different predictions about the nature of ridges, we will concentrate on ridges.

Let's start by considering the implications of the observation that spreading systems consist of stair-stepped ridges and transforms (Figure 4-11), individual segments of which vary in length from tens of kilometers to thousands of kilometers. If ridges always occur precisely above ascending limbs of convection cells, then the pattern of convection rising from the deep mantle would be the same as the pattern of offset ridges, each ridge segment corresponding to the rising limb of one convection cell. For very long ridge segments, this interpretation is consistent with studies of convection cells made in the laboratory or using computers. These show that the ratio of the width to the length of convection cells, termed the aspect ratio, is usually about one. Hence ridge segments with lengths of the order of a thousand kilometers would be consistent with convection cells with an aspect ratio of one originating deep in the mantle. However ridge segments as short as 10 kilometers, which are not rare on the seafloor, are another matter. It seems highly unlikely that in an intricate stair-step pattern of such short ridge segments, each segment would be caused by the rising limb of an individual convection cell with a very small aspect ratio rising from deep in the mantle.

The second model of convection provides a much simpler explanation of the origin of short ridge segments. When the initial boundary between two diverging plates is not perpendicular to the direction of plate divergence, the boundary becomes a stair-step pattern of transforms and ridges with transforms parallel to the divergence direction. The short offset ridge segments are simply cracks between the diverging plates. They are fed by magma coming not from a set of offset convective currents arising from deep in the mantle but rather from shallow sources in nearby parts of the asthenosphere.

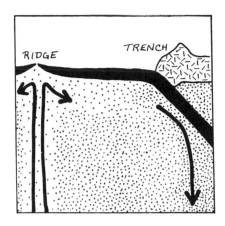

Second Test: Jumping and Propagating Ridges

Ridges sometimes jump to new positions for reasons that are not well understood but which appear to reflect changes in the direction of plate motion. Sometimes the distance jumped is short, sometimes it is long. Often a ridge with a new orientation propagates like a growing crack into a region near a dying ridge, replacing the old ridge. Generally the orientation of the propagating ridge is more nearly perpendicular to a new direction of spreading than was the old ridge.

The first model of convection requires that each ridge jump reflects a change in the pattern of convection: A convection current, rising from deep in the mantle, dies in an old location and a new convection cell forms in the new location. The observation that ridges commonly jump at time intervals of only a a few million years would require a physically implausible jerky mode of convection.

The second model of convection, in which a ridge is simply a crack between two diverging plates into which magma flows, offers an explanation that is physically more plausible. When a ridge jumps, the crack between diverging plates simply jumps to a new position. Propagating ridges are simply growing cracks which, as they propagate, tap nearby magma sources in the asthenosphere.

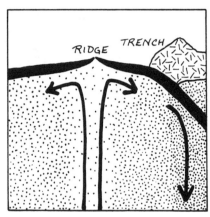

Third Test: Ridge Meets Trench

What happens when a ridge meets a trench? That ridges do, in fact, migrate into trenches is well established by the plate tectonic record. For example, 50 million years ago the Farallon plate was subducting beneath California and the Pacific-Farallon ridge lay on the far side of the Farallon plate about 500 kilometers from the subduction zone. The ocean floor record shows that during the next 20 million years the Pacific-Farallon ridge moved eastward rapidly until it was fairly close to the trench, at which time ocean floor spreading was still going on at a vigorous rate. When the ridge met the trench, seafloor spreading and subduction both stopped. What was going on in the mantle?

For the first model of convection (Figure 10-4), this boils down to the question: What happens when the rising con-

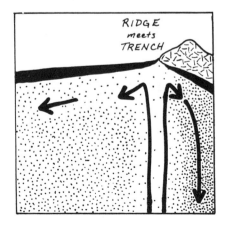

Figure 10-4.
First model: As ridge migrates to trench, hot ascending and cold descending limbs of deep mantle convection current converge.

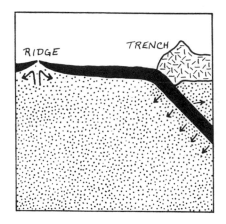

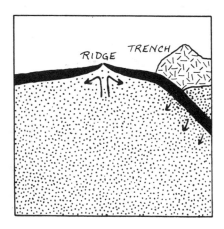

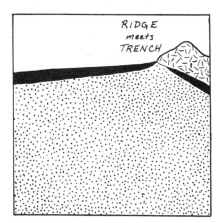

Figure 10-5.
Second model: Plate with ridge at its trailing edge simply sinks into trench.

vection current rising under a ridge meets the convection current descending under a trench? In the early days of plate tectonics, this issue had an air of mystery about it. Rising convection currents were seen as "sources" and subduction zones as "sinks"; when ridges and trenches meet, sources and sinks, being opposites, were thought of as annihilating each other. On a more physical level, as the distance from the ridge to the trench becomes smaller, the convection cell becomes an implausibly narrow one in which a hot current rising from depths of several thousand kilometers in the mantle is located only a few tens of kilometers from a descending cold current. Convection cells like this with extremely small aspect ratios are generally not observed in laboratory or computer experiments.

How does the second model of convection explain the same ocean floor observations? The main actor in this model is not the Pacific-Farallon ridge but rather the Farallon plate, so keep your eye on the plate and not the ridge (Figure 10-5). The Farallon plate is sliding into the trench at the rate of about 50 km/my along its eastern boundary with California. Along its western ridge boundary the plate is growing at a rate of about 25 km/my. As a result, the plate is getting narrower at a rate of about 25 km/my. The motion of the ridge toward the trench and the narrowing of the Farallon plate are simply the result of the Farallon plate sliding into the trench. When the last sliver of Farallon plate slides into the trench, the Pacific plate comes into contact with North America and the ridge and trench cease to exist. The Farallon plate then continues down the subduction zone, where it presumably is heated and absorbed into the asthenosphere. No source-sink paradox is involved, nothing more mysterious than a heavy object sinking.

These are some of the reasons why most geophysicists favor the second, active style of convection as the dominant one in plate tectonics. However these observations do not rule out the possibility of "two scale" convection, one scale being the one in which plates are active elements, the other being smaller-scale convection confined to the asthenosphere. The latter delivers heat to the bottom of plates but does not act as a significant driving force. The jury is still out on whether this second type of convection does, in fact, occur.

Return Flow in the Asthenosphere   What can be said about flow in the asthenosphere? First there is boundary layer flow. As a plate moves over the viscous asthenosphere, a boundary layer of asthenosphere forms next to the plate; within this layer, velocity decreases from that of the plate to that of the asthenosphere beneath the layer. Next let's consider vertical flow. Imagine a world-encircling horizontal surface located at a depth of 100 kilometers just beneath the base of the lithosphere and ask yourself: Where does flow take place across this surface? One place is subduction zones, where large volumes of lithosphere are obviously flowing downward, accompanied by a boundary layer of asthenosphere carried along by viscous drag. A second place is at rises: Asthenosphere must flow upward across the imaginary horizontal surface in the region where plates are growing near rises. For example, the volume of the entire plate that underlies the Atlantic basin has flowed upward from the lower asthenosphere at the Mid-Atlantic Ridge. Third, there is return flow through the asthenosphere, which carries asthenosphere displaced by subducting lithosphere at trenches all over the world to ridges wherever they are located. The resulting flow pattern is not the simple one of well-defined convection cells like those in Figure 10-2 but more like the complex flow in an aquifer which is being pumped at many wells and replenished at many sources.

Driving Forces

Two approaches have been used to analyze mantle convection. The first is to develop computer codes to solve the differential equations that describe convection in a fluid with distributed heat sources and with a rheology that varies as a function of temperature. The second approach, which is more attuned to the intuitive approach of the present book, regards a plate as a mechanical object subject to stresses acting along its surfaces (Figure 10-6). The stresses acting on different parts of the plate are then analyzed separately in order to assess the relative importance of ridge push, slab pull, and other potential driving forces. This approach, though less rigorous than mathematical analyses of the total convection process, nonetheless offers some interesting insights.

Mantle Drag Force F_{DF}   The force (or more properly, the traction) acting on the bottom of a plate, termed the **mantle drag force** or F_{DF}, is due to viscous coupling between the plate and the asthenosphere. The magnitude of the mantle drag force is proportional to the area of the plate and its

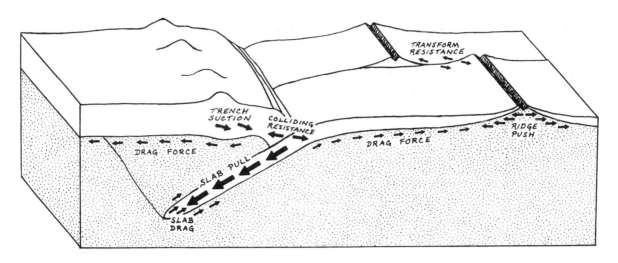

Figure 10-6.
Forces acting on plates.

velocity relative to the asthenosphere. If plates are passive and are driven by deep convection in the mantle, then presumably a given plate moves in the same direction as the asthenosphere beneath it, but at a slower rate, and F_{DF} is the dominant force. On the other hand, if the driving force comes from the plates themselves and the asthenosphere is passive, as we believe to be the case, the drag force F_{DF} is resistive and has a direction opposite to that of the motion of the plate.

An additional force can be expected for the part of a plate occupied by a continent if there is greater drag under continents. This would be expected if there is greater viscosity in the subcontinental asthenosphere or if a root extends beneath continents into the asthenosphere. This **continental drag force**, F_{CD}, which is added to the general mantle drag force F_{DF}, is also proportional to the velocity of the plate.

Ridge Push F_{RP}   The **ridge push force** or F_{RP} is usually described as an effective force per unit length of ridge that pushes plates away from ridges. This is more a metaphor than a physical description because this force does not act on the edge of a plate but is developed throughout the entire body of the plate. Its origin is the gravitational potential energy resulting from the height of ridges over abyssal plains. If ridges were composed of ice, the ice would flow laterally as glaciers. Similarly at ridges the lithosphere tends to slide off the sloping lithosphere-asthenosphere interface. At a given distance from a spreading center, the body force acting on a plate is proportional to the sine of the slope of the ridge at that distance. Thus ridge push forces are generated over the entire width of a rise, being greatest near the crest of the rise and decreasing laterally to zero where the rise meets an abyssal plain.

Slab Pull Force F_{SP} The **slab pull force** F_{SP} arises because subducting slabs are denser than the asthenosphere and therefore sink through it. The expression "pull" does not mean that the lithosphere adjacent to the trench is under tension in an absolute sense, but rather that as the slab sinks, the stress in the adjacent plate is less than the lithostatic stress that would exist in a static horizontal plate due to its own mass. The slab pull force is produced by the density contrast between a slab and adjacent asthenosphere. This density contrast is small at shallow depths because the temperature difference is small between the slab and adjacent mantle; it is greatest at depths of 200 to 300 km; and it dies out at depths of 700 to 1000 km as the slab heats to the temperature of the surrounding mantle. The slab pull force is greater for older than for younger lithosphere but appears to be independent of the velocity of subduction.

Slab Drag Force F_{SD} In addition to F_{SP}, which is independent of the rate of convergence, we can define a second velocity-dependent resistant force, F_{SD}, which is due to viscous drag as the slab descends into the mantle. If, as seems likely, the viscosity of the asthenosphere is much less than that of the underlying mesosphere, F_{SD} would be greatest near the bottom of slabs where they butt against or penetrate into the mesosphere.

Transform Fault Resistance F_{TF} Seismic activity along transform faults shows that friction is present as plates try to slide past each other, producing a resistive force. In most stick-slip models for the generation of earthquakes, faulting occurs when a critical yield stress is reached. This stress and therefore the magnitude of the **transform fault resistance** is independent of the relative velocity of the plates on either side of the transform.

Colliding Resistance F_{CR} Shallow seismic activity along zones of plate convergence indicates that friction between converging plates tends to resist the process of convergence. As was true for transform faults, in most stick-slip models for the generation of earthquakes, faulting occurs when a critical yield stress is reached, suggesting that the **colliding resistance**, F_{CR}, should also be independent of the velocity of convergence.

Suction Force F_{SU} Where ocean floor is being subducted beneath continents, as in South America and Asia, the continental plate appears to be moving in the direction of the trench while the trench itself retreats away from the continent. This suggests the possibility that the continent is being pulled toward the trench by a **suction force** F_{SU}. Several ideas have been advanced concerning the cause of this force. One is that the subducting slab induces an eddy-like flow in the asthenosphere that swirls up beneath the continent, dragging it toward the trench. Another idea follows from the observation that since Circum-Pacific continents are moving closer together, the trenches surrounding the Pacific must be moving toward the center of the basin. As a trench migrates, presumably the subducting slab producing the trench migrates through the asthenosphere broadside. The result is to displace a large volume of asthenosphere, creating a mass deficit on the side of the trench adjacent to the continent. This sucks the continent toward the migrating trench.

Motion Relative to the Mantle

Armed with these definitions, we are now ready to determine which (if any) of the forces is correlated with the observed velocity of plates. If ridge push is the dominant driving force, then plate velocities should be strongly correlated with the lengths of ridge and similarly for slab pull. If transform fault resistance is the dominant resistive force, plates with extensive transform boundaries should have low velocities. Since mantle drag (either resisting or driving) is proportional to the area of plates, if this force is important velocity should correlate with plate area. Three forces, F_{DF}, F_{CD}, and F_{SU}, should be proportional to the velocity of plates relative to the mantle.

Given a set of relative plate motions, the additional piece of information needed to determine velocities relative to the mantle is an "absolute" reference frame, by which we here mean one that is fixed relative to the sub-lithospheric mantle. Since the sub-lithospheric mantle itself is undergoing deformation, the best reference frame to use for our present purpose is one in which the mean velocity of the parts of

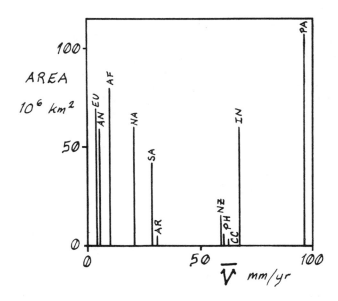

Figure 10-7.
Total area of plate versus absolute velocity. (Velocities are based on 1978 article by Clement Chase listed at end of chapter.)

the mantle that interact with the lithosphere is zero. Different ways of defining such a reference frame are discussed in the next section. The one used in the present analysis is essentially one in which it is assumed that hotspots rising through the mantle define a reference frame that moves with the mean velocity of the mantle. Values cited for mean plate velocities in the following discussion were found by averaging over the surface of individual plates the scalar velocity of points on the plate relative to this mantle reference frame.

Velocity Versus Plate Area   Plotting the areas of plates against their mean velocities (Figure 10-7), we see that there is no correlation between them. What conclusion can we draw from this? One that can (and has) been drawn is that the passive convection model discussed earlier is correct. The reasoning is as follows. The Pacific, Cocos, and Nazca plates are similar in the following ways: They are all completely oceanic, their boundaries all have about the same proportion of ridges, trenches, and transforms, and they are all moving at about the same velocity. In fact, the only way in which they are significantly different is in their areas, the ratios of which are 50:5:1 for Pacific:Nazca:Cocos. Since the mantle drag force F_{DF} is proportional to plate area, the drag on the bottom of the Pacific plate is 10 times greater than that on the Nazca plate and 50 times that on the Cocos plate. Therefore the driving forces of the three plates must be in the ratio 50:5:1 for the plates to all have the same velocity as observed. If one assumes that the magnitude of ridge push and slab pull forces is proportional to the total length of ridge and slab for a given plate, then the lengths must also be in these ratios. Yet the total ridge length and trench length of the Pacific plate is observed to be only 2 to 3 times larger than that of the Nazca plate. It is only 4 to 5 times larger than that of the Cocos plate. Thus (so the reasoning goes) the model in which ridge push and slab pull are the dominant driving forces and mantle drag the main resistive force is invalid. On the other hand, these data are consistent with a model in which the viscosity of the asthenosphere is high and plates are strongly coupled to convection currents in the mantle. Plate velocities then depend only on the velocities of flow in the underlying mantle, which happens to be fast under five plates and slow under six plates; trenches and ridges are merely secondary features.

However another interpretation is possible. Suppose the viscosity of the asthenosphere is extremely low, so that coupling between asthenosphere and the overlying plate is weak and the mantle drag force is negligible. Then the velocity of a plate is determined by the driving forces ridge push F_{RP} and slab pull F_{SP}, balanced against mantle drag acting on the subducting slab F_{SD} and the resistive forces that act on boundaries, namely, transform fault resistance F_{TF} and colliding resistance F_{CR}. Let's see if any of these show any relationship to the mean velocities of plates.

Velocity Versus Length of Transforms   Although transform faults by their nature must resist the motion of plates, Figure 10-8 shows no clear correlation between plate velocity and the fraction of a plate boundary occupied by transforms; apparently resistance to the motion of transforms is small.

Velocity Versus Length of Ridges   The story with ridges is less straightforward. In adding up the total length of ridge along a plate boundary, it is important to allow for the fact that ridges pushing on opposite sides of a plate tend to cancel. This happens in Antarctica. The **effective length** of a ridge is defined as the length that is not cancelled by an opposing ridge and thus is capable of driving the plate. In Antarctica, the total length of ridge is 20,800 km whereas the effective length of ridge is only 1,700 km.

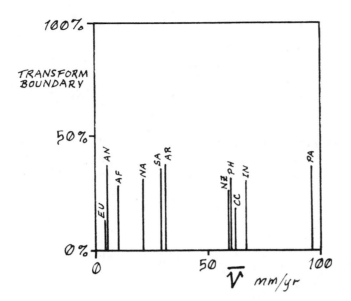

Figure 10-8.
Total length of transform boundary expressed as percentage of total circumference of plate.

In Figure 10-9 we see no correlation at all between velocity and total ridge length and at most a weak correlation between velocity and effective ridge length. If we pick 50 km/my as the boundary between slow and fast plates, most of the fast plates have lots of ridge. The one exception is the

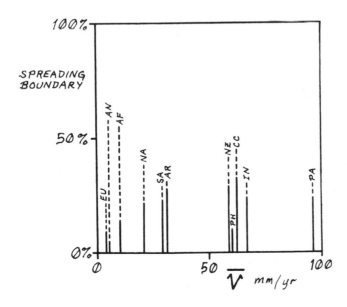

Figure 10-9.
Length of ridge as percentage of plate circumference versus absolute velocity. Dashed line is total length, solid line is effective length.

Philippines plate, which is bounded mostly by trenches. Even though it might be argued that the velocity of this plate is poorly determined and that back-arc spreading in the Mariana Islands provides this plate with a segment of ridge, this plate still tells us that plates do not need to have long ridge boundaries in order to be fast.

Velocity Versus Length of Subducting Slab This correlation (Figure 10-10) is a scientist's dream. The plates of the earth fall into two groups, one attached to subducting slabs over a large fraction of their boundaries, the other not. Those attached to subducting slabs, namely the Pacific, Nazca, Cocos, Philippines, and India plates, all have velocities relative to the mantle of 60 to 90 km/my. These are the fastest plates on earth. The other plates are not connected to downgoing slabs and have velocities of less than 40 km/my. These results leave no doubt that slab pull is an important driving force.

Velocity Versus Continental Area of Plates Plates with continental areas greater than 15 million square kilometers, including Eurasia, North America, South America, and Africa, all are slow. They have velocities less than 25 km/my. (Figure 10-11). Fast plates (those with velocities greater than 50

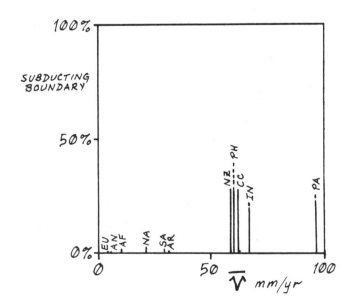

Figure 10-10.
Percentage of circumference of plate attached to downgoing slab. Dashed line is total length, solid line is effective length.

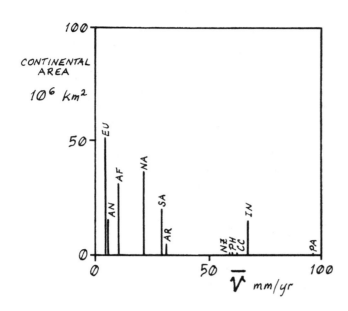

Figure 10-11.
Continental area of plates versus absolute velocity.

km/my) are completely oceanic. So in general, "continentality" appears to be associated with slowness. But let's look a little more closely. The India and Antarctica plates have the same continental area of 15 million square kilometers, yet Antarctica is slow and India fast. Moreover the oceanic Caribbean plate is slow. So the correlation is not perfect.

Several interpretations have been made of this correlation. One is that the mantle drag force is a resistive force that is weak beneath oceans and strong beneath continents. An alternative interpretation is the passive model of convection discussed earlier: Plates are coupled strongly to deep convection currents and are driven by them; convection currents beneath continental plates simply happen to be slow at the present time. A third interpretation is that plates with large continental areas are slow because they do not happen to be attached to subducting slabs at the present time. We tend to favor this last interpretation for reasons that are discussed below.

A Model for What Drives the Plates

The following model fits most of the above observations. Start with a plate resting on an asthenosphere with viscosity so low that mantle drag, the product of viscosity and plate area, is small; this feature of the model explains why large plates like the Pacific are able to have high velocities. Next let a cold, dense slab of lithosphere sink through the warm asthenosphere and into the underlying mesosphere. Recall that any object falling through a viscous medium arrives at an equilibrium settling velocity at which two forces are balanced. One force is gravity acting on the excess mass of the object over that of displaced medium. The other is viscous drag on the sides of the object. In other words, slab pull (F_{SP}) balances slab drag (F_{SD}). The viscosity in our model is layered: That of the mesosphere is high and produces a large amount of slab drag; that of the overlying asthenosphere is low and may itself be layered into an upper layer of low viscosity and lower layer of higher viscosity. So most of the slab drag force develops along the leading edge of the slab. The two quantities that determine the settling velocity of the slab are thus excess mass of the slab and the viscosity of the mantle. This settling velocity sets an upper bound to plate velocities of about 80 km/my.

Now attach the horizontal plate to the sinking slab. The term "slab pull" conjures up an image of a plate held together with strings under tension, being pulled by a heavy slab over a roller at the edge of a trench. The roller can be dispensed with because we know that when convection is produced in the laboratory or simulated on the computer, cold layers of fluid do, in fact, turn down and sink spontaneously. The tendency to turn down sharply is more pronounced if the cool, upper layer is more viscous than the warm layer beneath it. So a roller isn't needed to account for the bending of plates as they enter trenches.

The term "pull," as noted earlier, seems to imply that tension is required to transform the nearly vertical pull of a sinking slab into the horizontal motion of the attached plate. However, large bodies of rock are much too weak under tension to be pulled, especially when they are broken by faults as plates must certainly be. The explanation, as noted earlier, is that the stress transmitted to the horizontal plate by "slab pull" is not tension in an absolute sense, but rather is a reduction in the pressure that would exist in a horizontal plate at a given depth if the subduction zone were replaced by a continuation of the horizontal plate. So the horizontal plate is under compression, but less compression than it would be if the subduction zone were not there. If this explanation seems fishy to you, recall that even in water, which is weaker under tension than rock, when a cold layer analogous to the lithosphere turns down and sinks into the descending limb of a convection cell, a process analogous to slab pull is taking place.

To complete the model, put a dozen plates on the globe and attach subducting slabs to half of them. These become fast plates. The others are slow or stationary.

This model does not explain why plates like North America not attached to subducting plates are moving at all. Apparently slab pull isn't the whole story. Another driving force that may be significant is ridge push. Recall that a ridge is an enormous mass of rock rising several kilometers above the abyssal ocean floor and recall also the earlier analogy with an ice glacier of similar dimensions: The force exerted on an obstruction at the base of such a flowing body is substantial. Where ridges surround most of a plate, as in Antarctica and Africa, the forces of ridge push on opposite sides of the plate are nearly balanced and tend to compress the continent. The effective length of ridge, which is the unbalanced component of ridge push, probably imparts a velocity, albeit a low one, to the plate.

Most of the fast plates, including Pacific, Nazca, Cocos, and India, experience both pulls from subducting slabs and pushes from long ridges. Moreover the direction of the torque vector produced by ridge push in each of these plates is nearly parallel to the torque vector produced by slab pull. Since it is difficult to determine the contributions to plate motion of forces acting in the same direction, the relative magnitudes of ridge push and slab pull are poorly determined.

A process akin to ridge push occurs in other tectonic settings. The ridge force results from uplift of the ridge caused by thermal expansion, which is accompanied by an upwarp of the isotherm at the top of the asthenosphere. In essence the plate slides down this isotherm. This condition appears to exist in other areas such as the Basin and Range province of western North America, the Rift Valley of East Africa, and marginal seas behind many island arcs. The result is rifting like that at ridges, often accompanied by volcanism, but spread over a broader area, reflecting a broader upwarping of the isotherm at the top of the asthenosphere in these areas than at ridges. The rate of divergence across these rifts is less than that across most spreading ridges. In East Africa, the rate is slow because ridge push is exerted on Africa from both the Atlantic Ocean side and the India Ocean side. If in the future one of the oceanic plates attached to either side of Africa begins to subduct, the East African Rift will undoubtedly become a full-fledged ridge spreading at a rapid rate.

Ridge push may play another role in plate tectonics. The Pacific, like other large plates, presents an interesting problem in stability: It is 10,000 kilometers long, it is denser than the fluid beneath it, and it is riddled with mechanical imperfections. Yet this large, heavy object doesn't break up and sink except along its trenches. What holds it together? The answer may well be ridge push. In this regard it is interesting to note that the Philippines plate, the only fast plate with a long subduction zone but little or no ridge, is splitting apart along the Mariana Islands, where back-arc rifting is spreading at the remarkably fast rate of 60 km/my.

Plates are also undoubtedly driven and their motions modified by other forces, including drag exerted by flow in the asthenosphere, the suction force, colliding resistance, transform resistance, and compression transmitted across transforms. Many researchers have tried to identify the more important of these using different mathematical techniques to analyze known plate motions. No clear pattern has emerged from these studies other than the clear importance of slab pull, the probable importance of ridge push, and the possible importance of mantle drag as driving forces. Which of the resistive forces are most important is less certain, with viscous drag on subducting slabs being the main contender and drag on the bottoms of plates a distinct possibility.

Absolute Plate Motion

In the previous section, we used absolute rather than relative plate velocities in order to bring out important relationships like the correlation of plate velocity with length of trench. We're now ready to take a closer look at the idea of absolute plate velocity: How is it defined? How can we measure it?

At the risk of being childish, let's start by defining absolute motion on simple, cartoon-like planets where we can define relationships in a way that is crystal clear. We will then go on to Planet Earth, where matters aren't quite so clear.

Three Model Planets

Planet A This planet has a rigid mesosphere. Paint a set of latitude and longitude lines on the surface of the mesosphere. These define our absolute frame of reference. Add a layer of asthenosphere for lubrication, then cover the planet with moving plates. We define "absolute plate velocities" as plate velocities measured using the grid system painted on and moving with the mesosphere.

What do absolute plate velocites do for us on Planet A? First, some of the forces acting on plates, for example, the drag force, require that we know the velocity relative to the mesosphere; this is given by the absolute plate velocity. Second, let's assume that on Planet A most ridges have very high absolute velocities whereas most trenches have low absolute velocities or are stationary. Relationships like these, which also happen to be valid for Planet Earth, are obvious if we use absolute velocities but obscure if we use relative plate velocities.

Planet B   This is like Planet A except that the mesosphere is deformable like tar. Subducting slabs can penetrate into it, but the rate of deformation is sluggish, so after a time grid lines painted on the mesosphere will be distorted but still recognizable. Let's define our fixed reference frame on this planet by a perfect set of latitude and longitude lines which align themselves to fit the old, deforming grid system as well as possible in a least-squares sense. Absolute velocities defined in this reference frame would be almost as useful as on Planet A. The effect of deformation of the mesosphere would be to introduce noise into relationships based on the fixed reference frame and absolute velocities. For example, if two trenches remained fixed relative to the local parts of the mesosphere into which they were subducting and if these two parts of the mesosphere were moving relative to each other, then the positions of the two trenches would move in the fixed reference frame. However since on Planet B the rate of deformation of the mesosphere is much slower than the rate of plate motion, we would still be able to draw the useful conclusion that the rate at which trenches and their subducting slabs move laterally relative to the mantle is very slow.

Planet C   The mesosphere on this planet is undergoing lively deformation. Is the concept of absolute plate motion still useful? How can we define a fixed reference frame? Here's an alternative definition to the one we used on Planet B. Imagine different sets of reference axes originating at the center of the planet, each set rotating with a different angular velocity. When we calculate the instantaneous velocities at different points in the mesosphere, the value we get depends upon which set of reference axes we use. In Planet A, for example, if we select as our reference axes the fixed reference frame that moves with the mesosphere, the velocity at each point in the mesosphere is zero. On Planet C, we cannot find any reference axes for which this will be true because different parts of the mesosphere are moving relative to each other. However we can find one set of reference axes for which the sum of the velocities will be minimum in a least-squares sense. This reference frame moves with the mean angular velocity of some appropriate part of the mesosphere. Let's adopt this as our fixed reference frame for Planet C. Since we're interested in interactions between plates and the outermost part of the mesosphere, we specify that our fixed reference frame has the mean angular velocity of the outermost part of the mesosphere on Planet C. Absolute plate motion is measured relative to this reference frame.

By now we have the general idea. Let's see if we can apply what we know to a newly discovered celestial body, Planet E, which we will soon have a chance to visit. We created the cartoon planets ourselves, so we knew from the beginning exactly how their interiors worked. Planet E is more of a challenge: we will have to figure out how its interior works from observations made on the surface after we land there. We hope that Planet E is simple like Planet A but suspect it may be complex like Planet C. How do we proceed? To determine absolute velocities we need to establish a suitable fixed reference frame moving with the mesosphere. Here we hit a major problem: we can't see the mesosphere, so we can't determine its velocity directly. How can we measure something we can't see? It's time to visit Planet Earth, where geophysicists have been wrestling with this problem for years. In the next sections we will see some of the approaches they have used.

No Net Torque

Suppose Planet A has only one plate in the form of a single spherical shell (Figure 10-12). This plate is coupled by viscous drag through the asthenosphere to the rigid mesosphere. If you know the motion of the plate, what can you say about the motion of the mesosphere?

We begin by calculating the torque exerted by the lithosphere on the mesosphere. If $\mathbf{v}$ is the velocity at some point on a plate relative to the mesosphere, then the drag force $\mathbf{F}$ per unit area exerted on the mesosphere at that point is

$$\mathbf{F} = D\mathbf{v}$$

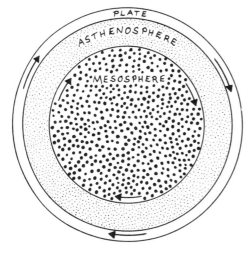

Figure 10-12.
Planet A: Mesosphere is rigid, entire lithosphere is one spherical plate.

where D is a drag coefficient. The corresponding torque **T** exerted on the mesosphere is given by the vector cross product

$$\mathbf{T} = \mathbf{r} \times \mathbf{F}$$

where **r** is the radius position vector. The total torque exerted by a plate is found by integrating **T** over the surface of the plate. For a spherical shell like that on Planet A, this integration gives a torque vector that is parallel to the angular rotation vector $_M\boldsymbol{\omega}_A$ of the shell relative to the mesosphere.

If the angular velocity of the mesosphere is different from that of the spherical shell, viscous drag through the asthenosphere will exert torque on the mesosphere, causing it to accelerate until it has the same angular velocity as the plate. Only then would there be no net torque exerted on the mesosphere by the plate and none on the plate by the mesosphere. This is the general condition sought in the **no-net-torque** approach to finding absolute motion: No net torque is exerted on the mesosphere by the set of plates that comprise the lithosphere and none is exerted on the total shell of lithosphere by the mesosphere.

Now suppose Planet A has two plates labeled N and S (Figure 10-13), each a hemispherical shell. The boundary between the plates is a transform along the equator. Motion across the transform is at the rate of 111 km/my, corresponding to a rotation vector with magnitude $_S\boldsymbol{\omega}_N = 1°$/my aligned with the rotation axis. What can you say about the motion of the mesosphere relative to either plate?

The torque exerted by the lithosphere on the mesosphere is the sum of the two torques exerted by the two shells. In order for this sum to be zero, the two torques must be equal and opposite. From the symmetry of the two plates, it follows that this condition will be met if the angular velocity of the mesosphere relative to Plate S is half that of Plate N relative to Plate S, that is, if

$$_S\boldsymbol{\omega}_M = {}_S\boldsymbol{\omega}_N / 2$$

Viewed from the absolute reference frame moving with the mesosphere, the angular velocity vectors of the two shells are equal and opposite in magnitude and are directed toward the two poles.

If Plate S is large and Plate N small, which plate will have the higher rate of rotation in an absolute reference frame moving with the mesosphere? Since drag force is the prod-

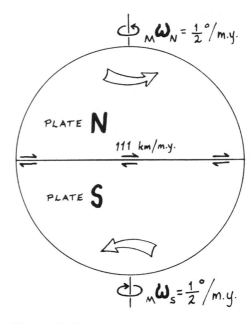

Figure 10-13.
Planet A with two equal hemispherical plates.

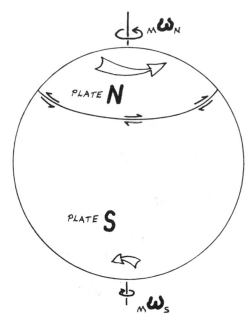

Figure 10-14.
Planet A with two unequal plates. In mantle reference frame, angular velocity of small plate N is greater than that of large plate S.

uct of velocity times area, the plate with the smaller area must have the higher velocity relative to the mesosphere in order for the two torques to balance (Figure 10-14).

Planet Earth On our own planet, about a dozen plates of different sizes are moving about different Euler poles. Each exerts a torque on the mesosphere which, for any given reference frame, can be calculated if we know the plate's angular velocity and the location of its boundaries. Using a straightforward mathematical technique, it is possible to find a unique **no-net-torque reference frame** such that the total torque exerted by all of the plates on the mesosphere is zero. We can think of this reference frame as moving with the mean velocity of the layer in the mantle immediately below the asthenosphere, namely, the mesosphere. In most contemporary analyses of global plate motions, absolute plate motions are given relative to a no-net-torque reference frame found in this way.

Hotspots

Planet A with Hotspots As before, Planet A has a rigid mesosphere and a single plate. But this time, instead of painting grid lines on the mesosphere to represent the reference frame, let's bury six spheres of highly radioactive material in the mesosphere to mark the locations of the $\pm\mathbf{x}$, $\pm\mathbf{y}$, and $\pm\mathbf{z}$ axes of the reference system (Figure 10-15). Now add the asthenosphere and the moving plates as before. The radioactive heat sources produce plumes of magma or hotspots that rise through the mesosphere, the asthenosphere, and the moving plate, forming chains of volcanoes upon the plate as it moves over the plumes. We could make Planet A less cartoon-like if we didn't place the rising plumes of magma at the locations of reference axes 90° apart but instead scattered them randomly throughout the rigid mesosphere. The result would be the same: Volcanic island chains would form above the plumes.

If a plate pivots around about an Euler pole that remains fixed relative to the mesosphere, the volcanic chain takes the form of a small circle centered on the Euler pole. This is shown in Figure 10-16 for the special situation where the plate exterior to the plate over the hotspot is not moving relative to the mesosphere, so that the hotspot track and the transforms are concentric—generally this is not the case.

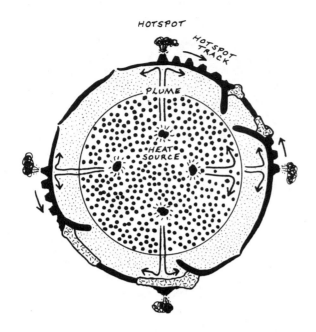

Figure 10-15.
Planet A: Radioactive heat sources are buried in rigid mesosphere, producing hotspot tracks on moving plates.

Plate reconstructions are easy on Planet A. The circular form of a hotspot track gives us the location of the Euler pole needed for a plate reconstruction. Ages of volcanoes along the track give us the angles Ω needed for reconstruction poles. As a check, if there are tracks from different hotspots on the same plate, they should give the same Euler pole and the same angles. Moreover because the meso-

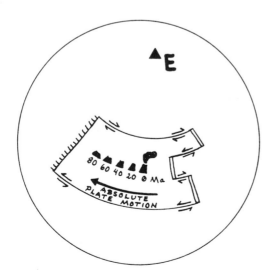

Figure 10-16.
Hotspot track forms along small circle centered on Euler pole **E** describing motion of plate relative to mesosphere.

sphere is rigid on Planet A, the hotspots retain the same coordinates in a common fixed reference frame and thus remain at fixed distances from each other. Therefore if we use tracks from two different hotspots to reconstruct positions of Plates A and B at an earlier time, the two plates will have the correct reconstructed positions relative to each other.

Planets B and C with Hotspots Heat sources are buried in a mesosphere undergoing deformation. Our reasoning and procedure are the same as on Planet A, but we have gnawing doubts about the validity of our reconstructions because we know that as the mantle deforms, the distance between hotspots is changing. Therefore when we used two different hotspots to restore two plates to their former positions relative to each other, we realize that these positions may be in error by as much as the amount of motion between hotspots.

Planet Earth with Hotspots Does anything on our own planet look like the small-circle hotspot tracks on Planet A? Yes indeed! The Hawaiian and Emperor seamount chains form the largest volcanic mountain range on Earth. Each chain is a sequence of coalescing seamounts that does, in fact, define a slightly wobbly segment of a small circle. The trends of the two chains are markedly different and they have distinctly different Euler poles. The age progression along this composite chain is remarkably systematic from the old end near Kamchatka, where the age of a seamount is 78 Ma, to the young end, where Hawaii is still active and a new submarine volcano is extending the chain at its tip (Figure 10-17).

The hotspot model explains this remarkable mountain range as follows. The Hawaii hotspot remained fixed relative to the mantle as the Pacific plate rotated first about an Emperor Euler pole that remained fixed relative to the mantle, then about a Hawaii Euler pole that also remained fixed relative to the mantle. Other somewhat fuzzier hotspot tracks on the Pacific plate have azimuths and age progressions that are fairly but not perfectly consistent with these Euler poles. Departures from a perfect fit are systematic enough to suggest that some of the Pacific hotspots are moving relative to each other.

Absolute plate motions relative to a hotspot reference frame are determined by a two-step process. In the first step, relative plate motions are determined for a global set of plates

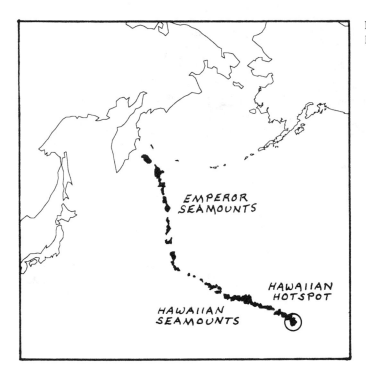

Figure 10-17.
Hawaiian-Emperor seamount chain.

by methods described in earlier chapters. In the second step, the mantle is given a velocity that produces as good a fit as possible to the azimuth of all known hotspot tracks and the progression of ages along them. This fit is made under the constraint that the motions between plates determined in the first step are not changed. Absolute plate motions determined in this way are given in most contemporary analyses of global plate motions. The absolute velocities used in the discussion of what drives the plates were determined in this way.

A global map of absolute velocity vectors found from hotspots by this method is given in Figure 10-18. Let your eye roam over and let your imagination soar. Examples of almost everything discussed in this book leap to the eye. The enormous Pacific plate is racing toward its trenches, as are the smaller Nazca and Cocos plates. The India-Australia plate is being pulled northward into the garland of trenches around southeast Asia, driving India into Asia and maybe even giving Eurasia a small northward shove. It's fun to run your eye along plate boundaries and mentally add vectors tail-to-tail to get relative plate motions. Try this along western North

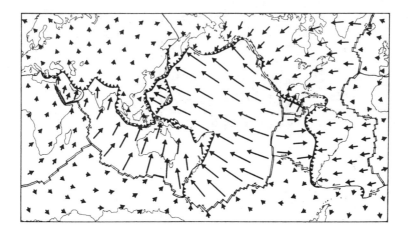

Figure 10-18.
Absolute velocity vectors found from motions of plates relative to hotspots. (Based on 1978 article by Clement Chase listed at end of chapter.)

America—it works! Where the Pacific and Nazca plates meet the Antarctica plate, the absolute velocity vectors are nearly parallel to transforms—do you see why? Compare the Chile-Peru trench and the Tonga-Kermadec trench on opposite sides of the South Pacific—do you see a difference in the absolute motion of the plates on the upper side of the subduction zones? Can you think of a possible connection between this observation and the fact the western zone has moderate earthquakes and marginal seas whereas the Chile-Peru subduction zone has enormous earthquakes and no marginal seas? Try to extrapolate the motions into the future. The Pacific basin is clearly shrinking as Circum-Pacific continents crowd in upon it and as the Atlantic basin grows. The arrows show the amount of motion in 15 million years. What will the map look like 100 million years from now?

As one of the more controversial topics in contemporary plate tectonics, hotspots are currently the subject of active research and lively debate. The following are some of the questions being asked. Are seamount chains produced by magma plumes rising from the mantle or by cracks propagating through the lithosphere and tapping local magma sources? If by plumes, where do these originate, in the asthenosphere, the mid-mantle, or at the core-mantle boundary? How are plumes deflected by flow in the asthenosphere? What is the rate of proper motion of plumes relative to each other? Have plumes existed throughout all of the Earth's history or only since the Jurassic, the age of the oldest preserved volcanic hotspot tracks?

A Consistency Test   Are absolute motions determined from hotspot and no-net-torque reference frames the same? If not, one might question the usefulness of the concept of absolute plate motion. Let's ask the question first for our cartoon planets. On Planet A, the answer is crystal clear: Since the mesosphere is rigid, the two approaches would give the same results. On Planet B, results should be similar but not identical. On Planet C, they could be either similar or vastly different, depending on the rate of deformation relative to the rate of plate motion. How about Planet Earth?

First the bad news. For slowly moving plates like Antarctica, the two approaches do not agree very well. However this isn't devastating because the uncertainties in both methods are about as large as are the low velocities themselves. Now for the good news. For rapidly moving plates like the Pacific, the two methods give essentially the same results. Using no information about hotspots, the no-net-torque approach yields absolute plate motions that agree quite well with the observed azimuths of hotspot tracks. It predicts correctly the azimuth of the Hawaiian seamount chain on the Pacific plate. It predicts the trends of the Galapagos Archipelago on the Nazca plate. It predicts the direction of the Yellowstone hotspot track on the North America plate fairly well. It is this internal consistency that suggests that both approaches are yielding plate motions in a reference frame that is more-or-less locked to a sluggishly moving sub-lithospheric mantle.

Single-Plate Torque Due to Slab Pull

Suppose that for a single plate, you know the location of its ridges, transforms, and trenches but you know nothing about its motion relative to other plates or to hotspots. Can you say anything about the absolute motion of that plate?

This question is interesting because, as we go farther and farther back into the geologic past, our knowledge about the global geometry of plates becomes more and more fragmentary. The no-net-torque approach requires knowledge of the locations of the boundaries and relative motions of all plates that existed at a given time. If we are willing to make educated guesses about plate geometry prior to the early Tertiary, we can use the no-net-torque approach to calculate absolute velocities; but it turns out that the results are quite sensitive to uncertainties in plate reconstructions. Hence the interest in trying to estimate absolute velocity from information about the boundary of a single plate.

An approach that has yielded surprisingly good results has been to simply calculate the torque that would be exerted on a plate by ridge push or slab pull. More exactly, a push or a pull is applied in a direction locally perpendicular to the ridge or trench boundary. For a given segment of boundary of one type, for example, a ridge not offset by a transform, if the magnitude of the applied force is one per unit length of ridge or trench, the torque vector has the orientation of and is proportional to the length of the chord connecting the ends of the segment. Other applied forces scale proportionately. For example, unit force acting perpendicular to a ridge of any shape running from the south pole to the north pole would exert a torque of magnitude 2 on the plate lying to the east of the great circle. The direction of the torque is parallel to the north pole.

When this calculation is made for the plates as they exist today, it turns out that for the present fast plates (Pacific, India, Cocos, and Nazca), the torques exerted by ridge push and slab pull are nearly parallel and both agree surprisingly well with the absolute motions found using the global no-net-torque and hotspot methods. An even better fit is obtained if, instead of unit pull per unit length of slab along trenches, the pull is weighted according to the age of the subducting slab. The observation that the torques exerted by ridges and trenches are nearly parallel suggests that when the direction of plate motion changes, the configuration of both trenches and ridges must also change.

This result also shows that to first order, the motion of fast plates is determined not by interactions with other plates but rather by some combination of ridge push and slab pull acting along the boundary of each plate considered by itself. A corollary is that when the absolute motion of a plate changes, this is due to a change in the configuration of trenches and ridges along the boundary of that plate.

The single-plate torque approach provides a neat explanation of the famous elbow where the Hawaiian and Emperor seamount chains meet, which occurs where the age of volcanoes is about 43 Ma. What can have caused this remarkable bend? If the chain is recording the motion of the Pacific plate over a hotspot, then the direction of the plate must have changed at 43 Ma. But why?

The traditional explanation has been that a global reorganization of plates took place in the early Tertiary and that the change in Pacific plate motion was part of this reorganization. Implied in this explanation is a vaguely defined global change in mantle convection currents. The single-plate torque

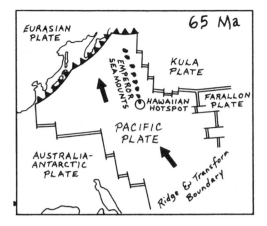

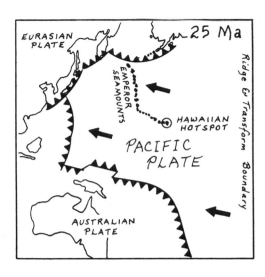

model provides a simpler and more explicit explanation. Today subducting slabs are attached to the Pacific plate along the Aleutian Trench in the north, along the Japan Trench in the west, and along the Tonga-Kermadec Trench in the south. Slab pull toward the north largely cancels that toward the south. As a result, when the total slab pull torque is calculated, the net result is to move the Pacific plate toward the Japan trench along the trend of the Hawaiian seamount chain. At 25 Ma the configuration of trenches and the direction of motion was about the same (Figure 10-19). The boundary of the Pacific plate was different prior to the time of the bend. Plate models for the north Pacific basin in the early Tertiary show the Pacific plate subducting into trenches only along its northern boundary. The slab pull torque for this plate model (Figure 10-19) moves the Pacific plate parallel to the trend of the 76 to 43 Ma Emperor seamount chain.

The success of the single-plate torque model in explaining the motion of the Pacific plate suggests the following alternative to global changes in mantle convection currents as an explanation of a global reorganization of plate motions: simultaneous changes in plate motion are simply due to simultaneous changes in the configuration of plate boundaries. The connection between the two is mainly a kinematic one. For example, a change in the motion of Plate A might convert a boundary between plates A and B from a transform to a trench; this change in boundary would cause the motion of Plate B to change.

Figure 10-19.
At 65 Ma, the motion of the Pacific plate was northerly toward trenches along its northern boundary. At 25 Ma, motion was more westerly as the result of vector sum of slab pull in several directions.

Paleomagnetic Euler Poles

Let's revisit Planet A and change it a little. Recall that we buried six spheres of highly radioactive material below the surface of the mesosphere to mark the locations of the $\pm\mathbf{x}$, $\pm\mathbf{y}$, and $\pm\mathbf{z}$ axes of the reference system. Let's get rid of five of these hot spheres and retain the one located along the $\pm\mathbf{z}$ axis, which corresponds to the North Pole. Now add the asthenosphere and moving plates as before. As a certain Plate P moves over the North Pole, the radioactive heat source produces a hotspot track on Plate P. If Plate P moves about an Euler pole that remains fixed relative to the mesosphere, the volcanic chain forms a small circle centered on the Euler pole. This volcanic chain always terminates at an active volcano located at the North Pole. Volcanoes with different ages along the volcanic chain correspond to former positions of the North Pole.

Now let's make a paleomagnetic study on Plate P of rocks spanning the age of the seamount chain. Recall that the final product of a paleomagnetic study is a paleomagnetic pole coincident with an ancient location of the geographic North Pole. The paleomagnetic pole path will coincide with the seamount chain, both lying along the same small circle. Paleomagnetic poles and seamounts of the same age will coincide. Therefore fitting the hotspot track and the paleomagnetic pole path with small circles will yield the same Euler poles. Both describe the motion of Plate P relative to the mesosphere of Planet A.

Now let's make Planet A more earthlike by displacing the hotspot from the rotation axis (Figure 10-20). Although the hotspot track and the pole path do not coincide, they are both concentric about the same Euler pole that describes motion of the plate relative to the rigid mesosphere of Planet A.

Now repeat the experiment on Planet A with the hotspot turned off. Again we can fit the apparent polar wander path defined by the set of paleomagnetic poles with a small circle and an Euler pole that describe the motion of Plate P relative to the mesosphere. This pole is called a **paleomagnetic Euler pole** or PEP for short.

Why bother with PEPs when we already have three other ways of determining absolute plate motion? One reason is that a hotspot track requires a plate to pass over a hotspot whereas a polar wander path does not require that a plate

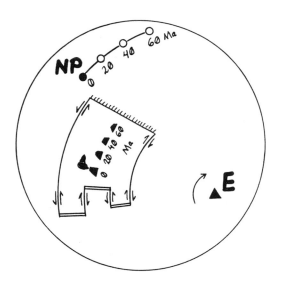

Figure 10-20.
Both the hotspot track and the paleomagnetic pole path are concentric about the Euler pole **E** for motion of the plate relative to the mesosphere.

pass over the geographic pole. So hotspot tracks are rarer than APW paths. Few hotspot tracks older than Jurassic have been recognized. A second reason is that in older rocks, it is difficult to use the no-net-torque approach because this requires information about plate boundaries not available prior to the Tertiary. On the other hand, paleomagnetic pole paths like the one shown in Figure 9-9 have been determined for most continental plates back into the Precambrian.

Some Concluding Thoughts

A final and perhaps more philosophical reason for concluding that absolute reference frames are useful is the following. The large and complex set of relationships that comprise the global set of relative plate motions fall into a simple pattern when considered in the absolute reference frame, and this pattern generally makes physical sense. For example, it turns out that the velocity of trenches is very low in the absolute frame. If a reference frame that supposedly was moving with the sub-lithospheric mantle gave the result that many trenches are moving fast, one would immediately smell a rat. Trenches are thousands of kilometers long. They mark the upper edge of slabs extending as much as 700 km into the mantle. Rapid motion through the mantle of such a "sea anchor" displacing large volumes of asthenosphere would be physically implausible.

What can we learn about the earth from the observation that fracture zones, hotspot tracks, and paleomagnetic polar wander paths are regular, gently curved, continuous features that often lie along small circles? One clear implication is that plate motions have great continuity: Once moving, a plate usually continues to move in the same direction for a long time. The circular form of many of these features reflects the mathematical fact that motion about a fixed Euler pole is the simplest way in which an object can move over the surface of a sphere. Just as constant force on a plane produces linear motion, constant torque applied to a plate on a sphere produces motion along a small circle. So the question of why hotspot tracks and pole paths have the form of small circles boils down to the question of why the torque acting on a plate should remain constant for a long time.

The torque that drives a plate is produced mainly by pushes and pulls acting at ridges and trenches along its boundaries. As long as the geometry of a boundary remains nearly constant, the torque acting on the plate will be nearly constant and the absolute motion will be along a small circle. Since the dominant driving force is probably slab pull, the torque would not remain constant if trenches were moving rapidly relative to the mantle. However as discussed above, the lateral motion of slabs through the mantle appears to be very slow. Therefore as long as a plate is attached to slabs subducting deep into the mantle along a nearly stationary array of trenches, the Euler pole describing its absolute motion relative to the mantle will remain nearly constant.

What about the three-plate problem discussed at the end of Chapter 7? Recall that if Plates A and B are moving relative to the mantle about Euler poles that remain fixed in a mantle reference frame, then the Euler pole for the motion of plate A relative to plate B will move relative to both plates and the boundary between A and B cannot remain a small circle. Yet the observation that many fracture zones are circles tells us otherwise. The resolution of this apparent contradiction (as discussed in Chapter 7) is that the boundary between Plates A and B can still be essentially a circular transform if either of two conditions is met: the absolute Euler poles of the two plates are nearly parallel, or their magnitudes are greatly different. Examples of the latter can be seen along the boundary between the Antarctica plate and the Pacific and Nazca plates (Figure 10-18), where transforms are nearly parallel to the motion vectors for the oceanic plates.

True Polar Wander

The distinction between true polar wander and apparent polar wander is analogous to that between relative and absolute plate motion. To an observer on a plate moving toward the pole, it seems like the pole is moving toward the plate. The movement of poles along a paleomagnetic pole path toward or away from an observer on the plate similarly reflects the motion of the plate. To emphasize the relative nature of motion between plates and poles, paleomagnetists describe such pole paths as *apparent* polar wander paths. Since all of our observations of poles are made on moving plates, isn't all polar wander apparent? How could one even define "true" polar wander?

A Thought Experiment

Let's return to our cartoon planets and see how we might define true polar wander there. Planet A has an equatorial bulge like the earth and a rigid mesosphere, which together rule out the possibility of polar wander. Planet B has a mantle capable of undergoing plastic deformation. This is more promising. Again let's paint a grid of longitude and latitude lines on the surface of the mesosphere of Planet B, centering the grid system on the North Pole, as defined by the rotation axis at time zero. Let's assume that Planet B's mantle is weak and will deform under the stress imposed by some new geologic feature like a mountain range or the Greenland icecap. Under these conditions, the entire mantle will deform in such a way as to move an excess mass, say the Greenland icecap, toward the equator. Deformational strain need not be large. Points that are close together before the deformation will be close together afterwards, so that the grid system painted on the mesosphere, while it may change shape a little, will still be easily recognizable. To an observer in space, the equatorial bulge does not seem to move but remains inclined at a fixed angle to the plane of the ecliptic. However, the entire painted grid system seems to move over the surface of the planet. An observer standing in Greenland, moving with the grid system, sees the geographic pole corresponding to the current rotation axis move further and further away as Greenland moves toward the equator. The mesosphere of Planet B is undergoing **true polar wander**.

Figure 10-21.
True polar wander has not occurred: Hot-spots stay at poles and equator; therefore all seamounts on the same track have the same paleomagnetic inclination (small arrows).

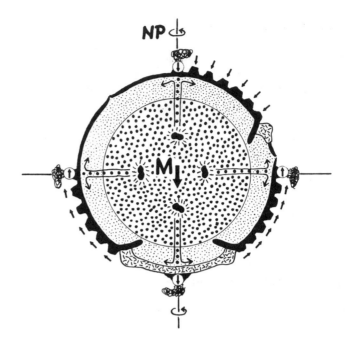

Now let's add a layer of asthenosphere and some rapidly moving plates. This makes it much more difficult to detect true polar wander. As an observer on one of the plates, the rotation axis will appear to you to wander just as it did to the observer on Greenland, even if the mesosphere is not experiencing true polar wander. How can you possibly determine whether the mesosphere is shifting relative to the rotation axis?

Try the following. Bury a sphere of radioactive material deep in the mesosphere at the North Pole of Planet B, producing a plume in the mantle and a hotspot track on a plate passing over the pole. If there's no true polar wander, the plume will stay at the North Pole and volcanoes forming there at different times will all be magnetized vertically downward (Figure 10-21). However with true polar wander, the entire mantle, including the plume, will shift relative to the rotation axis. As the plume shifts further from the pole, volcanoes forming above it will become magnetized at shallower angles (Figure 10-22). The test for true polar wander is thus very straightforward. Measure the paleomagnetic inclinations of volcanoes along a hotspot track from the young to the old end of the track. If the inclinations change, the latitude of the plume has changed and presumably true polar wander has occurred.

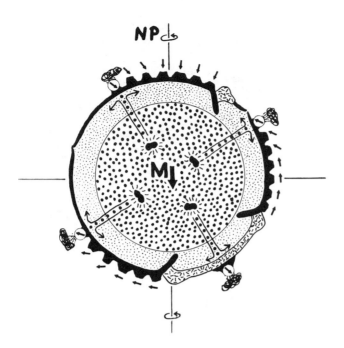

Figure 10-22.
True polar wander has occurred: Hotspots have all shifted about 30° relative to the rotation axis, and the paleomagnetic inclinations of seamounts changed at the time of the shift.

Observations on Planet Earth

On Planet Earth this experiment has been done in two different ways, the first using a single hotspot and the second using the global set of hotspots. The single hotspot experiment was done on the longest hotspot track, the Hawaiian-Emperor chain. The experiment consisted of measuring the paleomagnetic inclination at Suiko Seamount, a volcano with an age of 65 Ma, using a long core drilled from the seamount by the Deep Sea Drilling Project. The results were similar to those shown in exaggerated form for several hotspots in Figure 10-22: the magnetization of the seamounts, as shown by the arrow above each seamount, has the same inclination in successive seamounts until the hotspot shifts relative to the rotation axis. The paleomagnetic inclination of the core from Suiko Seamount is steeper than the inclination today at the active volcanoes on Hawaii. This experiment shows that during the past 65 million years, the Hawaii hotspot has shifted 7 degrees away from the geographic pole.

This experiment is not completely definitive because we know that hotspots shift relative to each other. One might argue that as the Hawaii hotspot was shifting away from the pole, other hotspots were shifting in other directions so that the mean motions of all hotspots cancelled. In this case, no

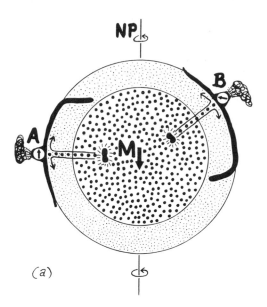

(a)

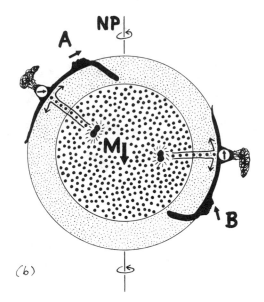

(b)

Figure 10-23.

Test for true polar wander. *(a)* Seamounts A and B form above hotspots. *(b)* The situation today: Seamounts A and B have moved toward trenches, and the hotspots have shifted relative to the rotation axis; poles calculated from the paleomagnetism of seamounts A and B would not agree. *(c)* The plates are reconstructed to place seamounts A and B over their respective hotspots; poles P_A and P_B calculated from the paleomagnetism of the two seamounts in their reconstructed positions now agree with each other but are displaced from the North Pole, showing that true polar wander occurred.

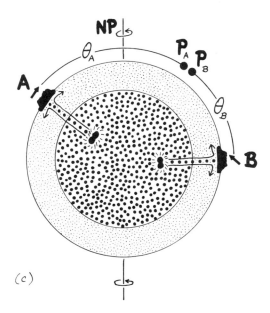

(c)

systematic shift of the mantle as a whole relative to the rotation axis is indicated.

Here's how geophysicists have tried to get around this problem. First, move plates back along hotspot tracks to their positions at earlier times (Figure 10-23). This restores the plates to former positions in a hotspot reference frame under the assumption that hotspots have not moved relative to each other. Paleomagnetic data are not used in this first step and no assumption is made about possible shifts between poles and hotspots. Next, with the plates reconstructed in a hotspot reference frame, use paleomagnetic data from the plates to find paleomagnetic poles. Without true polar wander, the poles will be grouped around the present geographic pole, the scatter in poles reflecting errors in the paleomagnetic data and shifts between the hotspots used for the reconstruction. If the mantle as a whole has shifted relative to the pole, carrying the global set of hotspots with it, the paleomagnetic poles will be displaced relative to the rotation axis.

Geophysicists have done this experiment, restoring plates to their positions in the late Cretaceous in a hotspot reference frame. The result is that late Cretaceous paleomagnetic poles are systematically displaced 5 to 10 degrees toward the central Pacific basin. True polar wander in the sense discussed earlier appears to have occurred.

The picture that emerges from all this is the following. Plumes rising from the mantle move with the parts of the mantle whence they come and through which they pass. During times of no true polar wander, the rotation axis is in a stable position relative to the density structure of the mantle. During these times, the mantle slowly deforms and plumes move relative to each other in a pattern that is random relative to the orientation of the rotation axis. Then, in response to some mass perturbation in the mantle, perhaps the initiation of a new plume or a new subduction zone, the mantle as a whole rotates relative to the rotation axis until it arrives at a new position that is stable relative to the new density structure in the mantle. An observer in space would see the entire global pattern of hotspots move, like the grid in our thought experiment, to some new position relative to the rotation axis.

Paleomagnetic and Hotspot Euler Poles

What implications does true polar wander have for the relationship between hotspot tracks and apparent polar wander paths? If a plume and the geographic pole both stay at the same position in an undeforming mantle, then, as noted earlier, hotspot tracks and paleomagnetic poles will lie along the same small circle. However this will not be true when a plume shifts away from the rotation axis, regardless of whether the shift is due to true polar wander or to a random process in a local part of the mantle. So we might expect apparent polar wander paths and hotspot tracks, while agreeing with each other in a general way, to display distinct differences. This is what is observed.

If both a hotspot Euler pole and a paleomagnetic Euler pole are available and if they are somewhat different, which provides the "best" measure of absolute plate motion? Generally the hotspot pole, since this is most closely linked to processes occurring in and moving with the mantle, and that's what we're usually most interested in. For example, during a time of rapid polar wander while the mantle as a whole was shifting relative to the rotation axis, a hotspot track on a plate and the trenches along the boundary of the plate would shift together with the Euler pole describing the motion of the plate relative to the mantle. Therefore the direction of absolute motion determined from slab pull for that plate would agree with the direction of motion determined from the hotspot track. On the other hand, during the time of true polar wander the paleomagnetic pole path would not coincide with the hotspot track, and the Euler pole derived from it would provide a less accurate description of the motion of the plate relative to the mantle than would the hotspot Euler pole. However as noted earlier, geophysicists rarely enjoy the luxury of choosing between several available estimates of absolute motion. For most of geologic time, paleomagnetic Euler poles are the main source of information about absolute motion.

Life Cycle of a Plate

Most geologic processes occur in regular or irregular cycles. Is this also true of plate tectonics? Do plates have typical life cycles? Do they all die? How long do plates live? Are the life cycles of oceanic and continental plates the same?

Are Continental Plates Intrinsically Slow?

Let's consider the last question first. One of the most striking results to emerge from analyzing plate motions in an absolute reference frame was the discovery of the great contrast in the speed of oceanic and continental plates. Most oceanic plates are presently moving at rates of 80 km/my or more. Most continental plates are moving at one-fourth that rate or less. If we applied the most famous proverb of geology,

"The present is the key to the past," then continental plates must have always moved slowly. If that is true, they must differ in some fundamental way from oceanic plates, perhaps by experiencing more drag force due to the presence of deep roots beneath continents.

Let's be contrarian and explore another possibility. Perhaps all plates alternate between times when they are moving rapidly and times when they are moving slowly. Perhaps the human race had the bad luck of discovering plate tectonics at a time when continental plates just *happen* to be moving slowly. After all, Nature has given us only a dozen large plates upon which to base our generalizations. An epidemiologist, given a sample this size and asked to decide whether, say, coffee causes cancer, would be amused if anyone were rash enough to try to draw a conclusion from so small a sample.

The obvious experiment to settle this is to look at the motion of plates at different times in the past. This has been done several times using absolute velocities determined by the approaches mentioned earlier. The most useful results have been obtained using hotspot and paleomagnetic Euler poles, since these require information from only one plate.

Tracks and Cusps   Hotspot tracks and paleomagnetic pole paths usually consist of two elements: Long, gently curved segments, termed **tracks**, and short segments of sharp curvature termed **cusps**, where the trend of tracks changes abruptly. The Hawaiian seamount chain is a well-defined track, as is the Emperor chain. The elbow where the Hawaiian and Emperor chains meet is a cusp. On either side of a cusp, the hotspot track or pole path commonly has the form of a small circle, the Euler pole of which describes the motion of the plate relative to the mantle. The Hawaiian- Emperor elbow is an example of a **fast cusp**: the volcanoes march around it without slowing down. On other plates there appear to be **slow cusps**, where the paleomagnetic pole or hotspot lingers for an interval of time before striking out in a new direction. Cusps obviously mark times when the motion of plates changed relative to the mantle.

Velocities of Continental Plates   A definitive answer to the question of whether continental plates have always had low velocities was obtained by the following analysis. First, Euler poles were found by fitting small circles to hotspot tracks and paleomagnetic poles for continental plates. Second, absolute velocities were found using these poles. Third, these absolute velocities were averaged over the surface of the plates. The results were clearcut. During times when paleomagnetic poles or seamounts were progressing rapidly along tracks, the mean velocities of North America and Eurasia were as great as those of the fastest oceanic plates. During other time intervals, these plates were nearly stationary. These studies show that plates carrying continents are not intrinsically slow. They simply happen to be slow today. The reason is probably that they are not presently attached to large, subducting slabs, as are the oceanic plates.

Life Cycle of Oceanic and Continental Plates

The following life cycle for plates is suggested by these results. Let's start with a typical oceanic plate like the Pacific. It is moving rapidly toward its trenches, where the plate bends downward and sinks into the asthenosphere. Trenches and their slabs move very slowly relative to the mantle in which the slab is embedded. The absolute velocity of the plate is thus essentially its velocity toward its trench system.

The direction of motion remains constant so long as the driving forces remain constant. These appear to be controlled mainly by the location of trenches along the boundary of the plate. When the driving forces change because of a change in the geometry of the boundary, the direction of plate motion changes. This happened at the time of the cusp in the Hawaiian-Emperor chain and at earlier times in the history of the Pacific plate.

While being destroyed along its trench boundaries, an oceanic plate like the Pacific is also growing along its ridge boundaries. Therefore the lifetime of the plate is not the time required for a piece of seafloor to travel from the ridge to the trench—it is much longer than this. We know from the age of magnetic stripes still preserved on the Pacific plate that it has been in existence for at least 150 my. If it stopped growing today and slid into its trenches at its present high velocity, it would take 80 my for the plate to disappear. So its expected lifetime is at least 230 my. Few geologic processes are active this long. However, eventually oceanic plates die when the last part of the plate is subducted. The once vast Farallon plate is in its death throes today off the coast of Oregon and Washington.

The life cycle of continental plates is more complex, alternating between times of rapid motion and times of slow motion. They move rapidly when attached to large subducting slabs. When not, they move slowly. This is the condition that exists today. The almost stationary Antarctica plate, for example, consists of a continent at the center of the plate surrounded by an apron of oceanic lithosphere growing outward. It is not subducting and it is standing still.

What causes a continental plate to start moving rapidly and what causes it to stop? Stopping is easy: continents are too buoyant to subduct. So when a continent arrives at a trench, the plate screeches to a halt. This explains an importance difference between oceanic and continental plates. Oceanic plates, lacking continental "stoppers," can go on growing along one edge while subducting along another for extremely long times. A continental plate can move rapidly only until the continent meets a trench, when it stops.

Another result of the buoyancy of continents is that continental plates never die in the sense that oceanic plates do. In the future, not a scrap of the Farallon plate will exist as an active plate. However the North American continent will always be part of some plate, so in a real sense the North America plate will always exist.

Why do continental plates, once stopped, start to move rapidly again? Let's be brave and speculate about the future of the Eurasia plate. Today it is the slowest plate of all. The key to the future is what happens to the part of the plate which consists of the segment of ocean floor between Europe and the Mid-Atlantic Ridge. This piece of oceanic lithosphere is growing along its western edge as the Atlantic basin grows and the Pacific basin shrinks. Eventually the Atlantic basin becomes as wide as the Pacific is today. The growing and aging oceanic plate attached to Europe becomes colder and denser. Eventually it begins to subduct along some zone of weakness, perhaps along an old fracture zone. A trench system develops and Eurasia moves toward the trench system at a rapid rate (Figure 10-24). During this time of rapid motion, volcanoes are widely spaced along hotspot tracks and paleomagnetic poles along pole paths. Eventually Eurasia arrives at the trench and its velocity decreases to a value typical of present continental plates.

One hundred million years from now, paleomagnetists will note paleomagnetic poles sweeping along an arcuate track centered on an Euler pole that describes the motion of the plate toward its trenches (Figure 10-25). The poles will then be observed to dawdle in the vicinity of a slow cusp. Geophysicists will conclude that this marks the time when Eurasia arrived at the trench and came to a screeching

Figure 10-24.
The future of Eurasia. At time t_1, Eurasia is attached to a short piece of oceanic plate. At time t_2, the oceanic plate has grown. At time t_3, the oceanic plate begins to subduct and Eurasia moves rapidly toward the trench. At time t_4, Eurasia arrives at the trench and its motion is stopped by the buoyancy of the plate.

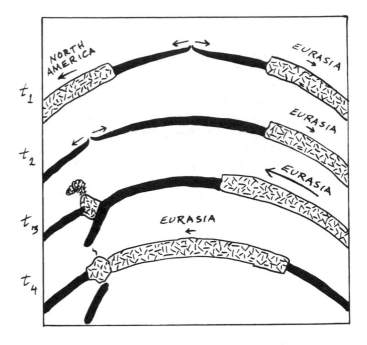

halt. Eurasia will then remain nearly stationary while growing another plate, and the cycle will repeat until Planet Earth runs out of heat and plate tectonics stops.

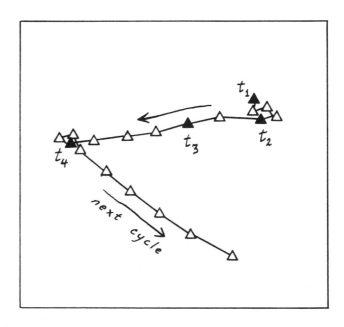

Figure 10-25.
Future paleomagnetic poles for Eurasia corresponding to previous figure. Times t_1, t_2, and t_4, located at slow cusps, are times like the present when Eurasia was not attached to a subducting slab.

Problems

10-1. The fracture zones on the southern margins of the Pacific and Nazca plates appear to be nearly parallel to the directions of absolute motion (Figure 10-18). Why is this so?

10-2. The Chile-Peru subduction zone on the east side of the South Pacific basin is characterized by extremely large earthquakes, the absence of a marginal sea, and a generally compressive tectonic style landward of the trench. The Tonga-Kermadec subduction on the west side of the basin has earthquakes of moderate magnitude, well-developed marginal basins, and an extensional tectonic style. Can you offer an explanation on the basis of the absolute motions (Figure 10-18) of the plates involved?

10-3. Using the paleomagnetic data from North America given in Table 9-1, try to find a paleomagnetic Euler pole for the track from 300 to 200 Ma. Using this Euler pole, try to estimate the direction and magnitude of the absolute velocity vector at 250 Ma for a central point in the North American craton.

10-4. The enormous volume of lithosphere flowing into trenches must go somewhere, presumably by some kind of return flow through the asthenosphere.

(a) Where does the return flow end up?

(b) Lay a piece of tracing paper over Figure 10-18 and try your hand at sketching the general pattern of return flow.

(c) Can you see a hemispherical or some other large scale pattern in the flow?

(d) Speculate on some possible tectonic consequences of the return flow.

Suggested Readings

Plate Driving Forces

Some Simple Physical Models for Absolute Plate Motions, Sean Solomon and Norman Sleep, Journal of Geophysical Research, v. 79, p. 2557–2567, 1974. *Classic paper defining plate driving forces and establishing no-net-torque method for determining absolute plate motion.*

On the Relative Importance of the Driving Forces of Plate Motion, Donald Forsythe and Seiya Uyeda, Geophysical Journal Royal Astronomical Society, v. 43, p. 163–200, 1975. *Classic paper establishing the importance of slab pull as a driving force.*

Flow in the Asthenosphere

Growth, Shrinking, and Long-Term Evolution of Plates and Their Tectonic Implications for the Flow Pattern in the Mantle, Zvi Garfunkel, Journal of Geophysical Research, v. 80, p. 4425–4432, 1975. *Classic paper showing that flow through the asthenosphere is required to account for the imbalance between the growth and consumption of plates in different parts of the globe. The patterns of plate growth, plate motion, and the return flow change considerably over periods of 100 Ma.*

Asthenosphere Flow and Plate Motions, J. F. Harper, Geophysical Journal Royal Astronomical Society, v. 55, p. 87–110, 1978. *Dynamical theory of flow in the asthenosphere. Torque analysis that incorporates effects of slab pull and drag forces due to flow in the asthenosphere provides good fit to observations.*

Asthenospheric Counterflow: A Kinematic Model, Clement Chase, Geophysical Journal Royal Astronomical Society, v. 56, p. 1–18, 1979. *Classic paper on counterflow.*

Geological Evidence for the Geographical Pattern of Mantle Return Flow and the Driving Mechanism of Plate Tectonics, Walter Alvarez, Journal of Geophysical Research, v. 87, p. 6697–6710, 1982. *Investigates the effect on return flow of deep roots beneath continents.*

Whole Mantle Convection

A Simple Model of Whole Mantle Convection, D. E. Loper, Journal of Geophysical Research, v. 90, p. 1809–1836, 1985. *Reviews geochemical and geothermal evidence for whole mantle convection; extensive references to earlier work.*

Thermal Evolution Models for the Earth, Ulrich Christensen, Journal of Geophysical Research, v. 90, p. 2995–3008, 1985. *A geochemical and geothermal model for whole mantle convection.*

Absolute Plate Motion from Single-Plate Torque

Absolute Motion of an Individual Plate Estimated from Its Ridge and Trench Boundaries, Richard Gordon, Allan Cox, and Clayton Harter, Nature, v. 274, p. 752–755, 1978. *Establishes the method of determining absolute plate motion from the ridges and trenches of a single plate. Explains the change in absolute motion of the Pacific plate at 43 Ma.*

Absolute Plate Motion from Hotspots

Plate Kinematics: The Americas, East Africa, and the Rest of the World, Clement Chase, Earth and Planetary Science Letters, v. 37, p. 355–368, 1978. *Global data are analyzed mathematically and with insight to obtain a set of relative plate motions and motions relative to hotspots.*

Present-Day Plate Motions, J. B. Minster and Thomas Jordan, Journal of Geophysical Research, v. 83, p. 5331–5354, 1978. *Classic determination of motions of plates relative to each other and to hotspots.*

Plate Motions and Deep Mantle Convection, W. Jason Morgan, Geological Society of America Memoir, no. 132, p. 7–22, 1972. *Seminal paper on hotspots.*

Hotspot Tracks and the Early Rifting of the Atlantic, W. Jason Morgan, Tectonophysics, v. 94, p. 123–139, 1983. *Classic paper on determining plate motions from hotspots.*

Hotspot Tracks and the Opening of the Atlantic and Indian Oceans, W. Jason Morgan, in *The Sea, 7,* edited by C. Emiliani, p. 443–487, Wiley Interscience, New York, 1981. *Classic paper.*

True Polar Wander

Early Tertiary Subduction Zones and Hotspots, Donna Jurdy, Journal of Geophysical Research, v. 88, p. 6395–6402, 1983. *Hotspots and subduction zones control the location of the earth's rotation axis.*

True Polar Wander: An Analysis of Cenozoic and Mesozoic Paleomagnetic Poles, Jean Andrews, Journal of Geophysical Research, v. 90, p. 7737–7750, 1985. *Demonstrates displacements between the paleomagnetic pole and the hotspot reference frame during the past 180 Ma.*

Index

Index of References

Irving, E., and G. A. Irving, "Apparent Polar Wander Paths Carboniferous through Cenozoic and the Assembly of Gondwana," 335

Isacks, Bryan, Jack Oliver, and Lynn Sykes, "Seismology and the New Global Tectonics," 217

———, and Peter Molnar, "Mantle Earthquake Mechanisms and the Sinking of the Lithosphere," 217

Jurdy, Donna, "Early Tertiary Subduction Zones and Hotspots," 381

Le Pinchon, Xavier, "Sea Floor Spreading and Continental Drift," 158

———, Jean Francheteau, and Jean Bonin, *Plate Tectonics,* 49, 260

Loper, D. E., "A Simple Model of Whole Mantle Convection," 380

McClain, James, John Orcutt, and Mark Burnett, "The East Pacific Rise in Cross Section: A Seismic Model," 218

MacDonald, Ken, "Mid-Ocean Ridges: Fine Scale Tectonic, Volcanic and Hydrothermal Processes Within the Plate Boundary Zone," 218

McDougall, Ian, and Don Tarling, "Dating of Polarity Zones in the Hawaiian Islands," 295

McElhinney, M. W., and D. A. Valencio (eds.), *Paleoreconstructions of the Continents,* 334

———, *Paleomagnetism and Plate Tectonics,* 335

McKenzie, Dan, and W. Jason Morgan, "Evolution of Triple Junctions," 84

———, and Robert Parker, "The North Pacific: An Example of Tectonics on a Sphere, 83, 157

Minster, J.B., and Thomas Jordan, "Present-Day Plate Motions," 158, 381

Morgan, W. Jason, "Hotspot Tracks and the Early Rifting of the Atlantic," 381

———, "Hotspot Tracks and the Opening of the Atlantic and Indian Oceans," 261, 381

———, "Rises, Trenches, Great Faults, and Crustal Blocks," 50, 83, 157

Munk, Walter, and Gordon MacDonald, *The Rotation of the Earth,* 49

Norton, I., and J. Sclater, "A Model for the Evolution of the Indian Ocean and the Breakup of Gondwanaland," 261

Parkinson, W. D., *Introduction to Geomagnetism,* 334

Patriat, P., and Vincent Courtillot, "On the Stability of Triple Junctions and Its Relationship to Episodicity in Spreading," 84

Phillips, Frank, *The Use of the Stereographic Projection in Structural Geology,* 176

Ragan, Donald, *Structural Geology—An Introduction to Geometrical Techniques,* 176

Ramsay, Johan, and Martin Huber, *The Techniques of Modern Structural Geology,* 176

Riddihough, Robin, "Recent Movements of the Juan de Fuca Plate System," 84

Snyder, J. P., "Map Projections Used by the U.S. Geological Survey," 125

Solomon, Sean, and Norman Sleep, "Some Simple Physical Models for Absolute Plate Motions," 379

Stauder, William, and Lalliana Mualchin, "Fault Motion in the Larger Earthquakes of the Kurile–Kamchatka Arc and of the Kurile, Hokkaido Corner," 218

Stein, Seth, and Richard Gordon, "Statistical Test of Intraplate Deformation from Plate Motion Inversions," 158

Sykes, Lynne, "Mechanisms of Earthquakes and Nature of Faulting on the Mid-Oceanic Ridges" 217

Tarling, Donald, *Paleomagnetism: Principles and Applications in Geology, Geophysics, and Archeology,* 334

Tau Rho Alpha and J. P. Snyder, "The Properties and Uses of Selected Map Projections," 125

Uyeda, Seiya, *The New View of the Earth,* 49

Van der Voo, R., and J. Zijderveld, "Renewed Paleomagnetic Study of the Lisbon Volcanics and Implications for the Rotation of the Iberian Peninsula," 335

Vine, Fred, "Spreading of the Ocean Floor: New Evidence," 295

———, and Drummond Matthews, "Magnetic Anomalies over Oceanic Ridges," 295

Wegener, Alfred, *The Origin of Continents and Oceans,* 321

Wilson, D., R. Hey, and C. Nishimura, "Propagation as a Mechanism of Reorientation of the Juan de Fuca Ridge," 84

Wilson, J. Tuzo, "A New Class of Faults and their Bearing on Continental Drift," 50